Medical Genetics

NOTICE

Medical Genetics

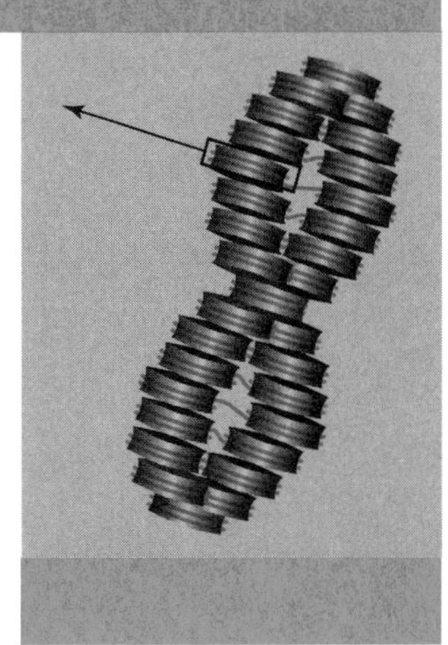

GEORGE H. SACK, JR., M.D., Ph.D., F.A.C.M.G.

Associate Professor
Departments of Medicine,
Pediatrics, and Biological Chemistry
The Johns Hopkins Hospital
Baltimore, Maryland

McGRAW-HILL
HEALTH PROFESSIONS DIVISION

New York St. Louis San Francisco Auckland Bogotá Caracas
Lisbon London Madrid Mexico City Milan Montreal
New Delhi San Juan Singapore Sydney Tokyo Toronto

McGraw-Hill 

A Division of The **McGraw-Hill** *Companies*

MEDICAL GENETICS

1234567890 QPK QPK 998

ISBN 0-07-057998-9 (Set)
0-07-134203-6 (Book)
0-07-134204-4 (CD-ROM)

This book was set in Times New Roman by Bi-Comp, Inc.
The editors were James T. Morgan III, Pamela Hanley, and Lester A. Sheinis.
The production supervisor was Richard C. Ruzycka.
The text and cover designer was Joan O'Connor.
The text illustrator was Mollie Dunker.
The cover illustrator was Joseph Gillian.
The indexer was Kathrin Unger.
Quebecor Printing/Kingsport was printer and binder.

This book is printed on acid-free paper.

*This book is dedicated to the honor of my parents,
Sophia and George Sack*

Contents

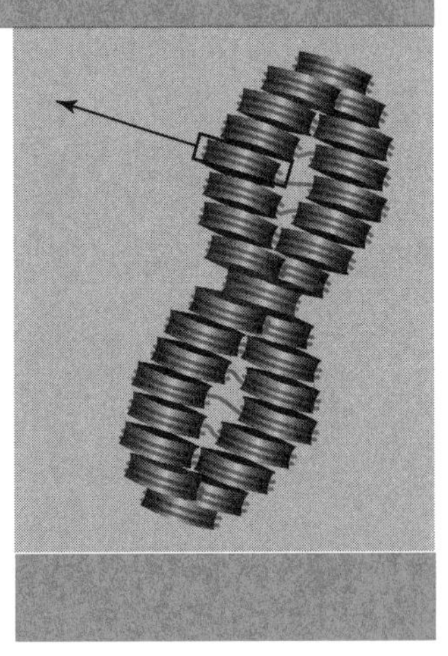

Preface

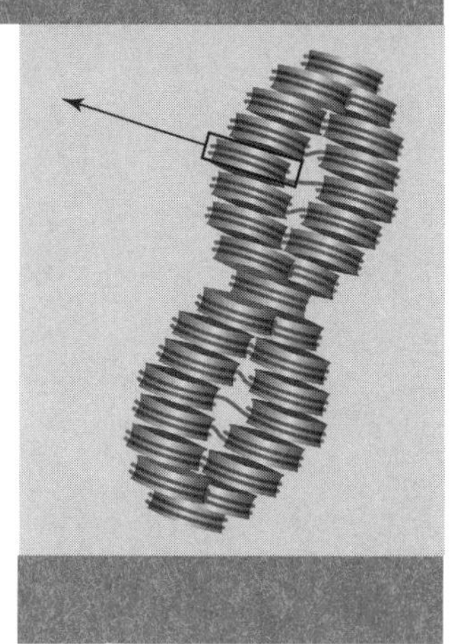

Rabbi, who sinned, this man or his parents, causing him to be born blind? (John 9:2)

Two thousand years have not lessened the poignancy of this question. It continues to express the concerns of many when genetic problems are encountered in the clinic. In fact, current developments in genetics have made similar inquiries quite frequent. Clinicians must be prepared to address these issues; those who are most effective will use their understanding of basic genetics, human biology, and individual personalities to appreciate the diversity of pathophysiologic variations in their patients.

Genetic aspects of medicine began with efforts to define and classify uncommon traits as well as attempts to establish relationships based on clinical evidence. Inevitably, such efforts often appeared esoteric and, seemingly, isolated from much of the daily practice of medicine. Nevertheless, many of the exciting recent developments are related to this extensive background of clinical observations and basic biologic data. This background has served as the starting point for applying the remarkable analytical power of molecular genetics. Gradually, genetics is achieving recognition as a clinical discipline.

The impressive progress in medical genetics reflects, in part, developments in several important disciplines including (but not limited to):

- mathematics—methods for mapping and linkage analysis as well as for managing large amounts of data for both storage and pattern recognition
- biochemistry and molecular biology—recombinant DNA technology, elucidating gene structure and action, study of recombination and mutations, and clarifying details of intermediary metabolism, membrane transport, and cell-cell communication
- cell biology—showing details of cellular and extracellular structures as well as of chromatin and chromosome structures in mitosis and meiosis
- virology—methods for isolation and study of nucleic acid fragments, models for recombination as well as possible vehicles for gene transfer
- obstetrics—details of early fertilization events and both invasive and noninvasive fetal monitoring
- general clinical specialties (particularly pediatrics and internal medicine)—disciplines through which increased prominence of genetics often confronts affected individuals for the first time
- pharmacology—new approaches to drug design and delivery as well as improving matches between drugs and individual patients
- veterinary medicine—useful animal mutations for comparison and study of inherited traits as well as possibilities for developing and testing treatments

Such a list of disciplines implies broad medical ramifications for genetics. An additional observation is that it has become unlikely that any individual will have the background needed for using all of the analytic and diagnostic tools as well as the new data. Until relatively recently, interactions of genetics with medicine have generally emphasized the obscure—conditions and problems so individually infrequent as to be encountered rarely in a lifetime of practice. Additionally, past approaches to genetics have assumed that one had to have considerable memory to recall all of the clinical presentation(s) of and diagnostic test(s) for any given syndrome. If this latter perceived requirement was a barrier in the past, it must appear even more formidable now. How can anyone remember the medical and biologic implications of changes in and interactions between 3×10^9 nucleotides?

Fortunately, despite this apparent (and often real) complexity, new approaches to data and information management are becoming an essential part of genetics in both research and clinical practice. Used and appreciated appropriately, these systems can provide access to timely, reliable information to assist in diagnosis, treatment, and counseling, thus enriching and improving the practice of medicine.

This book is not intended to be encyclopedic. It has been planned with the assumption that the students of today and the practitioners, researchers, scholars, and teachers of the future will have considerable comfort with computer-based information management and retrieval systems. I also have assumed a basic knowledge of molecular biology and biochemistry. Each chapter has been designed as a general introduction to essential information using several specific examples but should not be considered exhaustive. The chapters include references to conventional literature sources as well as an annotated guide to Internet sites that can enhance and extend the information presented. Most of these sites are updated regularly so that their information remains timely. An important concern about using such sites relates to their reliability and editorial quality control. This remains unresolved and the notion of *caveat emptor* ("let the buyer beware") still pertains. Thus, it is prudent to consider the sources and providers carefully. Except for the remarkable breadth and diversity of Internet resources available, however, this situation is not fundamentally different from past challenges, such as how to find the most valuable consultant or the best reference for a particular problem; these efforts often were managed through trial and error. With time, quality tends to become recognized, although the large amount of material currently available is intimidating.

Because of the rapid pace of research and the speed with which it is becoming clinically useful, it is an exciting time to consider genetics—with its many associated disciplines—in medicine. A considerable portion of this book is devoted to the discussion of so-called single gene disorders. In general, these follow the principles of transmission established by Mendel in the nineteenth century and often are called *mendelian* for that reason. These important disorders demonstrate the consequences of changes in single genes. Although individually rare, such conditions are not collectively rare and they have been important for detailed study and developing specialized care. Nevertheless, despite their heuristic value, learning only about these selected disorders will have relatively little relevance to general or even specialized clinical practice and certainly will provide a distorted view of the importance of genetics in medicine.

An entirely different perspective should be derived from Chapter 12, which considers aspects of genetics in understanding common diseases. For such conditions, which include some of the most frequent challenges in daily medical practice, individual gene contributions have been suspected often (from twin studies, family clusters, etc.) but have not been identified rigorously. Furthermore, available treatment options usually have not permitted much flexibility for the care of individuals with specific subtypes in each of the general categories. This situation is changing, however, and likely will become the area of greatest impact for genetics in many medical decisions. Predispositions to hypertension, diabetes, cancer, vascular disease, and dementia (to name only a few categories) must have at least some mendelian features. Using the extensive collection of molecular markers on the growing gene map in combination with large population studies and techniques such as mapping and linkage analysis should identify many individual contributions.

As these new genetic data become available, clinicians will be forced to accept the challenge of using this information to assist our patients. As described in Chapter 13, extensive amounts of genetic data have never been available for our patients before. Thus our diagnostic approaches and ethical perspectives have generally not had to consider such information. Concerns such as insurability, confidentiality (while maintaining appropriate access to data), and individual predispositions to specific conditions will have far more ramifications than before.

Using new genetic information in medicine will lead to changes in the notions of treatment. Current treatments for rare or "orphan" disorders receive considerable attention in scientific and lay press. Although they are important developments (even if only preliminary), these treatments affect so few individuals that they often can be ignored by conscientious physicians who do not encounter patients affected with these conditions. By contrast, consider "common diseases" for which future treatments may be individualized based on genetic studies; such disorders will necessarily be prominent in general medical practice. In such situations a textbook can serve only as an introduction. Continual awareness of new genetic discoveries and their implications will likely be most readily available through Internet resources and updated CD-ROM subscriptions. Thus, awareness of the implications of individual and collective gene changes will become central to continuing medical education and timely practice protocols.

It is important to remember that the practice of medicine must remain based on attending to an individual, often with a problem. Confronting such a person has always involved gathering data to answer two specific questions:

1. What is the problem?
2. What can I do about this problem?

These will remain essential in the effective practice of medicine. However, two additional considerations are becoming more important as new tools and genetic information become available. These questions are:

3. Why did *This* person develop *This* problem *Now*?
4. Are there ways to prevent or minimize a future problem (for this person or family members)?

These latter two questions necessarily involve increased knowledge of human biology and pathophysiology. Now and in the future they will involve genetics. Although genetics has finally achieved recognition as an individual clinical discipline, it has direct relevance to all other medical specialities. While it often is the case that certain subspecialties may not have to pay much attention to the detailed considerations of other specialities, the broad biologic implications of fundamental genetic distinctions cannot be ignored by anyone. The goal of this book is to emphasize the pervasiveness of inherited variations throughout medical practice. In fact, the current and prospects of future developments in this area lead me to paraphrase earlier advice from the distinguished clinician Dr. A. McGehee Harvey. The summary of this advice is "There is no such thing as medical genetics; there is only good medicine that you practice with an appreciation of genetics."

ACKNOWLEDGMENTS

This book would not have been written without considerable assistance from many people. I am particularly grateful to Mary A. Mix for all of her secretarial skills. Mollie Dunker did a marvelous job with the illustrations, and Fred Dubs and his colleagues provided excellent photographic support. Jim Morgan and Lester Sheinis made editorial interactions consistently pleasant. Several colleagues were kind enough to share photographs from their own collections including (in alphabetical order): Grant Anhalt, Gaylord Clark, Hal Dietz, Gail Stetten, and George Thomas.

Assembling a book requires remembering and integrating many important sources of advice and inspiration. However, naming these individuals should not require them to take responsibility for my various and peculiar interests. Mac Harvey, Philip Tumulty, and Samuel Asper remain my models for integrating clinical and laboratory information. I am particularly grateful to Dan Nathans, who did his best to help me appreciate the rigor of scientific thought, and also to Ham Smith, who consistently reminded me of how much fun it all could be. Margaret Abbott gave me fine advice on assembling pedigrees and then making sense out of them. Debbie Meyers helped considerably with understanding and applying linkage analysis. Linda Cork and her colleagues have provided an opportunity to examine the wealth of genetics in veterinary practice. Dan Lane and the entire Biological Chemistry faculty have given me treasured collegiality and consistent support. I cannot forget my many debts to my patients who continue to teach me so much.

Personal interactions are essential in maintaining one's direction and providing both tangible and intangible support. I am particularly grateful to my parents. Also, long-term support from Dan Kelly and Shirley and Bill Griffin has been essential. I remain in debt to Bill Baxter and the St. Thomas choirs, to Marylynn and John Roberts, and, especially, Elizabeth.

CHAPTER 1 The Basics

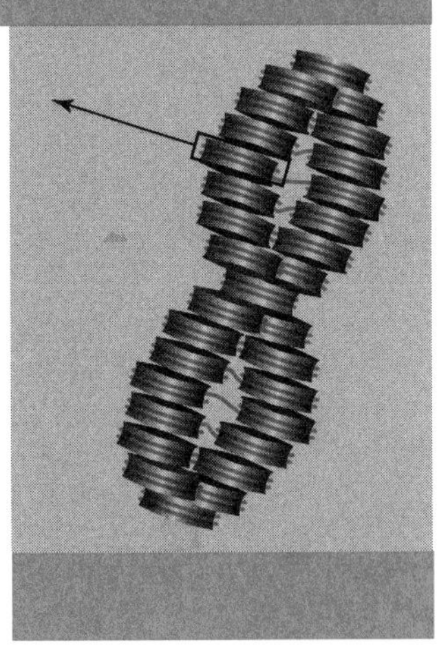

A. PROTEINS

Because proteins are determinants of cell structure and function, their variations frequently are the basis for appreciating genetic changes. Although the basic polymeric structure of proteins is rather simple, the three-dimensional structure of large proteins, as well as their physical and enzymatic interactions, become complex.

The linear structure of proteins is based on the formation of successive amide linkages between individual amino acids. As shown in Figure 1.1, such linkages permit the joining of any of the 20 regularly found amino acids in any prescribed order. Because amide linkages are common in all protein polymers, the differences among proteins are largely determined by the amino acids and their linear order. Because the different amino acid side chains have different chemical properties, ranging from acidic to neutral to basic and from hydrophobic to hydrophilic, the final properties of an individual protein represent the sum of a complex mixture of contributions.

The linear structure (also called the "primary structure") of a protein is only the initial determinant of its character. Proteins do not exist as linear polymers, however. Rather, they undergo complex folding so that various parts of the initial polymer may be closer or farther away from one another than suggested by the linear sequence alone. The three-dimensional structures of over 1000 proteins have been determined through a combination of x-ray crystallography and nuclear

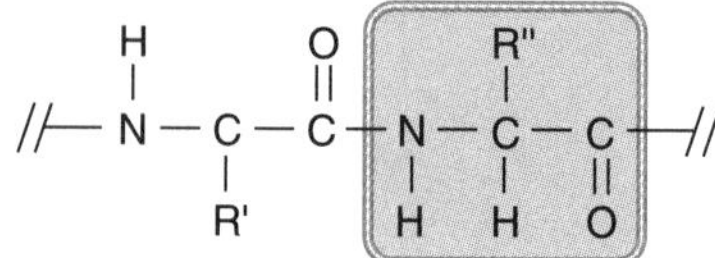

Figure 1.1 *Adjacent amino acids linked by an amide bond. R′ and R″ represent the substituents that define the specific amino acids. The shading for the second amino acid includes both the amide nitrogen and the carboxyl function. The substituents (R) can vary from a simple proton in glycine to imidazole (tryptophan) or a carboxylic acid (e.g., glutamic acid), with corresponding changes in size, configuration, and bonding characteristics.*

magnetic resonance spectroscopy; databases provide access to these structures. This three-dimensional information can then be used to predict the protein's location in the cell (or outside the cell) as well as its function(s) and relatedness to other proteins. Considerable emphasis justifiably has been directed toward understanding enzymes—proteins that function as chemical catalysts. Genetic variation(s) in enzyme function(s) underlie(s) many inherited disorders, as discussed in Chapters 5, 6, 7, and 10.

Many other proteins have structural rather than catalytic functions. For example, the intracellular polymers actin and tubulin are essential in determining three-dimensional internal support to establish, maintain, and remodel cell shape. Other structural proteins include the family of extracellular collagen molecules. Again, this is a group of support molecules characterized by physical strength and long, linear structures. Chapter 5 describes the array of clinical consequences resulting from alterations of the structural protein fibrillin in Marfan syndrome (OMIM #154700).

Still other proteins do not have enzymatic or structural functions. Globin combines with the prosthetic group heme to create the oxygen transport molecule hemoglobin. Other proteins, such as insulin, function as hormones.

Proteins may be modified after they are synthesized. Physical properties, three-dimensional structure, and function can be greatly affected by the addition of hydroxyl, phosphate, carbohydrate, and lipid groups. Such modifications also can help determine the protein's position in (or outside of) the cell. Most, but not all, protein modification is directed by specific enzymes. An example of nonenzymatic covalent modification is the formation of glycosylated hemoglobin. Other proteins form very tight but not covalent complexes with important functional groups. These groups include hemes (in hemoglobin and cytochromes), thiamine (vitamin B_1), and biotin.

Several contrasting three-dimensional protein structures are shown in Figure 1.2. These emphasize the remarkable differences among individual proteins. Because structure underlies function, any structural change has the potential to alter the function of the molecule and, ultimately, the health of the individual in whom it occurs. Considering such complex structures, however, also makes it obvious that some small changes might be tolerated because they fail to significantly perturb the overall structure. As will be considered below, such small changes can be sources of important biological variation(s) but may not greatly influence physiology and health. For example, hemoglobin, one of the best-studied human proteins, has many recognized structural variants. Some of these are responsible for important illnesses, such as sickle cell anemia (OMIM #141900) and hemoglobin C disease (OMIM #141900.0038), while others are largely laboratory curiosities with few or no known health consequences.

The linear structure of proteins is, of course, determined by informa-

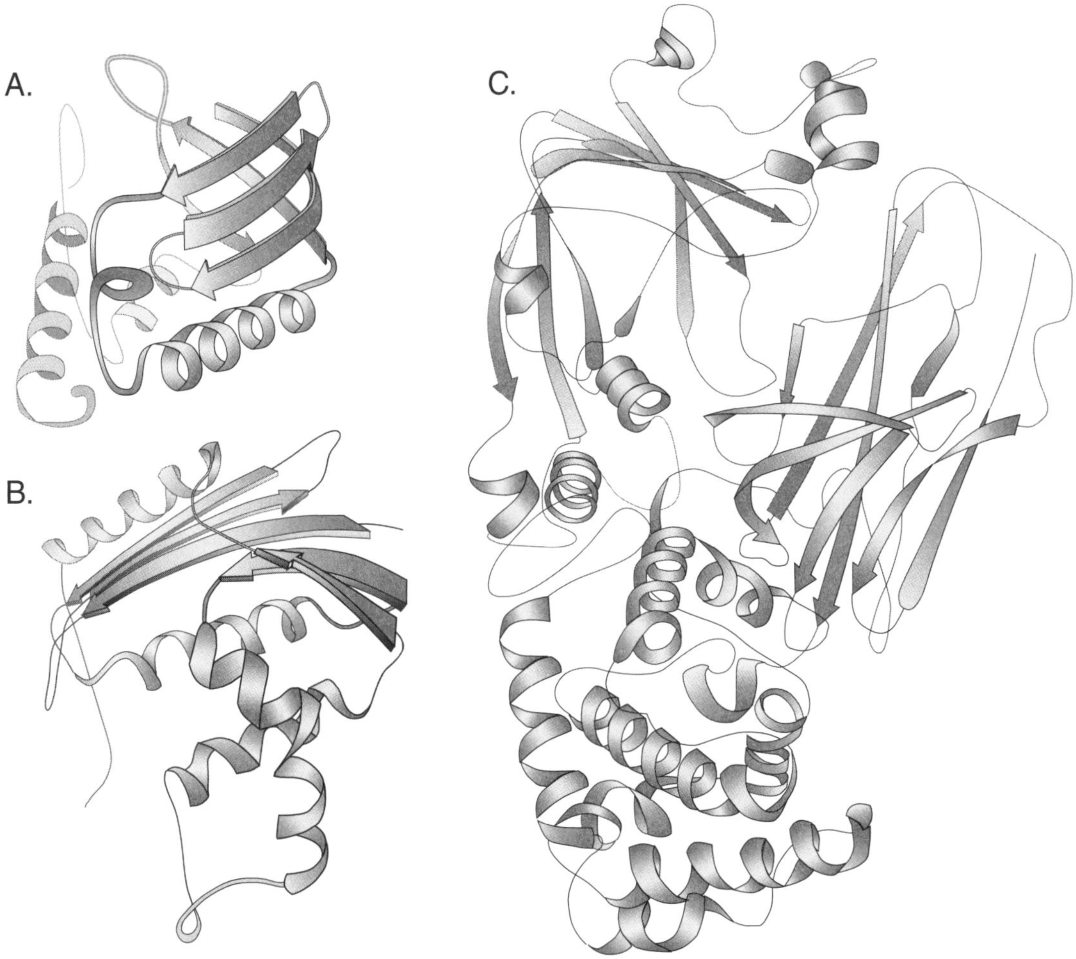

Figure 1.2 *Ribbon diagrams representing the three-dimensional structures of several well-studied proteins. Note the contrast between helices (shown as coils), β-sheets (flattened arrows), and random coil domains (thin lines). Such contrasting shapes underlie many of the different features of individual proteins. (A) Staphylococcal nuclease (Biochemistry 33:1063, 1994). (B) Ribonuclease H (Science 249:1398, 1990). (C) Diphtheria toxin (Nature 257:216, 1992).*

tion encoded in cellular DNA. The complex apparatus for synthesizing proteins on the basis of the genetic information within DNA involves activities within both the nucleus and the cytoplasm of the cell. Fundamental to this entire process is consideration of basic aspects of DNA structure and function.

1. GENE STRUCTURE

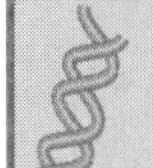

The double-helical structure of DNA serves as the repository for genetic information as well as the basis for DNA replication. These topics are addressed in detail in molecular biology texts

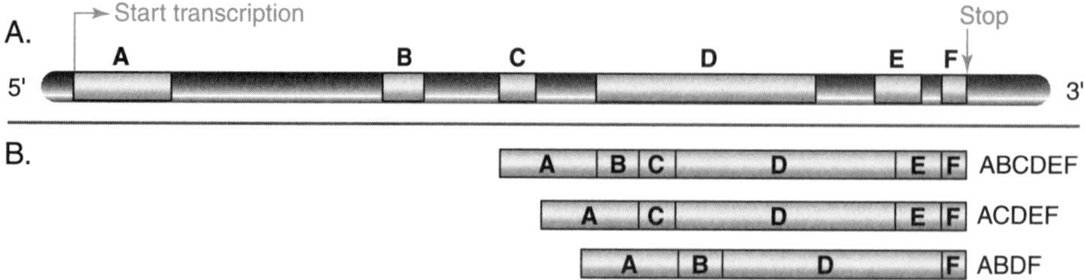

Figure 1.3 *(A) Representative linear arrangement of a eukaryotic gene. Exons are colored; the space between them defines introns. (B) After the entire region is transcribed into RNA, the process of splicing removes introns and assembles contiguous coding and control regions. As shown, it also is possible to eliminate exons during splicing, generating different final transcripts from one basic gene region.*

and will not be reviewed exhaustively here. Of particular importance in considering genetic contributions to medicine is an appreciation of the structure of individual genes. Genes represent discrete regions of DNA; they may be quite short or may extend over hundreds of kilobases (kb, 1 kb = 1000 individual base pairs [bp]). Individual regions of genes are defined by specific sequence features (see Figure 1.3). One of the most prominent features of most human genes is the presence of distinct segments, some of them responsible for protein-coding information and others separating such coding sequences. The former—coding—sequences are referred to as exons. The latter—noncoding—regions between exons are referred to as introns.

Noncoding sequences must be removed to assemble contiguous coding information. This is accomplished at the level of RNA by first transcribing the entire gene region from DNA into RNA and then subjecting the newly transcribed RNA to a "splicing" process. During splicing, introns are removed and exons are joined. In some complex gene systems, there may be more than one pattern of splicing. Such "alternative splicing" can generate a group of related—but still distinct—proteins from the same initial transcript (see Figure 1.3B). In addition to linear arrays of nucleotides, complex folded arrays can form, especially between single strands, as shown in Figure 1.4. Such arrays constitute additional control features and are intrinsic to the base sequence itself. Thus DNA sequence information encodes structural and control information in addition to specifying linear amino acid arrays.

Other critical features of gene structure include regions for controlling the initiation of transcription and signals for beginning and terminating translation. (All genes are transcribed and translated in the same direction [$5' \rightarrow 3'$].) These signals are necessarily based on nucleotide sequences, many of which must be recognized by proteins—transcription factors, enzymes, etc. It is important to realize that these nucleotide sequences have important topological features. It is instructive to examine a model of double-helical DNA in order to appreciate that the three-dimensional profile of the edges of the bases in the major groove permits clear distinction among base pairs. If you do this, you will find that the surface of the space occupied by a sequence of nucleotides can provide highly specific interactions with protein transcription factors, permitting strict regulation of gene expression.

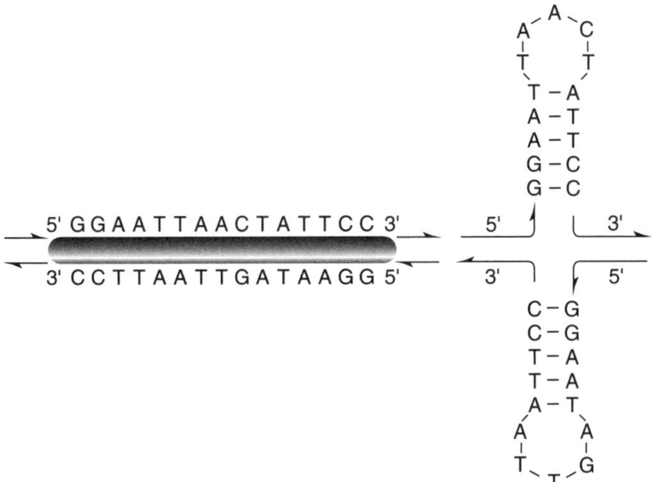

Figure 1.4 *Alternative structures for a stretch of 15 nucleotides, showing the transition between the conventional, linear helix (left) and a structure containing "stems" and "loops" (right). Although the unpaired bases in the loops cannot contribute to stability, increasing the length of the paired stem minimizes the effect of energy losses due to unpaired regions. Note that the stability of such a stem and loop structure depends on the presence of potential pairing regions—in this example, complementary groups of five nucleotides on each stem. Such alternative structures can serve as markers along a stretch of DNA that could be used for recognition, modification, and/or packing.*

There are approximately three billion (3×10^9) base pairs in human DNA. Not all regions of DNA are responsible for encoding proteins. We already have considered introns, whose sequences are removed from RNA transcripts prior to protein synthesis. Other regions of the DNA serve large-scale structural functions. These include areas such as telomeres at the ends of chromosomes and other sequences near centromeres that are essential for cell division.

Telomeres, centromeres, and other regions in human DNA are characterized by repeated nucleotide sequences. Some repeats are quite short. The dinucleotide CA is found in stretches of as many as 20 or more repeats in over 50,000 positions throughout cellular DNA. These are usually designated $(CA)_n$, where n is the repeat number. Repeats of longer sequences are also present. Generally, repeated sequences are not associated with protein coding information, and at least some may be evolutionary vestiges of no particular significance. Other repeated sequences are likely to have important roles in DNA structure, in the packing of DNA in chromosomes, and in recombination and replication. The specific locations of these repeated sequences in the DNA are very important, however. Because repeated sequences often are found within regions of DNA that also contain unique sequences, it is possible to define specific repeated segments on the basis of the unique DNA sequences flanking them. Thus, repeated sequences can

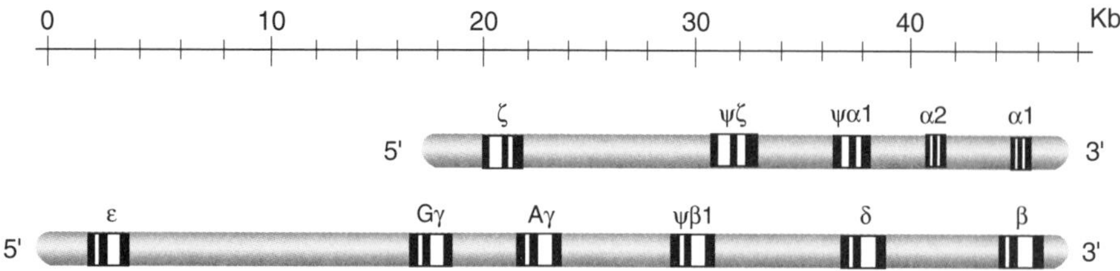

Figure 1.5 *The human globin gene families (α-globin above, β-globin below). These two groups of genes are closely related yet distinct. Each group includes genes expressed only during fetal development (ε, γ, δ, and ζ) and pseudogenes (ψ) in addition to the genes expressed in adult erythrocytes. The scale of this figure encompasses much longer distances than those in Figure 1.3. Dark regions correspond to exons.*

be assigned to unique positions in the linear array of human DNA based on their neighbors. As will be discussed below, this is a very important notion in terms of developing molecular markers for the gene map.

Many genes are related to one another. They make up so-called gene families. These are groups of genes of often similar structure and function. Some gene families are grouped as contiguous arrays. By far the best-studied gene families are those for the globin genes. The β-globin genes on human chromosome 11 and the α-globin genes on human chromosome 16 have been studied in great detail. They represent a group of related genes and include globin genes that are transcribed and synthesized at different times in development. In general, adults synthesize only adult globin genes but fetal globin genes remain present, although unexpressed, in adults. Figure 1.5 presents an outline of the structures of the globin gene loci.

Also present within many—but not all—gene families are "pseudogenes." Such sequences are no longer functional and cannot make proteins. Presumably, they have arisen as evolutional derivatives of their parent functional genes but cannot themselves be transcribed or successfully translated. Sometimes, as in the globin gene clusters (see Figure 1.5), the pseudogenes are present in the same region as their parent. Other pseudogenes may be dispersed in nonrelated areas of human DNA. Pseudogenes are examples of historical genetic remodeling and recombination events but appear to have no functional significance in themselves. They may be very similar to their parent genes, however, so that considerable study may be required to establish that they are, in fact, incapable of being expressed.

Although the α- and β-globin gene families are distinct, they also share many structural features. Together, they can be considered a "superfamily." Similar relationships exist among members of the immunoglobulin superfamily, as presented in Chapter 11. In the latter case, all related genes are not necessarily physically close or even on the same chromosome.

2. TOOLS OF MOLECULAR GENETICS

Analysis of DNA structure and its variations in health and disease uses specific analytical tools. A brief overview of their properties and use is essential for appreciating the analysis of normal and variant DNA in both laboratory and clinical testing. Because DNA is a very long polymer that appears rather monotonous from a distance, it is essential to be able to identify, isolate, and examine specific regions in close detail.

To isolate specific individual fragments of a larger DNA polymer, a series of bacterial enzymes are used. These are the enzymes mediating the bacterial process of "bacteriophage restriction," and the enzymes have thus become known as "restriction enzymes." Restriction enzymes operate by creating a break in the phosphodiester bonds of both strands of DNA. They do so by recognizing specific base sequences. Such base sequences are frequently palindromic, meaning that they read the same on opposing DNA strands. An example of one restriction enzyme recognition sequence is shown in Figure 1.6. The enzyme recognizes the topology of the bases in the intact double helix of DNA in order to establish specificity and then cleaves the two strands as shown. Significantly, and consistent with their physiological function in bacteria, DNA cleavage by restriction enzymes requires that the target DNA sequence not be modified. In general, restriction enzyme cleavage is prevented by the addition of specific methyl groups to base(s) within the target sequence. The base most frequently modified by methylation is cytosine, as shown in Figure 1.6. The addition of the relatively large and bulky methyl group onto cytosine is frequently sufficient to prevent cleavage by the enzyme by changing the surface topology and interrupting appropriate recognition and binding by the enzyme.

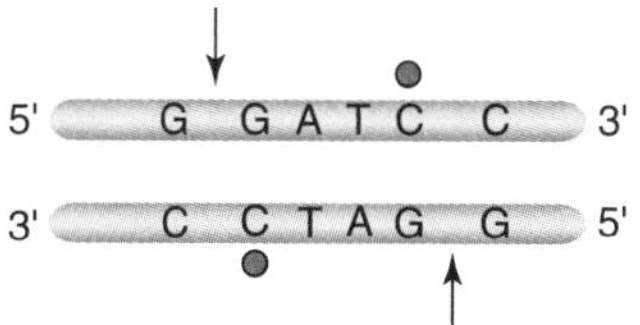

Figure 1.6 *The DNA recognition sequence for restriction endonuclease BamHI. Note that the sequence is palindromic—it reads the same on both strands. Sites of cleavage of the phosphodiester bonds are shown by arrows. The dots show the positions where methylation can block recognition of this sequence by the enzyme and thus prevent cleavage.*

A large array of restriction enzymes are available. Some cut DNA only rarely and thus form very large fragments. In general, such enzymes have long recognition sequences whose recurrence in the genome, simply on the basis of random base sequence, is very infrequent. Other enzymes, in contrast, cut quite frequently. For example, for the enzyme *Bam*HI, whose site is shown in Figure 1.6, it is simple to calculate the likelihood of finding similar sites in any stretch of DNA bases. The chance is $(1/4)^6$, because it is the product of having one specific base (out of 4) at each of six consecutive positions. This means that such an enzyme, on average, should cleave DNA once every 4096 bp. Table 1.1 shows examples of several different restriction enzymes, the sequences of their respective recognition sites, and an estimate of the frequency of their cutting based on the number of fragments formed by cleaving the DNA of adenovirus 2 (35.9 kb). It is possible to exploit the differences in specificity and base sequence requirements of restriction enzymes to permit analysis of many DNA variations, as will be considered below.

The ability to isolate a specific DNA fragment using restriction enzyme cleavage permits the joining of such a segment to a "vector" to produce a "DNA clone." Such vectors usually are derived from bacterial plasmids, extrachromosomal elements that can grow and divide independently in bacteria (see Figure 1.7). Because bacteria can

Table 1-1 ▶ **Examples of restriction enzymes**

Enzyme	Recognition Site (5′ . . . 3′)	Fragments from Adenovirus-2 DNA
Alu I	AG↓CT	158
Hha I	GCG↓C	375
Mbo I	↓GATC	87
Dde I	C↓TNAG	97
Hpa I	GTT↓AAC	6
BamH I	G↓GATCC	3
Not I	GC↓GGCCGC	7

be grown easily and plasmid numbers can become very high, "cloning" permits the preparation of large amounts of a specific DNA fragment and is the basis for much progress in the analysis of DNA and genes, as well as in biotechnology. It is generally possible to amplify and isolate large quantities of any DNA region of interest using cloning. Restriction enzymes and other DNA-modifying enzymes permit the creation of novel DNA rearrangements that may be useful for analysis as well as for forming recombinant molecules for biosynthesis of important proteins, drug development, and other uses.

A fundamental feature of double-stranded DNA is the complementarity of opposing single strands. This specific, complementary base pairing (A with T, G with C) is relatively weak for any individual base pair, reflecting only hydrogen bonding, but permits adding the strength of an array of many small interactions into a remarkably powerful and specific relationship. With a sufficient length of two sequences, it is possible to establish conditions under which a given test sequence will match only its specific complement. Such a procedure is referred to as "hybridization."

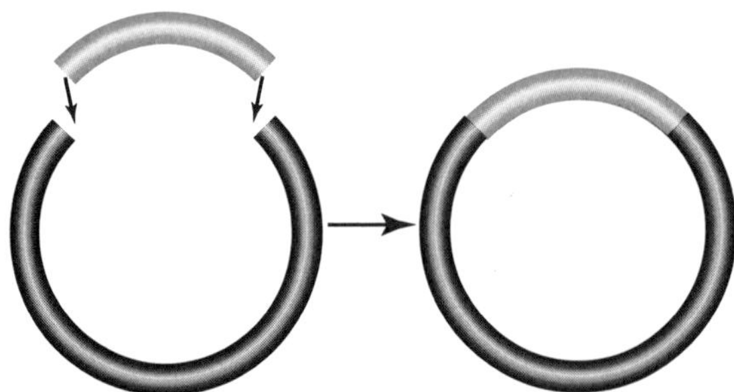

Figure 1.7 *A simple representation of cloning, in which a chosen DNA fragment is inserted into the DNA of a bacterial plasmid, usually at complementary restriction enzyme cleavage sites. Such a recombinant plasmid can then be propagated in host bacteria to reach very large numbers. The DNA can then be used for studies of sequence, transcription, or translation or for comparisons with other DNA. In principle, any DNA fragment can be recombined and amplified in this manner.*

To perform hybridization, the individual strands of the double helix must be separated (see Figure 1.8). This can be done with heat or denaturing agents, in a process called "melting." Isolated single strands can then be used to re-form hybrids. Depending on the strictness of

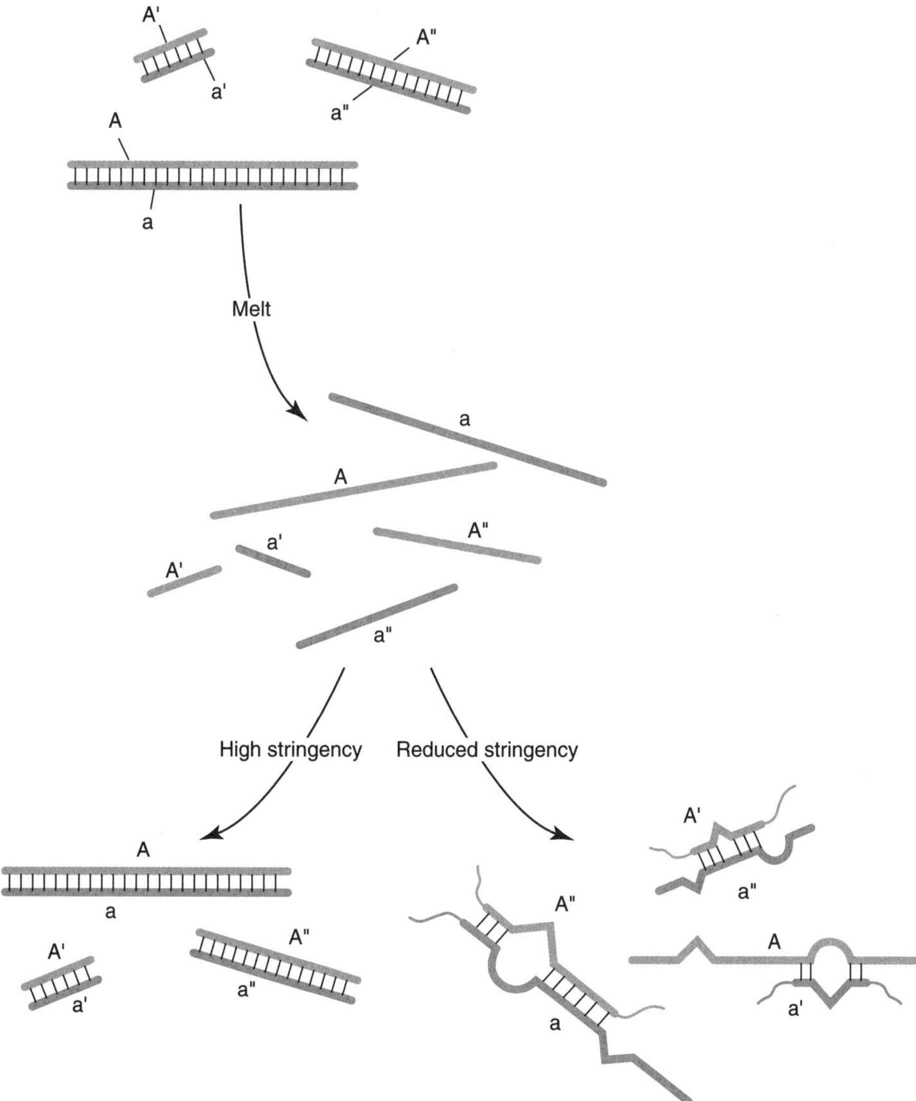

Figure 1.8 *Hybridization is a useful way of comparing the sequences of different pieces of DNA, indicated here as A, A', and A" with their respective complements a, a', and a". The paired strands are separated (melted) and then allowed to reanneal into double-stranded structures. If the pairing requirements during reannealing are high ("high stringency"), only perfectly matched hybrids will be able to form. Under reduced stringency, cross-hybridization can permit pairing at complementary regions, although gaps and loops will form at unpaired regions. The latter conditions are useful for comparing a specific DNA region with others that may differ in small or large degrees.*

the conditions under which hybridization is performed, absolute complementarity of every base pair may be required. Alternatively, the conditions of hybridization can be arranged so that some imperfect pairing or mismatches may occur. The spectrum of hybridization conditions is referred to as "stringency." Under highly stringent hybridization conditions, a given nucleotide sequence will hybridize only with its precise complement. Under reduced stringency, however, a given sequence may hybridize with other reasonably similar sequences. Hybridization is essential in studies of DNA variations and in establishing chromosome and gene identity. It forms the basis for many clinical diagnostic studies. It is important to note that hybridization can occur between DNA sequences, between RNA sequences, or between DNA and RNA sequences.

Because DNA has an electrical charge, it can move in an electrical field. When such a field is confined to a matrix, such as a carbohydrate or polyacrylamide polymer gel, electrophoresis occurs. DNA electrophoresis is very useful for separating fragments of different lengths. In conventional DNA electrophoresis, small fragments move faster than large fragments because there are fewer barriers to the movement of smaller fragments through the interstices of a gel. Because electrophoretic conditions can be standardized, it is possible to estimate the length of a given DNA fragment on the basis of its position in a gel following electrophoresis. The positions of DNA fragments within an electrophoretic gel can be determined by using a dye that binds to DNA. A particularly useful dye is ethidium bromide. This planar molecule slips between the stacked base pairs and shows a bright orange fluorescence when viewed under ultraviolet light. As shown in Figure 1.9, simple electrophoresis and ethidium bromide staining can permit direct comparison of different DNA molecules.

DNA or RNA fragments separated by electrophoresis can be transferred to special paper or other media for subsequent study. This is achieved by moving the fragments from the electrophoretic matrix directly onto the analytical matrix, which can be done electrophoretically or by mobilizing the fragments in solutions containing high salt concentrations. Once the fragments have become transferred to the new matrix, they can be immobilized there. Because specific fragments will have specific positions on the matrix, determined by their length, it is possible to identify them by hybridization. Such a transfer made with DNA fragments is called a Southern blot; a similar blot made with RNA is called a northern blot. Figure 1.10 shows the production of a Southern blot using DNA fragments produced by restriction enzyme cleavage. In such a situation it is possible to identify the specific fragment that contains sequences complementary to the test sequence, or "probe," by hybridization with the DNA in the blot matrix. Southern blot experiments have been essential in the analysis of genes and their variations.

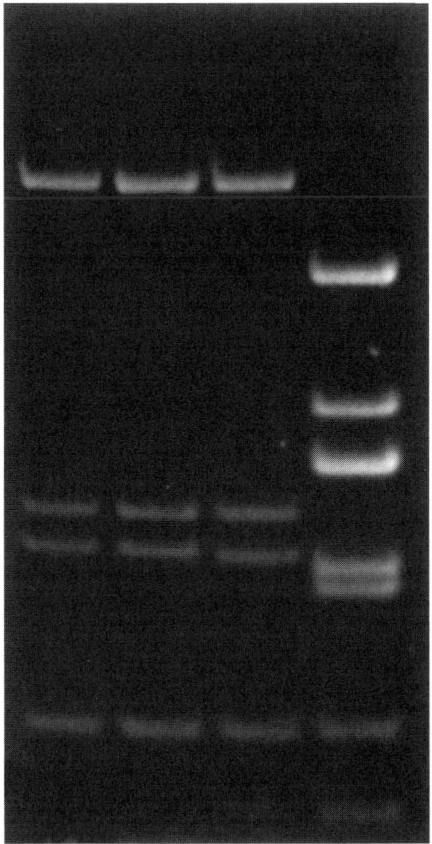

Figure 1.9 *Individual fragments of double-stranded DNA produced by restriction enzyme cleavage of four recombinant clones separated by electrophoresis, stained with ethidium bromide, and viewed with ultraviolet light. Electrophoresis was from top to bottom, as shown in this figure; smaller fragments migrate faster under these conditions. What conclusions can be drawn about the DNA in these four clones? (Sack, Gene 21:19, 1983.)*

The use of the polymerase chain reaction (PCR) has changed genetic analysis enormously. The principle of PCR is quite simple and is outlined in Figure 1.11. It takes advantage of both hybridization and complementarity by adding small, well-defined, short (~18 or more bp) pieces of single-stranded DNA (called "primers") to the double-stranded DNA mix. This mix is then heated and cooled, permitting the original strands to separate and the primers to hybridize with their complementary sequences. An enzyme is added to elongate the DNA, using the primer as a beginning for the new strand and the original DNA strand as a template. When this procedure is repeated, the DNA sequence between the primers is amplified geometrically. In theory, this could result in DNA amplification from a single initial piece of DNA, but it is usually begun with a small DNA sample. PCR has permitted many of the current mapping, mutation, and population

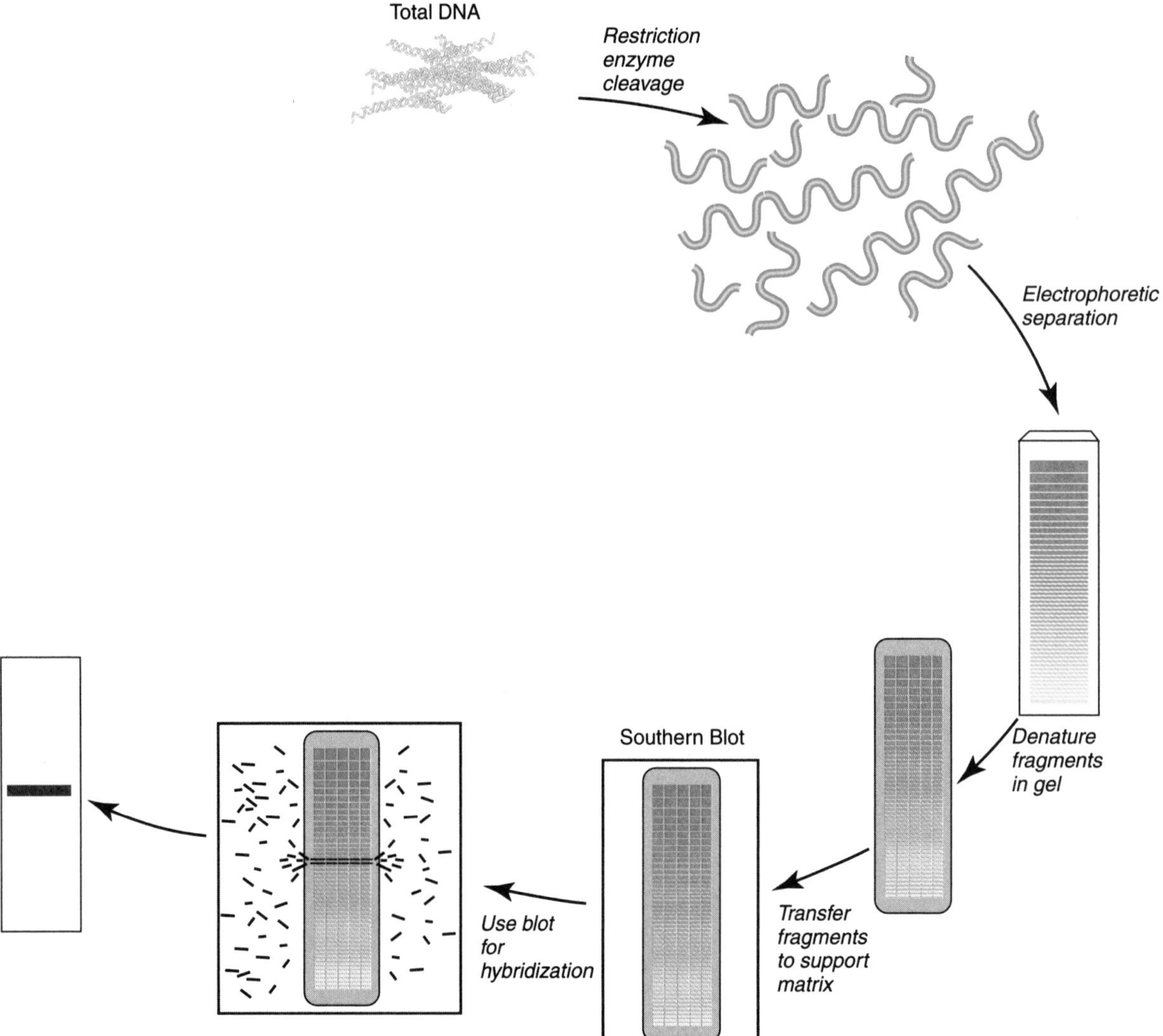

Figure 1.10 *Producing a Southern blot. High-molecular-weight DNA is cleaved into reasonable lengths by a restriction enzyme, and the fragments are separated by electrophoresis on a gel matrix. As shown in Figure 1.9, smaller fragments move ahead of larger ones. The final array of fragments can then be denatured and fixed to a specialized paper matrix where hybridization can detect the position(s) of a sequence(s) identical to or resembling the probe, depending upon the stringency (recall Figure 1.8).*

genetic studies to move rapidly and will undoubtedly remain the basis for considerable clinical information and testing in the future.

Because PCR depends on amplifying the DNA between defined primers, the sequences of these primers can be used to define a unique DNA region. Thus, by knowing the sequences of the PCR primers flanking a gene or other DNA region of interest, any laboratory can study the same region simply by making new copies of those primers.

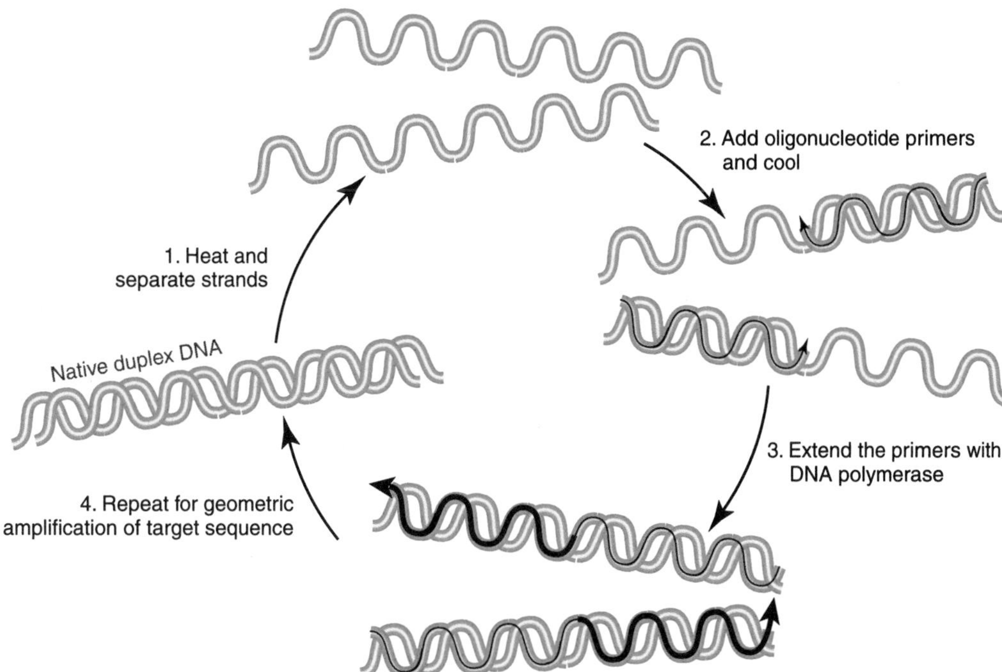

Figure 1.11 *The principle of PCR amplification. Individual strands of intact double-stranded DNA (left) are separated by heat (melting). Short single-stranded primers are allowed to hybridize with the separated strands and are then extended by DNA polymerase to synthesize complete complementary strands. The process is then repeated to produce the desired amount of DNA. In principle, this approach can begin with a single DNA molecule. There are numerous variations on this basic PCR scheme, but all are based on the same underlying principles.*

Current compendia of genes and sequences include flanking PCR primer sequences to facilitate this; a site thus identified has become known as a "sequence tagged site," or "STS".

The actual base sequence of a given region of DNA is the ultimate basis for individuality. DNA sequencing has matured considerably in both speed and fidelity and now is entering a stage of clinical usefulness. When combined with PCR amplification and the other tools described above, direct DNA sequencing permits the gathering of a great deal of information.

DNA sequencing is frequently done by sequencing machines. This technique is based on extending a primer using a mixture of normal (deoxy) and dideoxy nucleotide triphosphates. When the latter are incorporated, chain elongation ends. Adjusting the ratios of deoxy to dideoxy nucleotides produces an array of products representing termination at all possible positions, leading to a sequence "ladder" after chains are separated according to size. Automated sequencers are capable of reading about 500 bp in a single reaction, and both the sequencing reactions and the data output can be automated. Figure

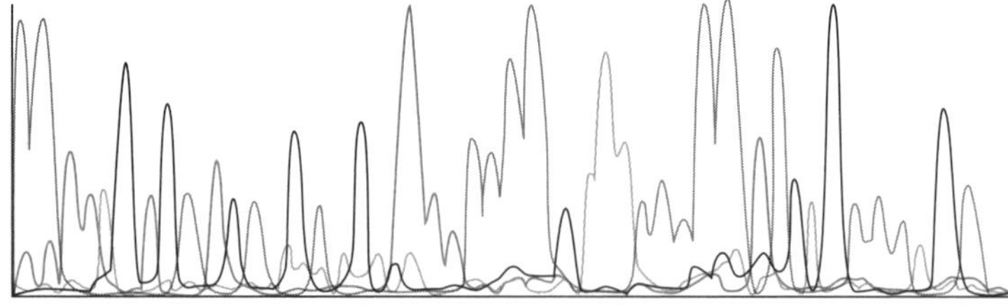

Figure 1.12 *Output from a DNA sequencer. Each of the four bases is tagged with a distinguishable dye, and the positions of successive bases are determined by spectroscopy. When the scans at the wavelength of each dye are superimposed, the sequence can be determined by comparing peak heights. Such sequence information can be stored directly for computer analysis.*

1.12 gives an example of sequencing data as derived from an automated sequencer. It presents a linear array of the sequence ladder, so that each nucleotide gives a characteristic, linearly ordered peak, permitting direct reading of the sequence. As will be discussed below, the efficiency of this approach is being increased in order to generate extensive stretches of human DNA sequence. In theory and in practice, sequencing permits the analysis of any genetic information. As of this writing, several viruses and bacteria have had their complete DNA sequences determined. There also are long stretches of human DNA for which the sequence is known (for example, mitochondrial DNA; see Chapter 8), and it is a reasonable goal and expectation that the entire human DNA sequence will be available in several years.

It is important to remember that the management of DNA sequence information is complex. It requires considerable data storage and refined search and analytical techniques. The computer methods and algorithms for storing and searching large DNA sequences are currently being perfected and will be essential for clarifying and making more useful the emerging array of sequence data.

C. MUTATIONS

Mutations represent differences in DNA organization or sequence in an individual with respect to some standard sequence. Many differences lead to observable amino acid changes in proteins, but some do not. Nevertheless, identifying different mutations has proved useful both for diagnostic studies and for determining the locations of genes and their alterations.

The simplest mutations represent local DNA base changes. These can include the substitution of one purine for another (A for G or G for A) or one pyrimidine for another (T for C or C for T); these are called transitions. Alternatively, mutations may exchange a pyrimidine for a purine or vice versa (C for A, T for G, etc.); these are called transversions. Such changes may lead to a change in the protein derived from that DNA sequence because one amino acid's codon is substituted for another's. Sickle cell anemia (OMIM #141900) is an example of a single base change that causes a single amino acid change (see Chapter 6). On the other hand, much of the coding information in DNA is said to be redundant, because there often are multiple triplet codes for the

Table 1-2

Three-letter codons

First position (5' end)	Second position U	C	A	G	Third position (3' end)
U	Phe	Ser	Try	Cys	U
	Phe	Ser	Tyr	Cys	C
	Leu	Ser	Term	Term	A
	Leu	Ser	Term	Trp	G
C	Leu	Pro	His	Arg	U
	Leu	Pro	His	Arg	C
	Leu	Pro	GluN	Arg	A
	Leu	Pro	GluN	Arg	G
A	Ileu	Thr	AspN	Ser	U
	Ileu	Thr	AspN	Ser	C
	Ileu	Thr	Lys	Arg	A
	Meth	Thr	Lys	Arg	G
G	Val	Ala	Asp	Gly	U
	Val	Ala	Asp	Gly	C
	Val	Ala	Glu	Gly	A
	Val	Ala	Glu	Gly	G

NOTE: U (for uracil) is used in place of T (thymine) because this is based on RNA sequence.

same amino acid (see Table 1.2). This situation can permit a base change to be invisible at the level of the protein because the "mutation" led to another codon for the same amino acid. Another consequence of base changes can be formation of a triplet that signals a stop to protein synthesis (see Table 1.2). Such a stop or premature termination results in a shorter protein, very frequently with aberrant properties. Still another result can be the substitution of a similar amino acid (e.g., alanine for valine). As described earlier, many such minor variations cause no meaningful changes in the protein or problems for the organism and are sometimes referred to as "conservative" changes.

Another possible DNA alteration is the loss of one or more bases. Deletions can cause serious problems. The loss of three contiguous bases can either lead to the loss of a single amino acid codon (as occurs in the most common mutation for cystic fibrosis [OMIM #219700]; see Chapter 6) or affect two contiguous codons (see Figure 1.13A,B). Nevertheless, with the loss of three bases in a row (or any multiple of three), the reading frame of the gene remains intact. The loss of different numbers of bases destroys the triplet reading frame and leads to complete aberrancy in the protein produced (Figure 1.13C). Deletions also can occur on a larger scale, such that entire DNA regions can be lost. In some situations these losses are large enough to be seen as chromosomal changes (see below); in other situations they are recognizable only with DNA studies. Large deletions generally have significant biological effects.

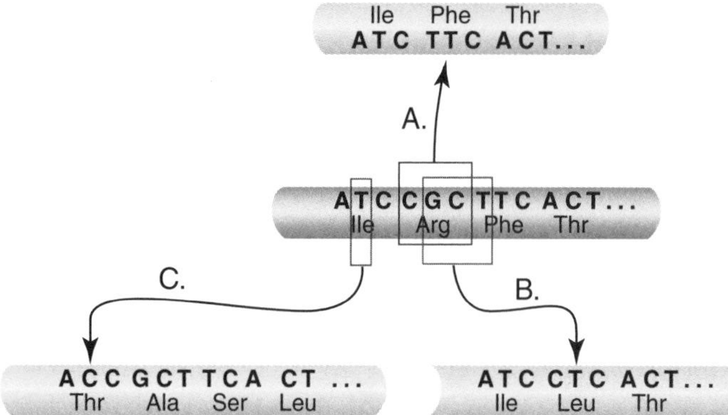

Figure 1.13 *The original DNA sequence with its corresponding encoded amino acid sequence is shown in the center. Deleting three contiguous bases within the reading frame (A) neatly removes a single amino acid from the coding region. In contrast, deleting three contiguous bases outside of the reading frame (B) leads to the insertion of a novel amino acid in place of the original two, but the correct reading frame is reestablished. Deleting a single base (C) puts the reading frame completely out of alignment and directs the polymerization of an incorrect chain of amino acids that continues until a stop signal is encountered. It is valuable to consider the consequences of the opposite situation of adding instead of deleting bases.*

The opposite of a deletion is an insertion. In this case, one or more bases are added to the DNA strand. The consequences are predictable based on the same reasoning presented above for deletions. Current sequence studies indicate that the rate of local differences in DNA sequence (polymorphisms) between individuals is about one variation in 500 bases; about 15% of these variations are insertions or deletions.

Local regions of DNA also can be duplicated. This may occur as part of the replication process for DNA or may be a result of errors in recombination (see below). Duplications may add a region of amino acid sequence to a protein or may cause changes in the reading frame and/or a termination, as discussed above.

The division of most mammalian genes into introns and exons means that splicing is required to achieve a usable transcript, as discussed earlier. Errors in splicing have important effects on gene and protein structure. Because intron sequences are not coding sequences, their retention in messenger RNA causes an aberrant gene product. The signals for splicing are found in the base sequences at the junctions between introns and exons. Thus, mutations in these regions can cause absent or incorrect splicing, possibly with retention of the intron. Another change can involve the mutation of a base sequence distant from the normal splicing site into a new splicing site; this also causes large problems for the fidelity of gene transcription by changing the sequences in the final spliced product.

The movement of a small or large piece of DNA from one position to another, a process referred to as transposition, is another source of DNA variation. In bacteria and many less complicated organisms, transposition of DNA sequences from one position to another occurs relatively readily. Although this is less frequent in mammals, it has been recognized as a source of mutation in humans. The movement of DNA sequences can have important effects on the sequence that is moved. It may have lost its appropriate controlling elements or may be only a fragment of the mature gene. Such a movement also may affect the region into which the sequence is moved—the movement of a stretch of foreign DNA into a structural gene may disrupt that gene by inserting anomalous information or introducing inappropriate control sequences.

As mentioned earlier, having repeated DNA base sequences establishes a propensity for variation in their number. This now has been documented in several important instances and is being recognized more frequently. The process of amplification (which may be considered an extension of the process of duplication) can result in very large increases in the number of small repeated regions of DNA. Two types of sequences are particularly recognized as subject to amplification. As described earlier, the so-called dinucleotide repeat is typified by $(CA)_n$ repeats. With 50,000 of these in noncoding regions of the human genome, individual variations in their length turn out to be useful position markers for mapping (see below) but do not generally cause diseases because they are not in coding regions.

In contrast, the category of "triplet repeat" disease is medically significant. In these conditions, amplification of adjacent groups of three base pairs may lead to aberrant gene products, aberrant gene control, or other pathological effects. With triplet repeats, the problems usually develop from an increase of the repeat number over a baseline (or threshold) level. Most clinically unaffected individuals have a relatively low number of repeats. Rarely, individuals have a higher repeat number, creating an unstable situation. Beyond this unstable intermediate level, further amplification leads to overt disease; thus, the intermediate level can be considered a "pre-mutation." This is an important category of genetic illness and will be considered in more detail below (see Chapters 5 and 7). Triplet repeat disorders frequently have neurological manifestations.

D. CHROMOSOMES

With the exception of mitochondrial DNA, all cellular DNA is found in chromosomes in the nucleus. The DNA in mitochondria is a double-stranded circular molecule and will be discussed in more detail in Chapter 8. Chromosomes are complex arrays of both the linear DNA polymer and its associated proteins. The length of DNA packed into individual chromosomes far exceeds the linear length of the chromosomes. If it were fully extended, the DNA of all the chromosomes in a single cell would be 1.5 to 2 m long! This obviously

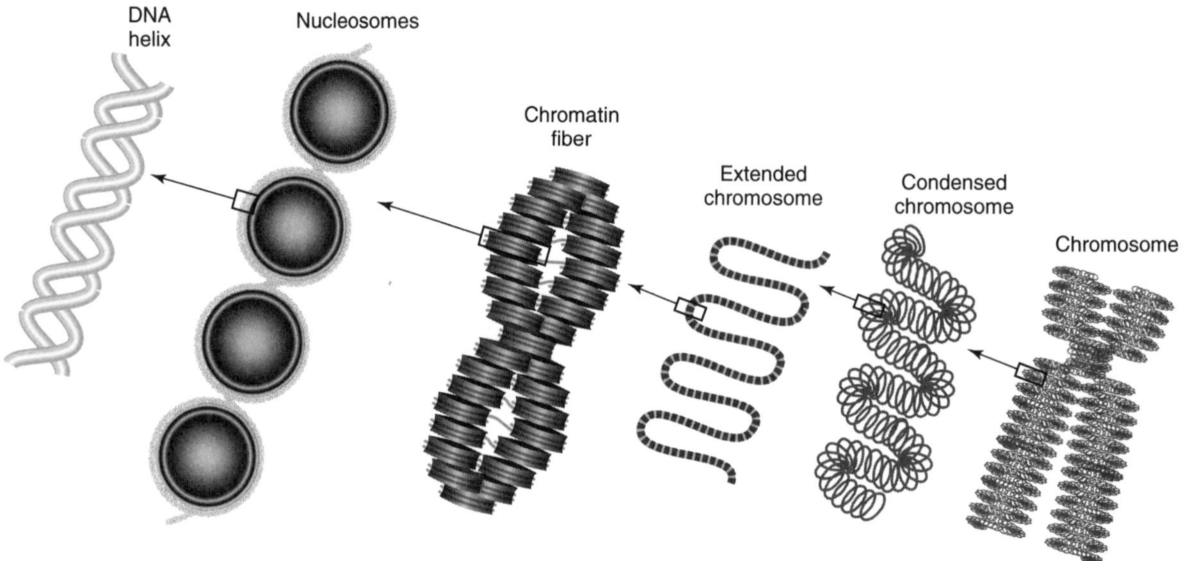

Figure 1.14 *Progressive compaction of DNA into chromosomes moves through several identifiable levels of organization. The nucleosomes are the first level, followed by folding into chromatin fibers and progressive condensation into chromosomes. All details have not been established, but regional changes in structure are often related to individual gene expression.*

implies substantial physical reorganization of the DNA when it is packed into chromosomes. Not all details of the packing are yet understood, but it is clear that DNA associates with the histone proteins as a first step. Wrapping DNA around histone proteins, approximately 210 bp per histone cluster, forms a "nucleosome" and serves as the basis for the next phase of DNA packing. As shown in Figure 1.14, additional folding and compaction steps are needed before the DNA can be fully packed into a chromosome. The chromosome has defined ends (telomeres) that represent repeated units of short hexanucleotide sequences and, at the very least, protect the free ends of chromosomes. Each chromosome also has specialized sequences and domains in the region of centromeres; these permit chromosome movement mediated by microtubules during cell division.

1. LONG-RANGE GENE RELATIONSHIPS AND GENE MAP

Because chromosomes contain such long lengths of DNA, they necessarily also contain large numbers of genes. The linear relationships among genes on the DNA also are related to the linear relationships of these same genes when packed into chromosomes. However, because the packing of DNA is still not completely understood and also because its density likely varies at a local level, the length relationships between DNA position and chromosome position for individual genes vary.

Despite these local variations, the relative positions of individual genes have been established to develop a linear gene map. For convenience, the map is divided into sections for individual chromosomes. This has facilitated studies and also has divided the amount of information that needs to be managed into smaller parcels. Establishing positions of specific, reliable markers along the length of chromosomal DNA has led to the current gene map. Dividing unmanageable lengths of DNA into shorter segments also is proving essential to solving problems of DNA sequencing and analysis, as discussed earlier. The obvious long-term goal of such studies is to be able to integrate the DNA sequence, the physical position of a gene or a marker on the chromosome, and the identity of genes throughout the length of all individual chromosomes; this is now a feasible vision.

The human gene map is growing rapidly. It consists of a series of fixed positions in a linear order. The positions are sites of defined DNA sequences, which may be of several types. The most obvious sequences are those of individual genes, including exons and introns, as well as control information. Another, sometimes more useful, type of positional genetic marker sequence comprises short, repeated sequences (such as the $[CA]_n$ repeats discussed earlier) bounded by unique DNA sequences. Establishing the relative positions of these repeated sequences along chromosomes is an important part of developing the current gene map. The gene map serves as the framework for integrating DNA sequence information as it becomes available. The map is essential for linkage studies and other clinical uses, as discussed below.

2. PHYSICAL STUDY TECHNIQUES

Because there are regional variations of DNA, packing, and associated proteins along the length of individual chromosomes, it is possible to distinguish chromosomes from one another. There are two particularly useful techniques for this approach, and they differ both in theory and in practice. The first approach, the elder, relies on the observation that different chemical stains have greater or less avidity for different chromosomal regions. This is related to combinations of local base sequences, histones and other proteins, and probably local packing organization. This variation in avidity for stains leads to the appearance of "bands" when stained chromosomes are examined microscopically. For a given chromosome, the pattern of bands is consistent; this has provided both an identification for individual chromosomes and a rough set of physical measurements for defining different regions. The current standard chromosome banding pattern, known as an idiogram, is shown in Figure 1.15. Examining chromosomes in this way requires growing cells in vitro and then arresting cell division with colchicine. The cells are then fixed to a microscope slide, where the chromosomes can be stained and examined. Chromosomes also can be counted and compared to identify variation(s). The array of stained individual chromosomes is referred to as a "karyotype." Because of the cell culture and laboratory techniques involved, the process of preparing a conventional karyotype can require several weeks.

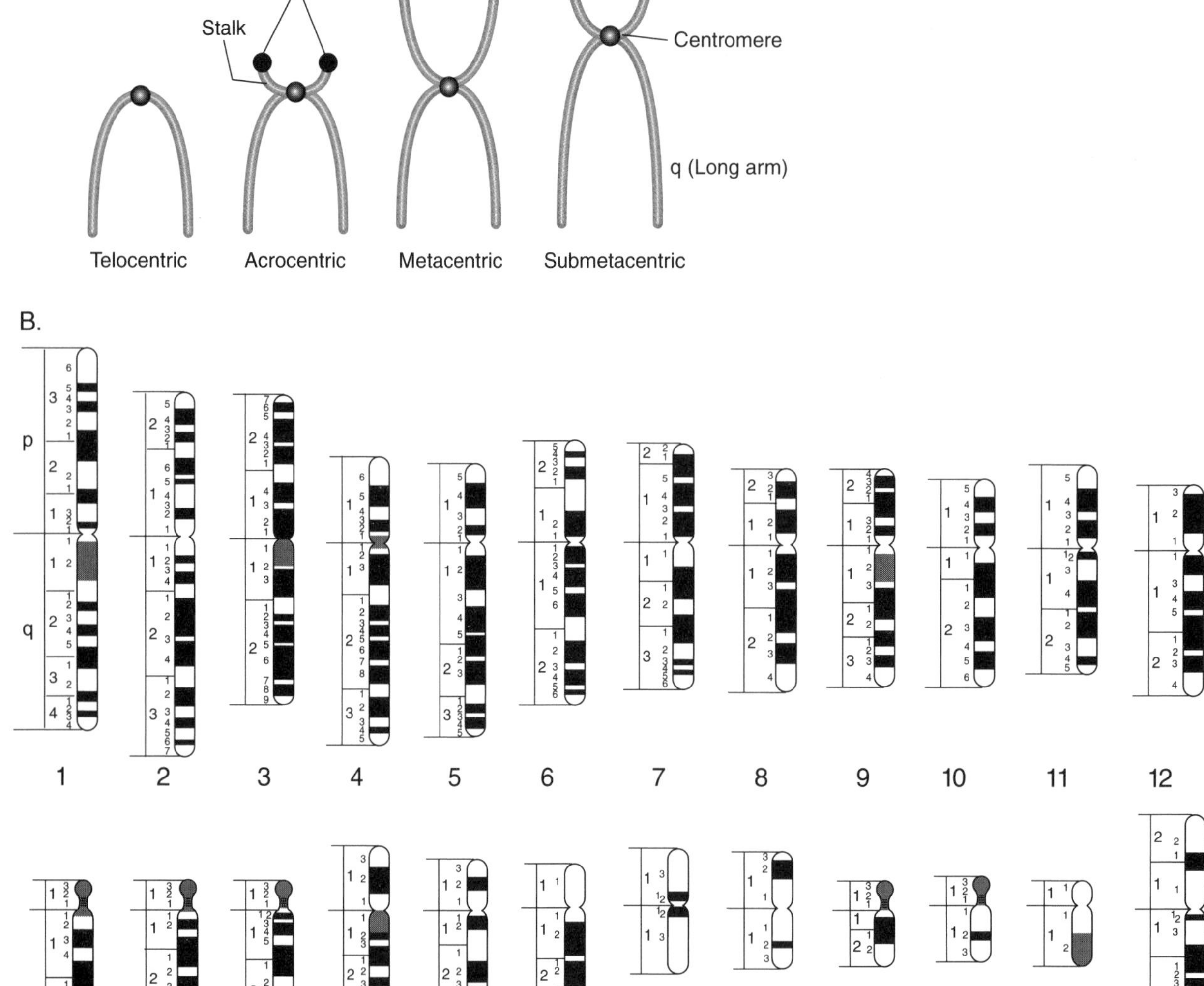

Figure 1.15 *(A) Basic structural features of chromosomes. (B) Idiogram showing current standard chromosome banding pattern using Q-, G-, and R-staining methods. Constrictions indicate centromere locations. The bands are reproducible for each chromosome, permitting their numbering as shown. In numerical designations, bands on the shorter of two arms are designated "p" (for "petite"); those on the longer arm are designated "q." Numbering can then move to major regions and identifiable subbands in those regions. Thus, the second subband in the second band region of the long arm of chromosome 6 would be designated 6q22 (International System for Human Cytogenetic Nomenclature, 1995).*

A contrasting and more recent development for chromosome studies has resulted from the joining of DNA hybridization techniques, as discussed earlier, with fluorescence microscopy. A given test DNA sequence or "probe" is labeled with a fluorescent molecular dye tag. This probe can then be hybridized to intact chromosomes. Under appropriate conditions, the probe will hybridize to a unique chromosomal location that can then be detected using fluorescence microscopy. An example of this is shown in Figure 1.16. Such hybridization is done in situ on preparations of entire chromosomes. The process is known as "fluorescence in situ hybridization" (FISH). Using FISH, it is possible to identify specific regions of chromosomes, to assign specific DNA sequences to physical locations, and to search for chromosomal rearrangements, which are large-scale reflections of local DNA rearrangements.

In addition to studies of single gene locations using FISH, it is possible to assemble a collection of molecular probes spanning an entire chromosome using information from the gene map, as discussed above. When all of these are labeled and hybridized to a chromosome preparation, fluorescence microscopy quickly identifies a specific chromosome without the need for banding or other studies. This has been called "chromosome painting." Combined with FISH, painting techniques hold considerable promise for rapid chromosome analysis, detailed study of chromosomal anomalies, chromosome counting, and, most likely, the application of computer technology to interpreting chromosome patterns. More detailed examples of FISH and painting will be presented in Chapter 2.

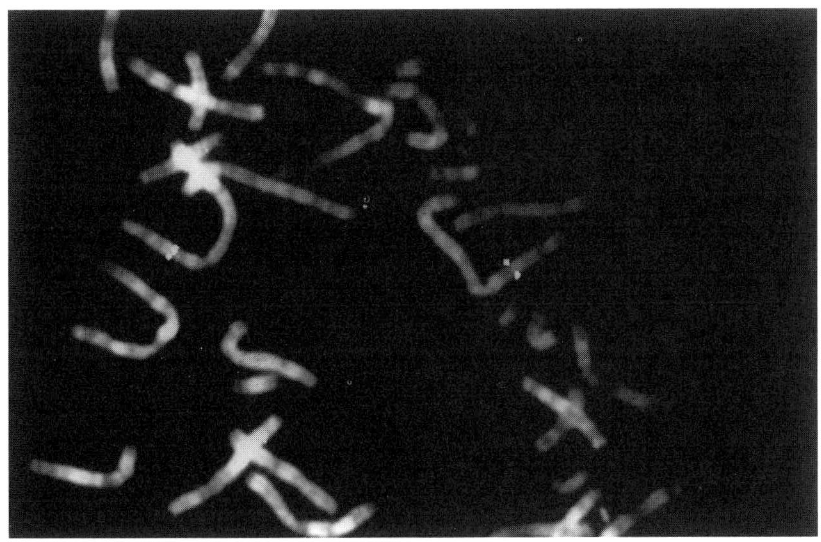

Figure 1.16 *Identifying single genes by FISH hybridization. (Only a portion of the field is shown.) Note that two sites are identified, because this is an autosomal gene mapping to 2p15 and both parental contributions appear. (Photograph courtesy of Dr. Gail Stetten.)*

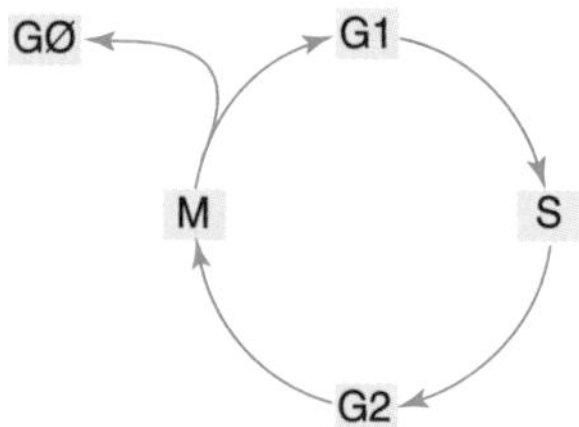

Figure 1.17 *A simplifed view of the cell cycle. Transitions between parts of this cycle are critically dependent on completion of previous steps, as explained in the text.*

3. MITOSIS

During the course of its growth, the single fertilized cell must undergo many divisions. Divisions continue until all the cells of the mature individual have been formed; this obviously requires considerable coordination. For cell division to be successful, there must be duplication of all essential cell constituents. Of particular interest in terms of genetics, inheritance, and mutation are the process and fidelity of duplication of the cell's DNA and chromosomes. For regular cell division, the process of duplicating chromosomes is called "mitosis." Mitosis is carefully regulated and occurs during a relatively brief but coordinated time. Mitosis is one of the critical stages in the "cell cycle," a diagram of which is shown in Figure 1.17. This diagram specifically refers to the events occurring within the cell nucleus. Most of the lifetime of most cells is spent in the G1 phase. During this phase, the cells can produce their specialized proteins and accomplish most of their essential functions.

When the signal is received for cell division, the cell enters the S phase, during which the DNA in all chromosomes is duplicated (see Figure 1.18). This duplication copies all of the DNA in each chromosome, so that each of the DNA strands can serve as a template, using the base pairing described earlier, for the synthesis of a corresponding new ("daughter") strand. At the end of the S phase, each of the chromosomes has been duplicated but they remain physically attached to each other through protein interactions at the centromere. The time delay called "G2" then separates the events of the actual separation of individual chromosomes from their duplication. During the M, or mitosis, phase, the two members of each pair of chromosomes go to opposite ends of the cell, moved by microtubules attached through the kinetochore and centromere. This separates 46 chromosomes into two clusters on each side of the original cell. Finally, a cleavage furrow develops, separating the original cell into two daughter cells. Each of these cells contains a copy of all of the genetic information of the parent.

4. MEIOSIS

While mitosis is used for cell duplication throughout the body in the so-called somatic tissues, one group of cells—"germ" cells—uses a different method for cell division, called "meiosis." The differences between meiosis and mitosis have important consequences. Meiosis, the process of germ cell division, also has been called "reduction division." This means that while the normal number of human chromosomes is 46 per cell, the final number of chromosomes in each germ cell is 23. Thus, the chromosome number has been reduced by one-half as a result of meiosis. This reduction permits the joining of male and female germ cells during fertilization to reconstitute an individual

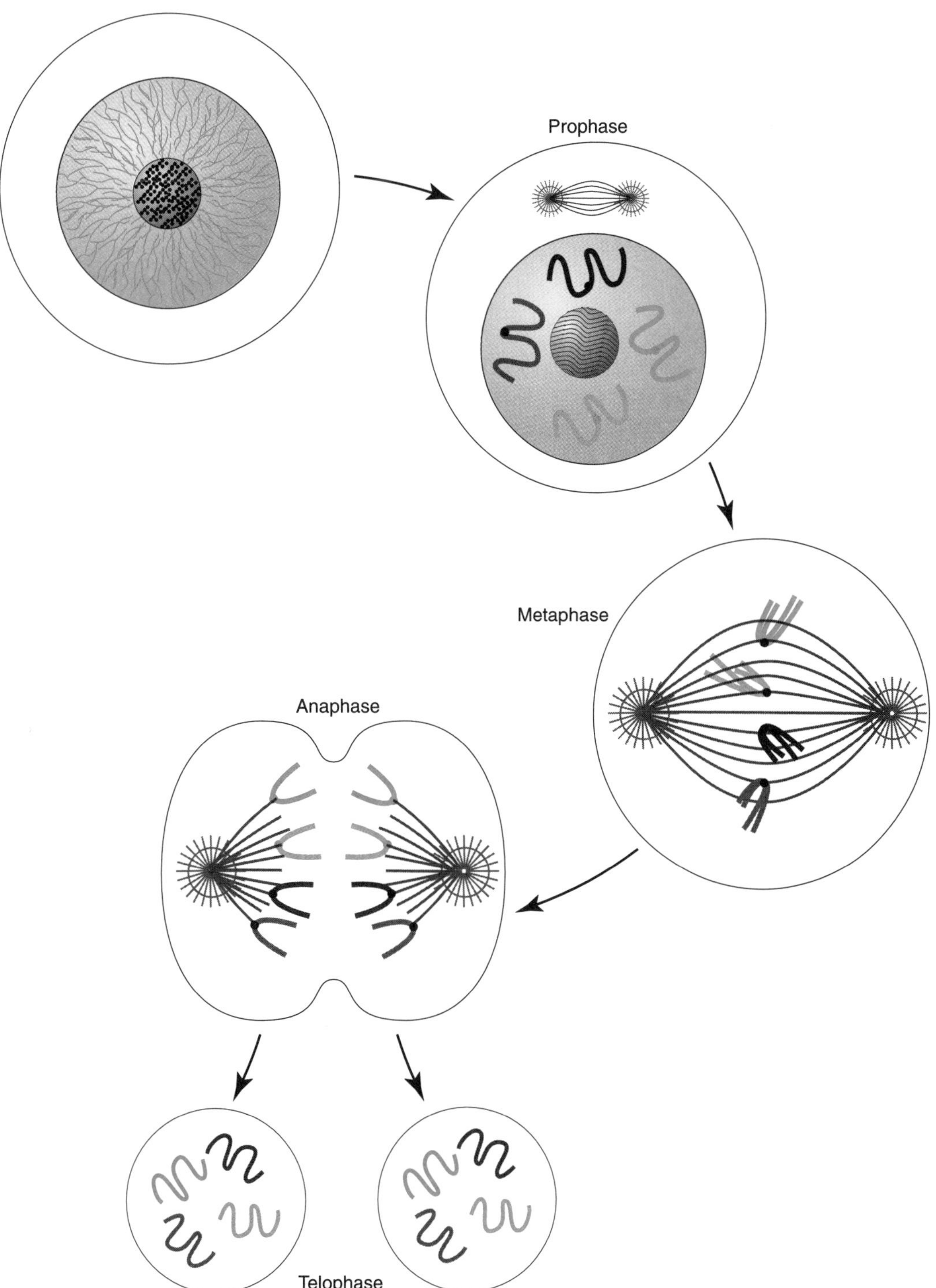

Figure 1.18 *Stages of mitosis. Only four chromosomes are shown for clarity. The phases are described in the text.*

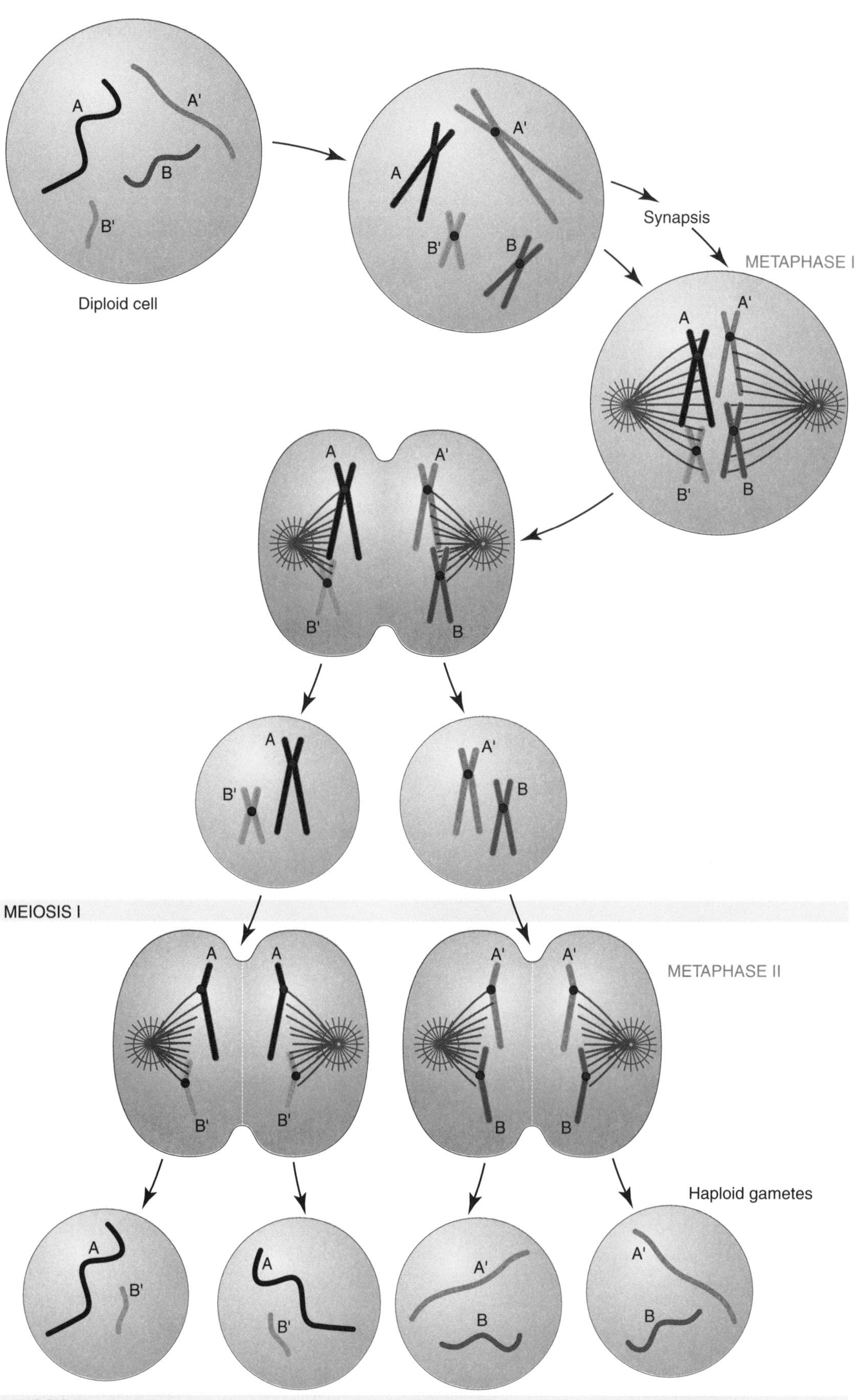

A
A'
B
B'
Diploid cell
A
A'
B'
B
Synapsis
METAPHASE I
A
A'
B'
B
A
A'
B'
B
A
B'
A'
B
MEIOSIS I
A
A
B'
B'
A'
A'
B
B
METAPHASE II
Haploid gametes
A
B'
A
B'
A'
B
A'
B
MEIOSIS II

cell with 46 chromosomes. The other important difference between meiosis and mitosis is that the goal of mitosis is to preserve the linear integrity of each chromosome. Meiosis, on the other hand, permits the exchange of genetic information between pairs of related chromosomes, a process referred to as crossing-over. This physical exchange of genetic information is essential to the reassortment of genes prior to their transmission to new individuals. We will consider the implications of that below.

Meiosis can be considered two processes—meiosis I and meiosis II (see Figure 1.19). Meiosis I begins very much like mitosis: Each individual chromosome serves as the template for synthesizing two sister chromatids. At this point in mitosis, the sister chromatids would separate to opposite poles of the cell. In meiosis I, in contrast, the newly replicated sister chromatids contact one another in a precise pairing fashion referred to as "synapsis." During synapsis, physical exchange between individual chromatids is possible. This process, referred to as "crossing-over," permits the exchange of regions of chromosomes between sister chromatids. In this way, a chromosomal segment with its attendant genes, markers, and possibly mutations originally received from one parent can become joined to an adjacent chromosomal segment received from the other parent. The process of crossing-over during meiosis I is essential to establishing linkage relationships, which will be discussed below. As a result of crossing-over, each chromatid can become a mosaic of segments derived from the two parents. Figure 1.20 shows an example of staining of different segments of chromatids following this process of sister chromatid exchange.

Meiosis II follows meiosis I and allows the centromeres of each chromosome pair to move apart, permitting separation of the sister chromatids. This process, also referred to as "disjunction," results in the production of four final cells, each containing 23 chromosomes. While under normal circumstances these chromosomes are identical in gene complement and order to those of the parent cell, they nevertheless can contain various regions of paternal and maternal genetic information because of crossing-over. The only exception to this is that germ cells produced by a male will be of two types—one type has 22 autosomes and an X-chromosome, the other type has 22 autosomes and a Y-chromosome.

In males, meiosis occurs in the process of forming sperm (see Figure 1.21). The primary source of sperm—the spermatogonium—leads to cells that ultimately undergo both first and second meiotic divisions to form first spermatids and, after maturation, mature sperm. Each sperm contains 23 chromosomes but has had a remarkable reduction in cytoplasm. A large number of sperm are produced (approximately 2×10^8 per ejaculate). Sperm are produced continually throughout the life of the mature male. For this reason, sperm produced by older men have undergone many more divisions of the spermatogonium prior to meio-

Figure 1.19 *Stages of meiosis. Only four chromosomes and no sister chromatid exchanges are shown for clarity. Stages are described in the text.*

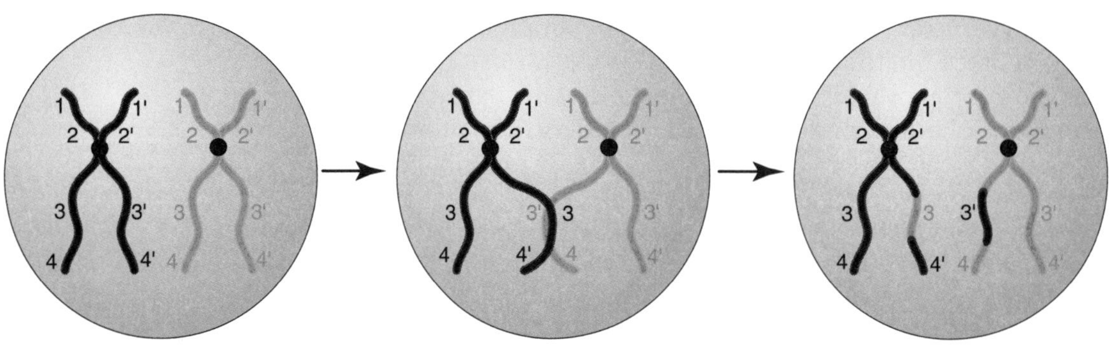

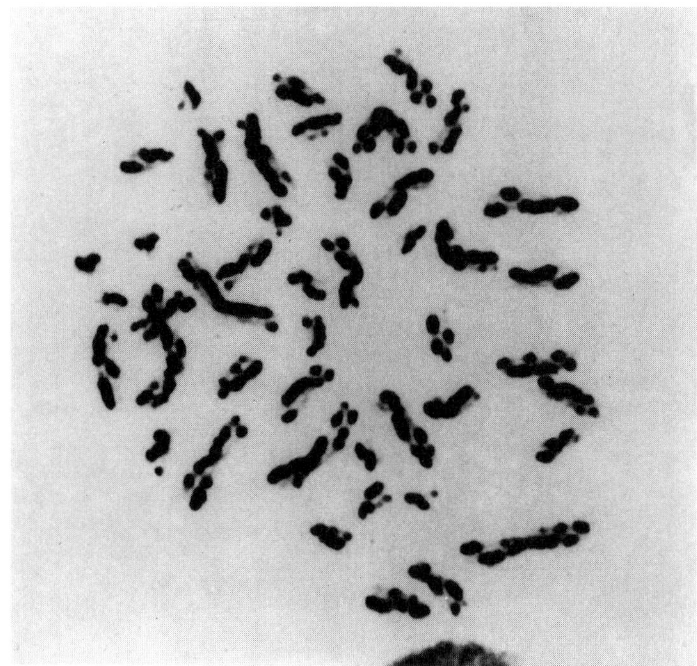

Figure 1.20 *Sister chromatid exchange. (A) A simplified diagram showing a single event on one arm. The physical point of exchange is known as a "chiasma" (pl, chiasmata). (B) Photograph showing human lymphocytes that have undergone two cycles of cell division in the presence of bromouracil deoxyribonucleotide. The lymphocytes have been stained with the fluorescent dye 33258 Hoechst and examined by fluorescence microscopy. Note alternate chromatid arm staining, giving evidence of past meiotic exchanges. (Photograph courtesy of Dr. Gail Stetten.)*

sis. As will be considered later, this presents the possibility of introducing additional mutations.

Meiosis in females differs from that in males in several ways (see Figure 1.22). However, most of the differences are related to the timing of the events and not to the meiotic process itself. Obviously, because female cells contain two X-chromosomes, all of the meiotic products

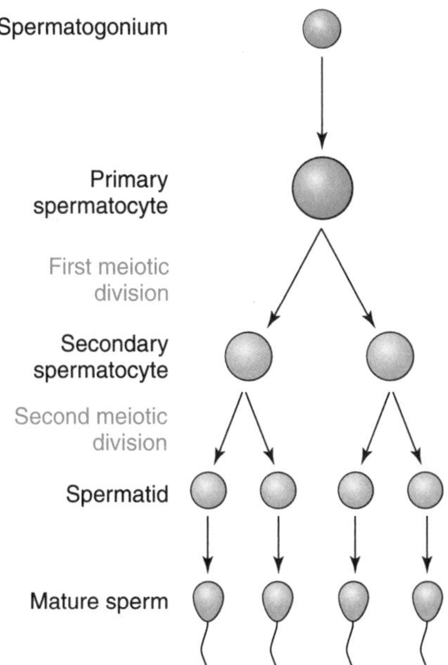

Figure 1.21 *Sperm formation. Note the positions of the first and second meiotic divisions. Formation of mature sperm is a continuous process in the adult male.*

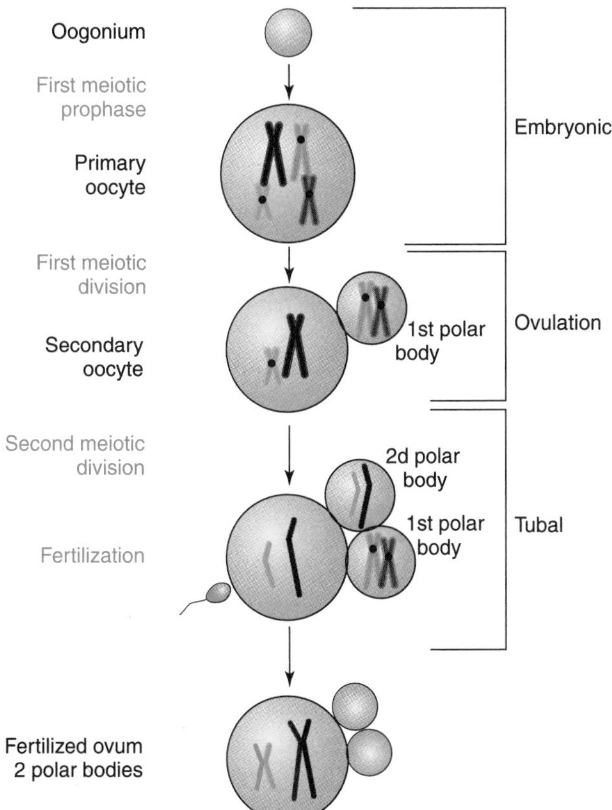

Figure 1.22 *Oocyte formation. Note that the second meiotic division occurs very late in the maturation process. Meiosis in these cells is strikingly asymmetrical with respect to the cytoplasm.*

will have 22 autosomes and a single X-chromosome. During embryonic life, the oogonia begin to develop into primary oocytes. These gradually enter meiosis but remain in the first meiotic prophase until ovulation; this may require 40 or more years. After this interval, an individual maturing oocyte is stimulated to complete meiosis I. However, meiosis I is asymmetrical, because, although the chromosomes are divided, one cell contains most of the cytoplasm. This asymmetry results in the formation of both a "secondary oocyte" and a "polar body." The second meiotic stage proceeds after ovulation, usually during passage of the ovum into the uterus. Meiosis II is not completed until fertilization occurs, at which time a second polar body is formed, containing a tiny amount of cytoplasm and one meiotic chromosome complement. In general, the polar bodies are discarded and the haploid 22-X ovum is joined with the haploid 22-X or 22-Y sperm to form a diploid fertilized ovum. From this point, mitosis continues as described earlier.

Because the commitment of female germ line cells to gamete production is completed prior to birth, females are born with a limited number of germ cells. Not all of these germ cells survive until puberty, but it is likely that about 10^5 cells are still viable at the onset of puberty. Even these will not all survive into ovulation. This long quiescent interval means that oocytes may remain for 40 or more years before completing meiosis. This also has implications for chromosome segregation (see below, particularly Chapter 3).

5. MITOCHONDRIAL DNA

As alluded to earlier, there is one other location of DNA in human cells. The mitochondria contain a double-stranded circular DNA molecule. This is replicated within each mitochondrion and transmitted to its progeny as the mitochondria divide. The mitochondrial DNA is limited in length, comprising only 16,569 bp; its entire sequence is known. Because mitochondrial DNA is relatively short, it encodes only a few proteins. These are important in mitochondrial function, however, and are not represented in nuclear DNA. The genetic information from mitochondria is transcribed into RNA, which is then transported to the cytoplasm where the proteins are synthesized. These proteins must then be transported back into the mitochondria.

While in meiosis and mitosis have similar final effects in males and females and transmit an identical number of chromosomes from both parents to their progeny, the genetic transmission of mitochondria and mitochondrial DNA genetic information is different. As noted above, sperm have lost most of their cytoplasm and the part of the sperm that enters the ovum on fertilization rarely contains mitochondria. Thus, the mitochondria of a fertilized egg are usually derived exclusively from the ovum. This establishes the unique pattern of mitochondrial inheritance that will be discussed in detail below (see Chapter 8), but the general implication is that the transmission of mitochondrial DNA and of mitochondrial genetic traits occurs through the female.

6. LINKAGE

On the basis of the discussion of physical crossing-over during meiosis I (see above), it is clear that during the production of germ cells there is physical exchange of maternal and paternal genetic contributions between individual chromatids. This exchange necessarily separates genes in chromosomal regions that were contiguous in each parent and, by mixing them with retained linear order, results in "recombinants." The process of forming recombinants through meiotic crossing-over is an essential feature in the reassortment of genetic traits and is central to understanding the transmission of genes and the development of a useful gene map.

Recombination generally occurs between large segments of DNA. This means that contiguous stretches of DNA and genes are likely to be moved together. Conversely, regions of the DNA that are far apart on a given chromosome are likely to become separated during the process of crossing-over. Because each pair of sister chromatids undergoes at least one physical exchange or recombination during crossing-over, there is likely to be some independent movement of maternally and paternally derived sequences. These exchanges can be seen microscopically as chiasmata (see Figure 1.20A). Sometimes chiasmata can be counted during the study of karyotypes, but it is difficult to find large numbers of meiotic cells in normal human tissues. Nevertheless, this means that the chance that any two markers on a chromosome will remain together through meiosis is 50% at most.

Based on the techniques discussed above, it is becoming possible to use molecular markers to clarify the recombination events that take place during meiosis. As discussed earlier, many small, noncoding DNA variations are present along the chromosomes of all individuals. This frequently occurs in situations such as the presence of $(CA)_n$ repeats of different lengths. Because these repeats are dispersed throughout human DNA and because, presumably, there is little selective pressure on their length, they can be used as position markers and as regional identifying characters along chromosomes. When markers are sufficiently different (usually in length due to variation in the number of repeats "*n*"), they can be used to distinguish paternally derived from maternally derived gene regions. This implies that an individual can show DNA marker patterns along autosomal chromosomes that represent regions derived from maternal and paternal counterparts. Because this is the case, it becomes possible to analyze the transmission of these markers to offspring.

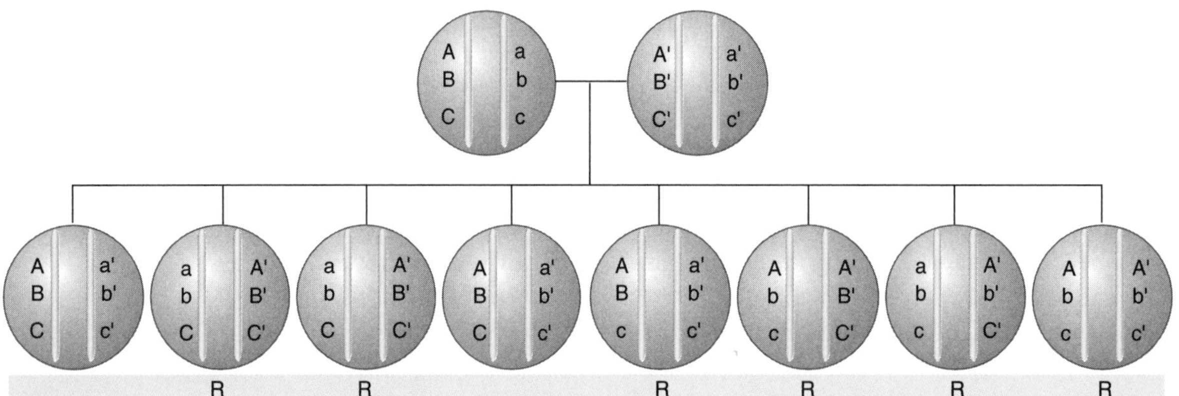

Figure 1.23 *Comparison of the haplotypes of eight siblings with those of their parents identifies those with recombinant chromosomes (R). Such identifications can be made only with a sufficient number of variable (i.e., polymorphic) markers of known position. In this idealized case, the phase relationships for the parents were known (see also Figure 1.24).*

The pattern of a set of markers along a chromosome is referred to as a "haplotype." By analyzing the haplotypes in a series of offspring of parents whose haplotypes are known, it is possible to establish which parental segment of which chromosome was transmitted to which child. This is shown for an idealized kindred in Figure 1.23. With a sufficiently dense array of chromosome markers, it becomes possible to trace any chosen individual section of parental chromosomes through subsequent generations. The result of such studies establishes the transmission pattern of discrete regions of individual human chromosomes (hence, specific genes) through a kindred. Once the pattern of transmission is established for a group of markers, it becomes possible to use their distributions to estimate the relative distance between individual marker segments. These distances are called "recombination distances"; the likelihood of their separation is the "recombination frequency."

Because, as noted, relatively large segments of chromosomes are involved in meiotic exchange, markers close together on a given chromosome are more likely to be transmitted together (not separated by recombination) through meiosis than those that are far apart. This is the intuitive basis for linkage analysis. "Linkage" refers to the likelihood of having one marker transmitted with another through meiosis. Markers that are transmitted together frequently are said to be "closely linked," while those that are transmitted together infrequently, or between which recombination is very likely to occur, are "loosely linked." Because, as noted above, at least one crossing-over occurs between sister chromatids for each chromosome, the maximum frequency with which any two markers can be transmitted together is 50% per generation. All of the markers on the same chromosome are said to be "syntenic." It should be obvious, however, that not all syntenic markers, particularly those that are physically far apart on a chromosome, are likely to be closely linked.

It is possible to transform the physical and empirical observations of sister chromatid meiotic exchange into mathematical expressions. These turn out to be very useful. The calculations yield the likelihood, based on studies of large numbers of meiotic events, that any two given markers will be transmitted together during meiosis.

In the kindred shown in Figure 1.23, three genetic markers, all on the same chromosome, are shown. Two of the markers are close together; one is far apart from them. The maternal and paternal copies of each of these markers can be distinguished from one another. In this case, they can be distinguished by different electrophoretic gel mobilities due to different lengths of $(CA)_n$ repeats. It also is clear that several individuals received all three markers transmitted without recombination (these are said to be "nonrecombinant"), while others show a mixing of the patterns. This prototype kindred is shown as an example of using markers to determine relative chromosomal positions. It is obvious from examining the pattern that two of the markers were transmitted together frequently, while the third often becomes separated from them. This is a physical statement of the simple notion that two are closely linked and the third is nearly unlinked. Extending this sort of analysis to grouped markers in the same region can establish which are most consistently associated with one another and distinguish them from those that frequently are separated at meiosis. Because we know that all these markers were chosen from the same chromosome to begin with (i.e., they are syntenic), it is possible to establish a relative linear order for the markers based on the frequency with which they become separated or remain associated through several generations. This is the fundamental principle of linkage analysis.

The most useful approach to expressing these relationships mathematically is to use a ratio called the "likelihood ratio." The likelihood ratio is based on a "recombination fraction," conventionally expressed as θ. The calculations are expressed as the ratio of the likelihood that a given pattern of marker segregation would be seen if a certain recombination fraction is present—that is, if the markers are in fact linked—to the likelihood that the same pedigree pattern would be seen if the markers are not linked at all. Several computer programs can be used to calculate different recombination fractions (i.e., values of θ). These ratios are then expressed as their base 10 logarithms—the logarithm of the odds that the data represent a real relationship. Obviously, the higher this ratio, the more likely a presumptive marker position is to be correct. Because logarithms are used, the term for this calculation is the "lod score" ("*l*og of the *od*ds"). By convention, a lod score of 3 or higher is taken as evidence for significant linkage. Obviously, because of the logarithmic nature of the score, a lod score of 3 means that the likelihood is 1000 (10^3) to 1 that this is a legitimate linkage; the higher the score, the better the evidence.

Because the lod score is a logarithm, scores calculated on the basis of data from different kindreds can be summed. Thus, relatively small families can be studied and their data combined to establish a likelihood ratio. It often happens with small families or groups that data from an individual family will not be able to resolve a linkage question. However, combining data using the same set of linkage markers in several families that have the same clinical problem or novel genetic marker can frequently yield results that achieve statistical significance. Here it is essential that the problem be *identical* in each kindred, or combining the data would be meaningless or misleading.

An important requirement for linkage studies is that the phase relationship between the two segregating entities (markers, disease, phenotypes) be known. "Phase relationship" refers to whether the specific alleles are on the same (*cis*) or different (*trans*) chromosomes. Figure 1.24 shows this distinction. Coupling (or *cis*) places two alleles on the same chromosome (i.e., they are derived from the same parent). Repulsion (or *trans*) means the opposite. Obviously, the more variation(s) at a given locus (or marker), the easier it is to distinguish paternal from maternal contributions. (Such markers are said to have a high "polymorphism information content," or "pic," value.) This has made repeat sequences, such as $(CA)_n$, particularly helpful in these studies. Also, having data from multiple generations often helps identify phase relationships. When these calculations are made for different likelihoods of recombination (different values of θ), one can get a notion of the likelihood of recombination between any two segregating entities. The range of possible values of θ is from 0.0 (no recombination between the two markers) to 0.50 (random assortment). Whether one entity is a genetic DNA-based marker and the other a disease or whether they are two different markers, the calculations give an estimate of the maximum likelihood; this is the value of θ at which the lod score is greatest. The smaller this maximum likelihood value θ is, the more likely the two segregating entities are to be close to one another.

An example of the usefulness of this approach comes from a study of the transmission of Huntington chorea (OMIM #143100). This study took advantage of a large kindred from Venezuela in which transmission of the disorder was clearly established. The work began by looking for changes in a group of DNA markers that might accompany the presence of the disease. A useful DNA marker called "G8" showed

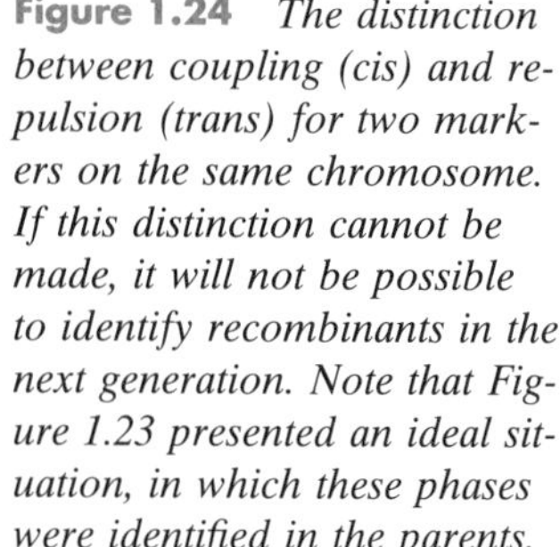

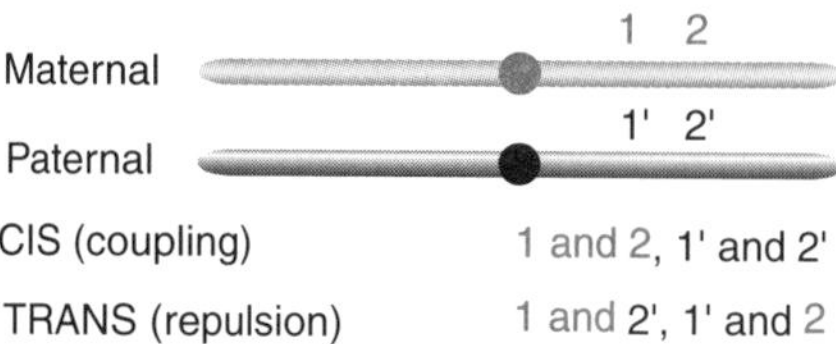

Figure 1.24 *The distinction between coupling (cis) and repulsion (trans) for two markers on the same chromosome. If this distinction cannot be made, it will not be possible to identify recombinants in the next generation. Note that Figure 1.23 presented an ideal situation, in which these phases were identified in the parents.*

LOD scores for chromosome 4 markers versus Huntington disease ◀ **Table 1-3**

Test	Recombination fraction (θ) 0.00	0.05	0.10	0.20	0.30	0.40
Huntington disease versus G8	6.72	5.96	5.16	3.46	1.71	0.33
Huntington disease versus MNS	$-\infty$	−3.22	−1.70	−0.43	−0.01	0.07
Huntington disease versus GC	$-\infty$	−2.27	−1.20	−0.32	0.00	0.07

* Data from Gusella et al (*Nature* 306:234, 1983)

changes (polymorphisms) that followed the phenotype through the kindred. This marker was localized to chromosome 4. This observation led to studies of additional DNA markers on chromosome 4 to try to improve the estimate of the map position.

Table 1.3 shows the data. Note the striking differences between the lod scores for three different markers on chromosome 4. In this example, the G8 marker was so close to the gene that the highest score came at $\theta = 0.0$. This means that the best (maximum likelihood) estimate of the position of the gene was sufficiently close to that of the G8 probe that no crossing-over occurred between the gene and the marker. Compare the data for G8 with those for two other markers: The latter's scores begin as negative values and indicate a high likelihood of having a crossing-over between those markers and the disease gene's location. Such studies can now be performed with many more markers, because the marker map is now more complete but the basic principles are unchanged.

Genetic distance is defined in units called centimorgans (cM), a term chosen in remembrance of the initial mapping studies by Thomas Hunt Morgan. A distance of 1 cM implies a 1% chance of recombination in any meiotic event. Mapping evidence has placed 3300 cM in the human genome. Because we know that this corresponds to approximately 3×10^9 bp, this gives roughly 1×10^6 bp/cM.

The serial comparison of multiple genetic markers through extended kindreds will lead to a series of linkage relationships. These can be ranked in order of distance based on recombination frequencies. Such a set of relationships establishes a "recombination map." The recombination map gives a set of mathematical likelihoods that given markers are close together or far apart and allows some measurement, albeit indirect, of the relative positions of the markers. It is important to emphasize that a recombination map is not the same thing as a physical map, although recombination positions and distances bear some relationship to physical distances between markers. Nevertheless, the recombination map is very useful in relating different marker clones and positions to one another and, as it becomes based on more markers, it is becoming progressively more useful clinically. We will consider some of the uses of the recombination map below.

E. PEDIGREE ANALYSIS

1. SYMBOLS

Pedigrees are diagrammatic representations of family structures. They provide a quick picture of the location of an affected individual within a larger kindred and/or of the spread of a problem (or any genetic trait or marker) through the kindred. Because pedigrees contain so much information, it is useful to use a common set of symbols to simplify studies and comparisons. The conventional symbols are shown in Figure 1.25. A given family is represented by a series of symbols, the oldest individual in a generation, or sibship, being on the left. By convention, males in an individual mating are on the left and females are on the right.

Sometimes it is useful to simplify a kindred and present only an outline. This can be particularly helpful if one is tracing a long lineage and if all of the details are not known. Several examples of this are shown in Figure 1.26. The inclusion of a carefully assembled and drawn pedigree in a clinical note can communicate a great deal of information quickly. It also can be simpler than long verbal descriptions of the same relationships.

Determining the degree(s) of relatedness of members of a kindred can be confusing. Figure 1.27 presents a five-generation kindred showing these relationships to the proband III.7. First-degree relatives include sibs, children, and parents. Second-degree relatives include grandparents, grandchildren, aunts, and uncles. First cousins are third-degree and second cousins are fourth-degree relatives. Note the confusing status of individuals IV.3 and IV.4. They are fourth-degree relatives

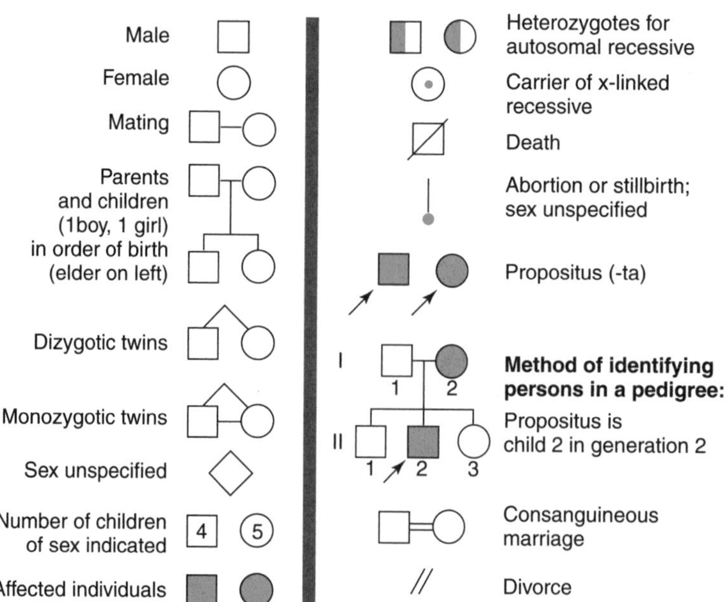

Figure 1.25 *Standard symbols for pedigrees. See also Figure 1.26 for examples of pedigree drawings; others will be presented in later chapters.*

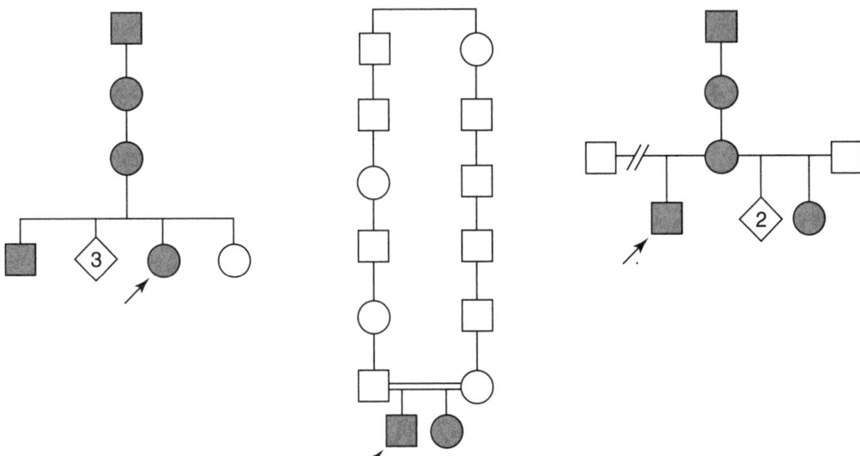

Figure 1.26 *Three multiple-generation kindreds showing gene or marker transmission using the symbols defined in Figure 1.25. Note that not all members of these kindreds have been studied. However, members essential for identifying the transmission patterns have been ascertained. Such a pedigree often serves as the beginning for more extensive kindred analysis.*

of the proband through their father (III.3) but second-degree relatives through their mother (III.4). The degrees are determined by the number of meiotic events separating the individuals. These relationships define the proportion of genes in common between kindred members (see also Table 12.1).

2. ASCERTAINMENT

Assembling a pedigree may be complicated. While it is usually simple to determine the structure of a nuclear family, larger families, especially those with secondary relatives, may provide a challenge—particularly when not all of the information is readily available. Often it can be useful to ask a patient or a family member to gather the family information. Sometimes this can be a useful opportunity to take advantage of the resources of older family members for whom it might not be easy to come to the clinic.

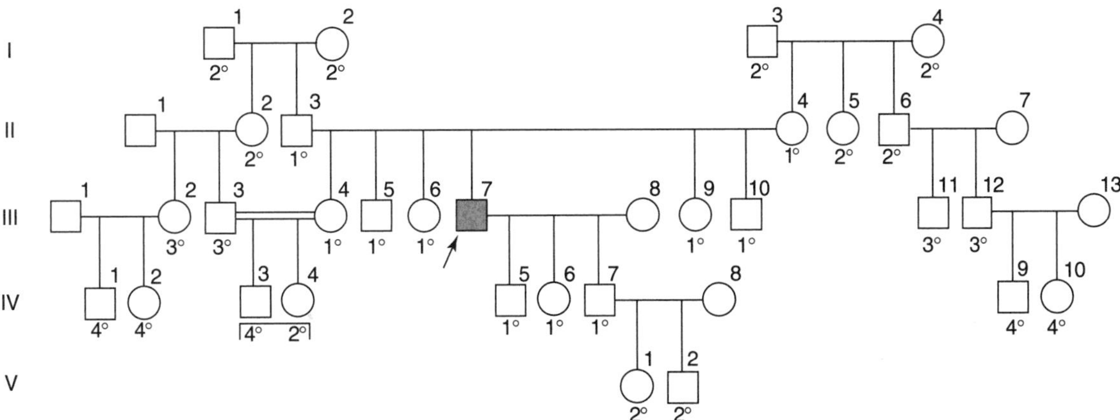

Figure 1.27 *Degree relationships in an extended kindred. All are defined with respect to the proband (III.7). The degree numbers are determined by the number of meioses separating the individual from the proband. See text for details.*

When assembling a kindred in which a trait occurs, it is particularly important to attempt to establish any possibilities of having various manifestations in other family members. For instance, when dealing with neurological problems, it can be useful to inquire about older family members. If people died "in a hospital," it is helpful to know how long they were there and whether any diagnosis is available. In particularly critical situations, it can be helpful to contact the hospital to obtain more diagnostic information. Another aspect of family analysis is the recognition that a given problem may not present in the same fashion in every affected family member. We will consider this in more detail below (particularly in Chapter V), but recognizing the possibility that different individuals may have different expressions of the same basic genetic change can make a pedigree far more valuable. This often is particularly helpful in the analysis of dominantly inherited conditions.

Ascertaining pedigree information may require much time. When it is particularly valuable and essential for establishing the diagnosis, sometimes families themselves can arrange family gatherings or reunions. These can be ideal opportunities for getting more information. Often a pedigree is not complete the first time a patient is met, and follow-up information becomes essential. This additional information is valuable for several reasons:

- It makes the pedigree more complete and gives a better picture of the pattern of this condition.
- The pedigree pattern may identify others who are at risk for a particular problem and who might benefit from further study.
- Identifying other affected family members may provide information critical for establishing the diagnosis in a given individual.
- Considering the manifestations of a given problem in some family members may help explain the prognosis for others.
- Assembling large kindreds can be essential for performing linkage studies and other diagnostic testing.

When complete pedigree information is especially important, it is valuable to seek other sources of data. Such sources include church or synagogue records, courthouse records, birth records from local hospitals, and other community information. Even immigration records can be used. In addition, physicians caring for other members of the family may be able to offer important additional information.

While obtaining the information about members of an extended kindred, it is important to respect the legitimate desires of individual members to withhold information, and it is essential to maintain confidentiality. This often becomes the case where there is anxiety or discomfort about the diagnosis. We will consider examples of such situations below. Ironically, extending a pedigree around an individual who is either unavailable or reluctant to cooperate can sometimes provide important information about that person and his or her relatives even though direct testing was not performed.

STUDY QUESTIONS

1 Discuss the consequences of the following mutations:

a Deletion of the dinucleotide AT in the middle of the first intron
b The transition T → C in the codon TTA
c Increasing n from 4 to 6 in a $(CA)_n$ cluster in intron 2

2 Explain the base changes leading to these protein sequences. Reconstruct a plausible nucleotide sequence.

native ...Arg Phe Phe Ser Pro...
a ...Arg Phe Ser His GluN...
b ...Arg Phe Phe
c ...Arg Phe Phe Pro...
d ...Leu Leu Phe Leu Thr...

3 A colleague has come to you excited that she may have isolated a DNA clone containing the gene for YoYo disease. However, she has been frustrated because both her Southern blot and FISH studies show multiple hybridization positions. You tell her:

a Her DNA clone may contain a repeated sequence that pairs with many others in the genome.
b Try the FISH and Southern blot hybridizations again at higher stringency.
c Sequence the clone and see how many matches there are in the databases.
d She has picked a member of a large gene family.
e She should go back to the clinic.

4 Your friend from Question 3 returns disappointed. Her DNA clone contains a repetitive sequence, and sequencing does not suggest that it contains coding information. From all that was known about YoYo disease, however, the clone should be close to it on the chromosome. You tell her:

a Try to modify the clone, eliminating the repeats.
b Isolate a new clone.
c Find a large kindred segregating YoYo disease and test your clone as a linkage marker.
d Consider getting an MBA.

5 Your friend reappears after 2 months. She took your advice and used her original clone as a marker for DNA studies of the kindred below, with the results shown in Figure 1.28. You tell her:

a She has established linkage between YoYo disease and her polymorphic marker.
b Marker allele 3 segregates with the phenotype through 7 of 8 meiotic events.

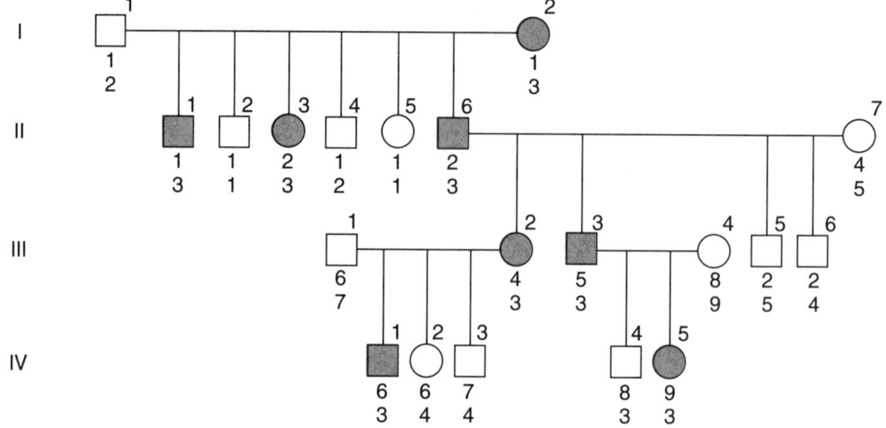

Figure 1.28 *Haplotype transmission for the variants of the new clone versus the segregation for YoYo disease in the study kindred. The highly polymorphic nature (high "pic" value) of this clone as a hybridization probe permitted considerable detail to be realized.*

c Individual IV.4 is a recombinant between the marker and the phenotype, so her clone is some distance from the actual responsible gene.

d Maybe she should work in genetics.

6 You and your friend are impressed with the linkage data but disappointed by the data for individual IV.4. Recombination here means that she has a potentially long distance to move from her clone before she gets to the gene for YoYo disease. Could there be another explanation for this picture?

FURTHER READINGS

Alberts B et al. *Molecular Biology of the Cell*, 3d ed. New York, Garland, 1994.

Caskey CT. Molecular medicine: A spin-off from the helix. *JAMA* 269:1986, 1993.

Hort MA and Geiser JR. Genetic analysis of the mitotic spindle. *Annu Rev Genet* 30:7, 1996.

Miyazaki WY and Orr-Weaver TL. Sister chromatid cohesion in mitosis and meiosis. Annu Rev Genet 28:167, 1994.

Sack GH. Clinical applications of molecular genetics, in Stobo JD, Hellmann DB, Ladenson PW, et al (eds). *The Principles and Practice of Medicine*, 23d ed. Stamford, CT, Appleton and Lange, 1996, pp 690-698.

Terwilliger JD and Ott J. *Handbook of Human Genetic Linkage*. Baltimore, Johns Hopkins, 1994.

USEFUL WEB SITES

DICTIONARY OF CELL BIOLOGY

http://www.mblab.gla.ac.uk/~julian/Dict.html
Internet edition of reference book of same name. Permits searching of 5450 entries with links to other, related entries.

GENE CARDS

http://bioinformatics.weizmann.ac.il/cards/index.html
Presentation of integrated data from large genetics databases in a concise screen of data. Data include gene name, synonyms, gene locus, protein product(s), associated disease(s). Searching can be extended from basic cards to other sites.

NATIONAL CENTER FOR GENOMIC RESOURCES

http://www.ncgr.org
Genetics and public issues pages, including legislative proposals. Section on continuing medical education about genetics. News updates. Access to Genome Sequence Database.

NOMENCLATURE SEARCH ENGINE

http://www.gene.ucl.ac.uk/cgi-bin/nomenclature/searchgenes.pl
Data on gene identification and methods for searching gene database.

INDIANA UNIVERSITY BIOTECHNOLOGY

http://biotech.chem.indiana.edu
Basic introduction to principles of genetics and molecular biology.

NATIONAL HUMAN GENOME RESEARCH INSTITUTE HOME PAGE

http://www.nhgri.nih.gov
Review of basic genetics. Genetic maps. Discussion of "How to conquer a genetic disease." Background on Human Genome Project. Discussion of ethical and social issues.

INSTITUTE FOR GENOMIC RESEARCH

http://ftp.tigr.org
Extensive genetic database collection and management software. Includes Human Gene Index.

GENLINK

http://genlink.wustl.edu/
Software and linkage information to integrate physical and genetic data. Very useful cross-references.

HOWARD HUGHES MEDICAL INSTITUTE

http://www.hhmi.org
Summaries of sponsored research in genetics. Timely news updates. Useful section on structural biology.

MOLECULAR CYTOGENETICS AND POSITIONAL CLONING CENTER (BERLIN)

http://www.mpimg-berlin-dahlem.mpg.de/~cytogen/
Describes a family of yeast artificial chromosomes (YACs) cytogenetically and genetically anchored, spread evenly over the entire human genome. A resource for FISH mapping. There also is a collection of YAC probes for all human chromosomal ends, averaging several cM from actual ends. This is a reference panel for studies of telomere rearrangements.

CYTOGENETIC RESOURCES

http://www.kumc.edu/gec/geneinfo.html
A particularly valuable entrance to multiple sites, including: Cytogenetics Images Index, Cytogenetic images and animations, Gene Map of human

Chromosomes, Human Cytogenetics Database, Chromosome Databases, Chromosome empiric risk calculations, and Karyotypes of normal and abnormal chromosomes. Excellent cross-references to other sites.

RESOURCES FOR HUMAN MOLECULAR CYTOGENETICS

http://bioserver.uniba.it/fish/Cytogenetics/welcome.html
Libraries of partial chromosome paints, recognizing a definite region of a chromosome. >900 fragments characterized from both normal and radiation-induced somatic cell hybrids.

CHAPTER 2 Genetic Testing

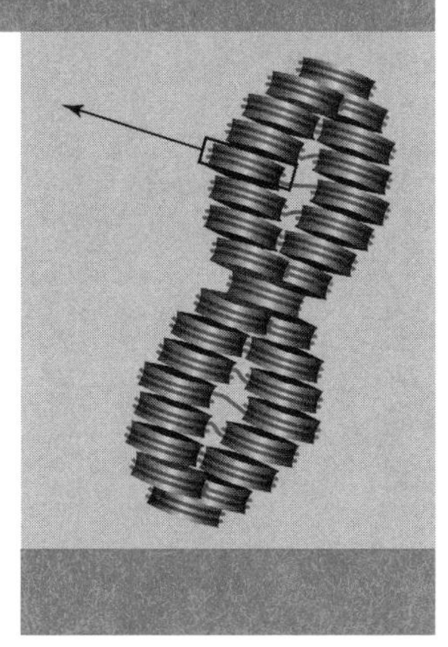

A. TESTING METHODS

The goal of genetic testing is to establish whether a particular genetic change is present in an individual. This may be done using different methods, depending on the type of problem. Laboratory testing for genetic conditions generally emphasizes one of three areas: (1) the DNA itself, (2) the chromosomes, or (3) the protein product of the gene of interest. Linkage analysis, described in Chapter 1, is done in the context of studying multiple members of a family, usually with DNA markers. Most of the methods for testing were described in Chapter 1 and will only be summarized here. Nevertheless, it is important to keep them in mind as options when a particular clinical problem is encountered.

Study of the DNA itself can be done using several approaches. Despite the enormous number of bases in human DNA, it frequently is possible to detect a specific single base change by using restriction enzymes. As described earlier, restriction enzymes are useful because they recognize a specific DNA base sequence. Obviously, any change in this base sequence will prevent cleavage by the enzyme. Thus, single-point mutations occurring within the recognition site of a particular enzyme (recall Table 1.1) can be detected readily by observing the altered mobility of the DNA fragments after enzyme digestion (see Figure 2.1). When combined with specific hybridization or PCR studies, this approach can, in principle, detect a single base change in the entire

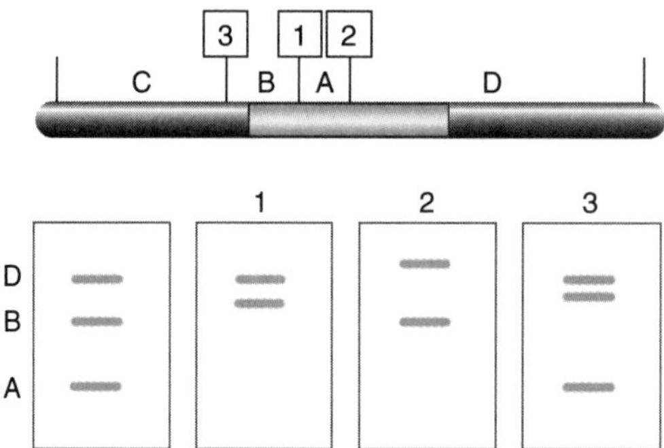

Figure 2.1 *The use of changes in restriction enzyme recognition sites for genome analysis. A molecular hybridization probe is available for the colored region of the DNA. Restriction enzyme cleavage sites are indicated by vertical lines. The four lower panels show patterns of hybridization using the probe. The panel on the left shows the pattern when cleavage occurs at all sites; the others show the patterns produced when site 1, 2, or 3 is changed to prevent cleavage. Recall that Southern blot patterns can be calibrated to determine fragment lengths; this helps distinguish bands of novel length that result from combining smaller fragments. Why is fragment C not detected by itself?*

genome, for example, in sickle cell anemia. (This remarkable sensitivity is 1 in 3×10^9.)

Sometimes it is useful to look at a larger region around a given gene. This may develop in a situation in which the mutation is not known or in which more than one mutation may explain the same clinical problem. (We will examine examples of this below.) In such cases, it may be simpler to determine the DNA sequence of a particular region of DNA directly. This is now usually performed with an automatic sequencing device and may use PCR to generate the DNA for analysis (see Figures 1.11 and 1.12). The use of PCR permits studies such as sequencing to be performed with only a small amount of material isolated from the patient (for example, a small blood sample that contains only the DNA in leukocytes).

Another useful approach takes advantage of physical changes in DNA due to sequence variation. By adjusting electrophoretic conditions, single strands of DNA containing a mutation(s) can be made to migrate at slightly different rates; such migration variants are called "conformers." Usually the single strands are produced by PCR. Detecting such a "single-strand conformational polymorphism" (SSCP) often is the first approach to finding new mutations. Figure 2.2A presents the principle of the method, and Figure 2.2B shows an example of its use in detecting mutation(s) in males with X-linked adrenoleukodystrophy (OMIM #300100). Study of X-chromosome mutations in males is simplified because there is only one X-chromosome (and hence only two strands in SSCP).

Chromosomes obviously represent an important area for analysis. As discussed earlier, obtaining a karyotype is particularly helpful in situations where chromosome abnormalities are suspected. Karyotype

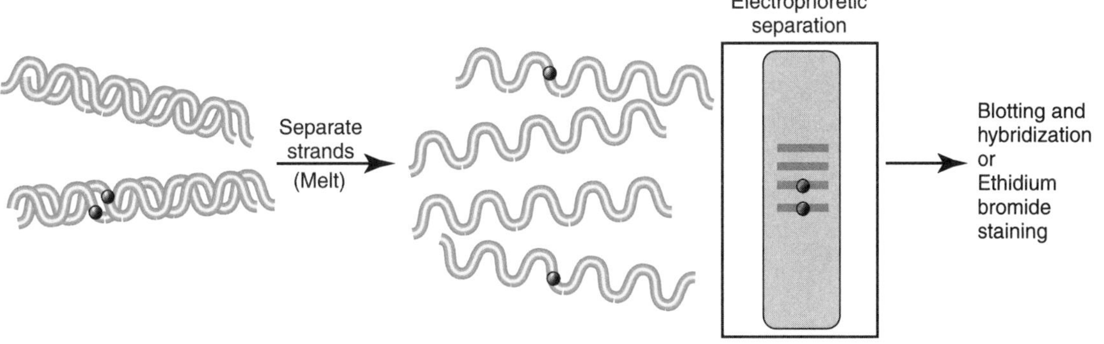

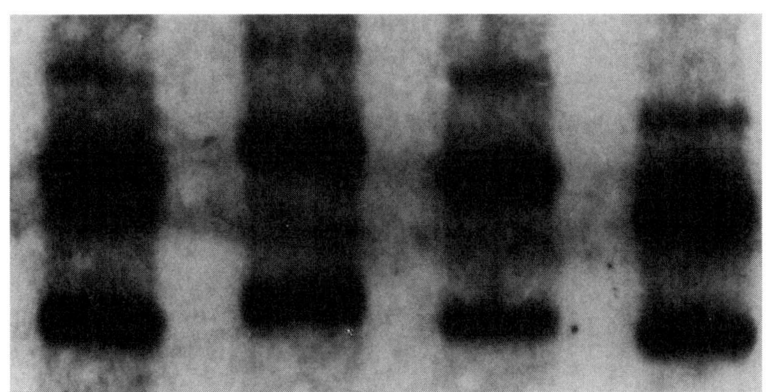

Figure 2.2 *Single-strand conformational polymorphism (SSCP). This separation technique often detects small DNA changes. (A) Fragments of defined length (generally produced by PCR amplification from known primer positions) containing a single-point mutation (shown by the small circles) can migrate with slightly different mobility when subjected to electrophoresis as single strands. The positions of the fragments can be detected by hybridization to Southern blots or by direct staining with ethidium bromide. (B) Data from SSCP study of three males with X-chromosome mutations. DNA was amplified from a single exon, and the strands were melted and separated by electrophoresis. Note that patient 8 was indistinguishable from normal (N) by this method, while the bands from patients 10 and 12 showed slower and faster migration, respectively. The two heavy bands represent the complementary strands of the single X-chromosome (from Kok et al., Hum Mut 6:104, 1995). Finding anomalous fragment migration permits direct sequencing of the affected region to determine the responsible base change(s). What can you conclude about patient 8?*

determination is becoming simpler and more rapid using FISH and painting techniques, as presented in Chapter 1.

An example of the usefulness of FISH can be found in studies of Williams syndrome (OMIM #194050). This condition is notable for hypercalcemia in infancy as well as supravalvular aortic stenosis, mental deficiency, facial dysmorphism, and a generally pleasant, outgoing disposition (see Figure 2.3A). Although few kindreds are known, the disorder has been mapped to the long arm of chromosome 7 (7q). Because the unusual form of aortic stenosis also exists as a separate entity and was linked to the elastin gene (7q11.23), this chromosomal region was studied in detail in individuals with Williams syndrome. Although microscopically visible chromosomal changes have not been found, FISH using an elastin probe has shown deletions in 96% of stringently defined patients (see Figure 2.3B).

Another type of testing is based on analyzing the altered protein product(s) of the gene. Such alterations may be detected by several methods. The protein of interest may be an enzyme whose chemical reactivity has been altered by the genetic change. Using standardized assay conditions, a laboratory can determine whether the enzyme has a normal substrate requirement and whether it produces the expected

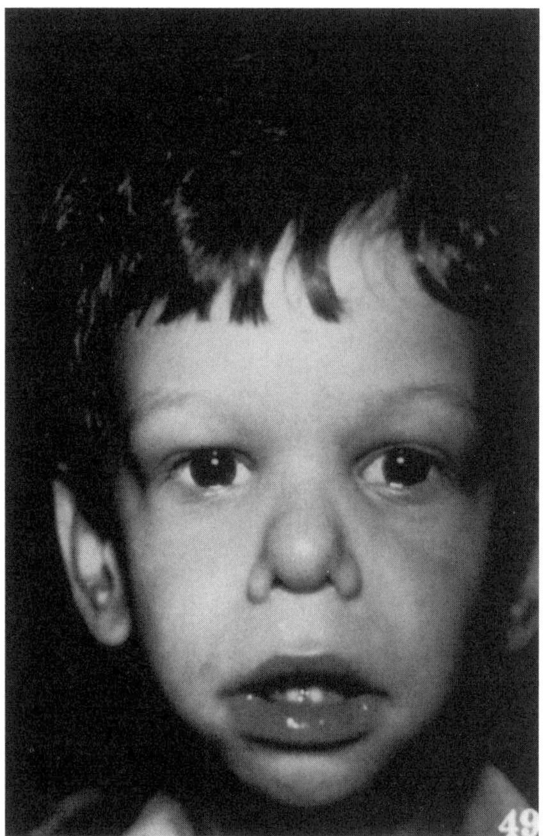
A

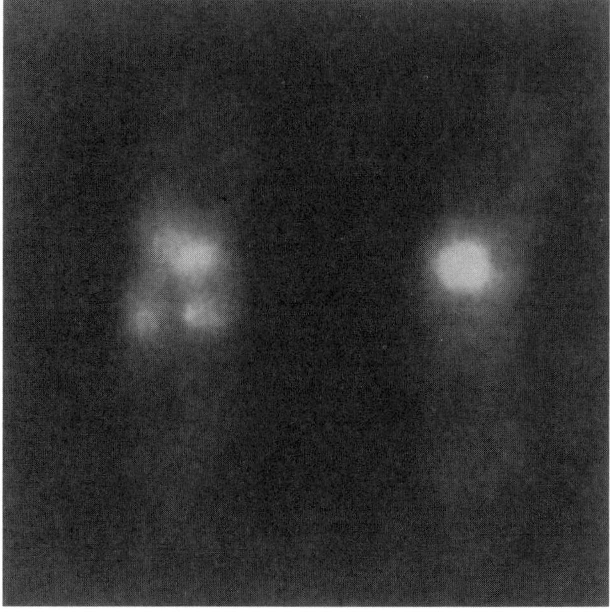
B

Figure 2.3 *(A) An individual with Williams syndrome. (Photograph courtesy of Dr. George Thomas.) (B) FISH study of chromosome 7 from a patient with Williams syndrome. Metaphase chromosomes are used. The upper signal comes from a repeat (α satellite) probe used as a control; the lower signal comes from the elastin probe. Note absence of the elastin signal on the chromosome on the right. (From Lowery et al., Am J Hum Genet 57:49, 1995.)*

type of product in the expected amount of time. Some enzymes will work with one substrate but show altered reactivity, depending on the presence or absence of mutations, with other substrates. Diagnostic laboratories have well-established conditions for performing these assays so that results can be interpreted clearly.

In other situations the gene product may not be an enzyme. Under such circumstances it may be useful to know the size (molecular weight) or the expected electrical charge for the protein. Both of these characteristics can be measured by electrophoresis. As discussed above, the fact that proteins bear an electrical charge makes them mobile in an electrical field. When the field is confined to a gel matrix, the movement of the protein can be detected easily. Under appropriately sensitive conditions, the change of a single amino acid often can be detected by the change in electrical charge and hence in the mobility of the parent protein. Of course, lengthening or shortening of a protein also can be detected by the electrophoretic migration position.

The notion of linkage analysis, introduced in Chapter 1, can be valuable in genetic testing. Here the clinician and the laboratory are confronted by a different question. Linkage analysis frequently is performed because the fundamental defect is unknown. Despite the fact that the defect itself may be unknown, other studies will have established the position of the responsible gene on a chromosome. The growing gene map then permits a choice of polymorphic DNA markers close to the region of the gene. Using a combination of several markers surrounding the presumptive location of the gene, a haplotype can be determined for affected and unaffected family members. As shown in Figure 1.23, if a sufficiently extensive family is available, the haplotype of the region containing the gene can be traced. Finding the haplotype that segregates with the genetic change in either an asymptomatic individual or an individual who has requested diagnostic studies strongly suggests that that person has the aberrant gene. Obviously, the main limitation of linkage analysis is related to the likelihood of having a recombination event occur between the actual gene change and the DNA markers used in haplotype analysis. As the number and frequency of markers along each chromosome increases, the gene map becomes more dense and it is possible to be progressively more accurate in the clinical use of linkage. The longer-term goal of much laboratory work, however, is to replace linkage analysis with a test that can precisely identify the underlying genetic change. Thus, testing for a particular genetic problem may be based on linkage studies for many years prior to the later development of a gene-specific assay. Sometimes, however, determining the exact gene mutation in the proband is very complicated and expensive even after the responsible gene has been identified; in such situations linkage using a tightly linked marker is a more practical choice. Linkage studies do require the cooperation of other family members, however, sometimes in several generations.

It is important to emphasize that much genetic testing is not performed in most clinical laboratories. Many of the tests are performed infrequently, so that it is inefficient, expensive, and impractical to have them available in all laboratories. In addition, the relatively infrequent need for some of the tests means that laboratories offering them also will be able to offer the benefit of larger numbers of test results so that reliability and quality control will be improved. Many tests are developed in individual research laboratories that have made a particular study of a certain problem. Slowly, these tests reach the stage where they can reliably be offered by commercial laboratories. It is likely that this will remain the pattern for many types of genetic testing simply because of the relatively low numbers of individuals affected with certain conditions.

The testing methods described above are applicable to many different clinical situations. Obviously, if a mature individual is being tested, more material (e.g., blood for leukocytes or preparing lymphoblasts, or skin biopsy for fibroblast studies) as well as additional clinical information may be available. By contrast, prenatal testing usually is done with only family history and clinical suspicion as a guide. However, the emerging importance of prenatal diagnosis has led to the development of several useful techniques. These will be discussed below.

B. PRENATAL DIAGNOSIS

Prenatal diagnostic testing has become particularly appealing for many conditions, and demand for it has grown rapidly. Here the obvious question is whether the fetus is affected with a given condition. Table 2.1 summarizes clinical indications for prenatal diagnosis and counseling; these will be discussed in more detail below. There are several uses for this information, as shown in Table 2.2. A broad range of disorders can be detected prenatally. Table 2.3 lists several relatively common single-gene disorders (many discussed in detail later in this book) for which prenatal diagnosis is possible.

The marked increase in risk for Down syndrome with increasing maternal age (see below and Chapter 3) has been the major indication for establishing an early diagnosis. The option of terminating a pregnancy has supported the development of these diagnostic efforts. Although this is currently the main option for parents, it is very important to emphasize that other therapeutic and ameliorative approaches are

Table 2.1 ▸ Indications for prenatal diagnosis/counseling

- Advanced maternal age
- Risk factors for neural tube defect
- Family history of known genetic condition for which diagnosis is possible
- Known chromosomal abnormality
  - De novo finding in previous child
  - Structural change in parent

Table 2.2

Uses of prenatal diagnosis

Informed parental decisions
Clinical management
- Prenatal treatment
- Neonatal care
- Obstetric considerations

being developed for specific clinical indications. It also is important to recall that, although elective early pregnancy termination may be the parental choice following prenatal diagnosis, such decisions account for a very small fraction (probably less than 2%) of elective abortions performed in the United States. Furthermore, the improved safety and reliability of prenatal diagnostic approaches often have led individuals who would not consider termination but who are at increased risk to have studies performed in order to reduce their anxiety during the course of pregnancy. The improvement of diagnostic techniques and protocols over 30 years has led to a useful array of approaches to different clinical problems.

1. ULTRASOUND

This approach to prenatal diagnosis is noninvasive. The improved resolving power of ultrasound now permits considerable information to be gathered about the developing fetus. Table 2.4 summarizes several of the more frequent uses of ultrasound in obstetrics. At a sufficient fetal age, sex determination is straightforward and other specific questions also can be addressed. The general rate of growth can be determined from ultrasound measurements. At the same time, assessments

Table 2.3

Single-gene disorders detectable by prenatal diagnosis

Disorder	OMIM Number
Autosomal dominant	
Achondroplasia	100800
Marfan syndrome	154700
Neurofibromatosis	162200
Myotonic dystrophy	160900
Huntington disease	143100
Autosomal recessive	
Sickle cell anemia	141900
Phenylketonuria	261600
Gaucher disease (I, II, III)	230800
Cystic fibrosis	219700
X-linked	
Hemophilia A	306700
Fragile X syndrome	309550
Duchenne muscular dystrophy	310200
Ornithine transcarbamylase deficiency	311250
Adrenoleukodystrophy	300100

Table 2.4 ▶ Prenatal uses of ultrasound

General obstetric considerations
Fetal age
Fetal sex
Multiple pregnancies
Fetal movement
Placental localization
Detection of specific abnormalities
Heart defects
Neural tube defects
Skeletal dysplasias
Polydactyly and limb defects
Ambiguous genitalia
Polycystic kidneys
Cranial malformations
Cleft palate/lip

can be made of physical changes in the fetus. These include changes around the spinal canal and the brain, as well as limb development. With future increases in its resolving power, ultrasound for monitoring pregnancies undoubtedly will increase in value but it will likely remain particularly valuable for establishing obstetric milestones.

2. AMNIOCENTESIS

For situations in which genetic analysis of the fetus is needed, other approaches are used; the best known of these is amniocentesis. With amniocentesis, amniotic fluid surrounding the developing fetus is sampled using an aspiration needle (see Figure 2.4 and Table 2.5). Amniotic fluid contains cells that are shed by the developing fetus; these cells

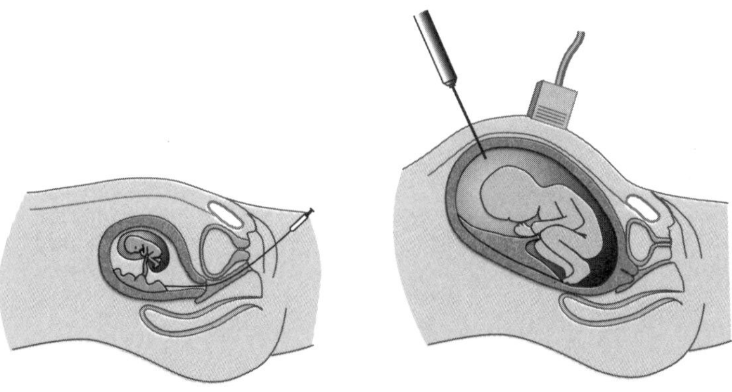

Figure 2.4 *Comparison of chorionic villus biopsy (left) and amniocentesis (right) for prenatal diagnosis. Note that chorionic villi (the fetal trophoblast) are accessible transcervically (although they must be approached transabdominally in certain situations). By contrast, amniocentesis samples fluid surrounding the fetus at a later gestational age. The fetal and placental positions are determined using ultrasound (as indicated by the transducer).*

Invasive prenatal diagnostic methods

Table 2.5

Method	Amniocentesis	Chorionic Villus Sampling	Fetal Blood Sampling
Performance time	16–20 weeks	8–12 weeks	18+ weeks
Fetal loss rate	<0.5%	≤1%	1–2%
Advantages	▪ AFP available ▪ More experience ▪ Fluid obtained for study	▪ Performed earlier, more time for parental decisions ▪ Time for later amniocentesis	▪ Useful when DNA study not possible ▪ Relatively rapid results
Disadvantages	▪ Little time for repeat ▪ Rare material infection ▪ Possible Rh sensitization ▪ Performed later ▪ Possible fetal injury (rare) ▪ Cells grow slowly	▪ No fluid obtained ▪ Minimally higher rate of fetal loss ▪ Occasional failure ▪ AFP data not available ▪ Slightly lower success of chromosome study ▪ Possible maternal contamination (mosaicism)	▪ Technically difficult

contain fetal DNA. The fluid itself may be assayed for various components (e.g., enzymes and α fetoprotein [AFP]) (see below). The cells can either be studied directly or have specific DNA regions amplified by PCR. An alternative approach is to grow the fetal cells in culture and perform additional measurements. These studies frequently involve chromosome analysis and sometimes the study of specific gene products. Amniocentesis is usually performed at 16 weeks of gestation but can be considered as late as 20 weeks. It has proved remarkably safe and has become the benchmark for assessing the accuracy of other testing methods. The choice of age 35 for recommending routine amniocentesis has been established by balancing the likelihood of carrying a Down syndrome conception versus the chance of fetal loss or complications of the procedure (see also Chapter 3).

3. CHORIONIC VILLUS SAMPLING

The second approach has been developed more recently and is based on evaluation of placental tissue. Placental tissue contains fetal trophoblast cells, and small villi can be removed by biopsy between 9 and 12 weeks of gestation. This involves establishing fetal and placental positions by ultrasound and then performing a transcervical biopsy. In certain situations (usually depending on placental position), a transabdominal biopsy is necessary. Thus, both the approach and the tissue assayed differ between amniocentesis and chorionic villus studies (see Figure 2.4 and Table 2.5). Chorionic villus biopsy provides both a source of

DNA and a source of cells for culture and further analysis, such as karyotype and protein studies.

4. FETAL BLOOD SAMPLING

A third source of fetal tissue for genetic study is the fetal blood supply. For this technique a very fine needle is used with ultrasonic guidance to obtain a blood sample from the umbilical vein of the developing fetus. This technique is performed at centers having specialized experience with the procedure and also serves as a useful source of cells for DNA or protein studies. Fetal cells grow rapidly, and hematological and chromosome studies are usually possible in less than 1 week (see Table 2.5).

5. MATERNAL BLOOD SAMPLING

A surprising observation has developed and holds considerable promise for future testing. A small number of fetal cells can be detected in the maternal circulation. At least in theory, this presents the possibility of analyzing fetal cells from a maternal blood sample. The major difficulty with this approach is the relative rarity of fetal cells in the maternal circulation. Enrichment techniques are being developed, however, and this may turn out to be a useful option in selected cases.

Because fetal sampling of any sort is not without risk (see Table 2.5), there have been studies to stratify the risks for the fetus and to direct prenatal studies to those most likely to have detectable abnormalities. A particularly important area (discussed in more detail below) relates to Down syndrome. The goal is to seek chemical changes in the maternal circulation that are associated with an increased likelihood of having a fetus with Down syndrome. If such relatively simple testing can be performed noninvasively (with no fetal, chorionic villus, or amniotic fluid sampling), it will simplify testing by concentrating on individuals who are at higher risk.

The best current combination of tests is to measure maternal serum levels of AFP (see also below), unconjugated estriol, and human chorionic gonadotropin (HCG). Finding low levels of the first two and an elevated level of the third, particularly when maternal age is considered, increases the likelihood of having a fetus with Down syndrome. Although these measurements are still only relative and additional studies are in progress, they hold the promise of helping to achieve the risk stratification that is needed.

6. ALPHA-FETOPROTEIN

Developmental abnormalities are important to consider in pregnancies. As has been described earlier and will be detailed later, not all developmental problems involve genetic changes. Nevertheless, identifying such changes can provide helpful guidance in pregnancy management and may permit presymptomatic treatment.

The most prominent developmental problems currently being assessed prenatally are related to abnormalities of the nervous system, particularly the spinal cord. In general, these are referred to as "neural tube defects." They include conditions in which the neural canal has not closed appropriately at some point along its length. Such a condition, often called "spina bifida," creates neurological deficits distal to the site of the abnormality. The consequences of these deficits are severe.

As noted above, some neural tube defects can be detected by ultrasound. However, it also is useful to perform screening for chemicals whose level(s) correlate with the presence of such defects. The most common current screening approach is to measure levels of AFP,which is produced by the fetus and reaches peak fetal levels early (Figure 2.5A). AFP also enters amniotic fluid; measurement of amniotic fluid AFP (AFAFP) follows a similar pattern (Figure 2.5B). AFAFP levels rise when neural tube defects are present. When AFAFP levels and ultrasound examination are combined at 18-20 weeks, they reliably identify open spina bifida and anencephaly.

Measuring AFAFP requires amniocentesis, however, and to avoid the risks involved with amniocentesis, the level of maternal serum AFP (MSAFP) has become more widely used. MSAFP levels rise later in gestation, as shown in Figure 2.5C, and a range of values is expected in a normal pregnancy. Figure 2.5D and 2.5E present MSAFP profiles for pregnancies with neural tube and abdominal wall defects. These profiles can be combined with ultrasound, allowing considerable information about these important developmental problems to be obtained noninvasively. The data in Figure 2.5F show that MSAFP levels in Down syndrome pregnancies often are lower than normal but have considerable overlap with normal levels. As noted above, additional testing is necessary to determine the reasonable level of suspicion.

It is important to recall that MSAFP levels can differ in different population groups and thus that a clear notion of the expected levels for the population at risk is needed. MSAFP measurements have become a regular part of obstetric monitoring. They have helped considerably to narrow the search for individuals at risk for neural tube defects and, combined with ultrasound studies, to assist in managing affected pregnancies. Other screening approaches are under development but have not yet reached the stage of wide clinical applicability.

7. PREIMPLANTATION DIAGNOSIS FOR IN VITRO FERTILIZATION

Genetic testing is becoming applicable to the technology of in vitro fertilization (IVF). IVF and its uses and indications will not be reviewed here, but it is important to realize that at early stages embryonic cells

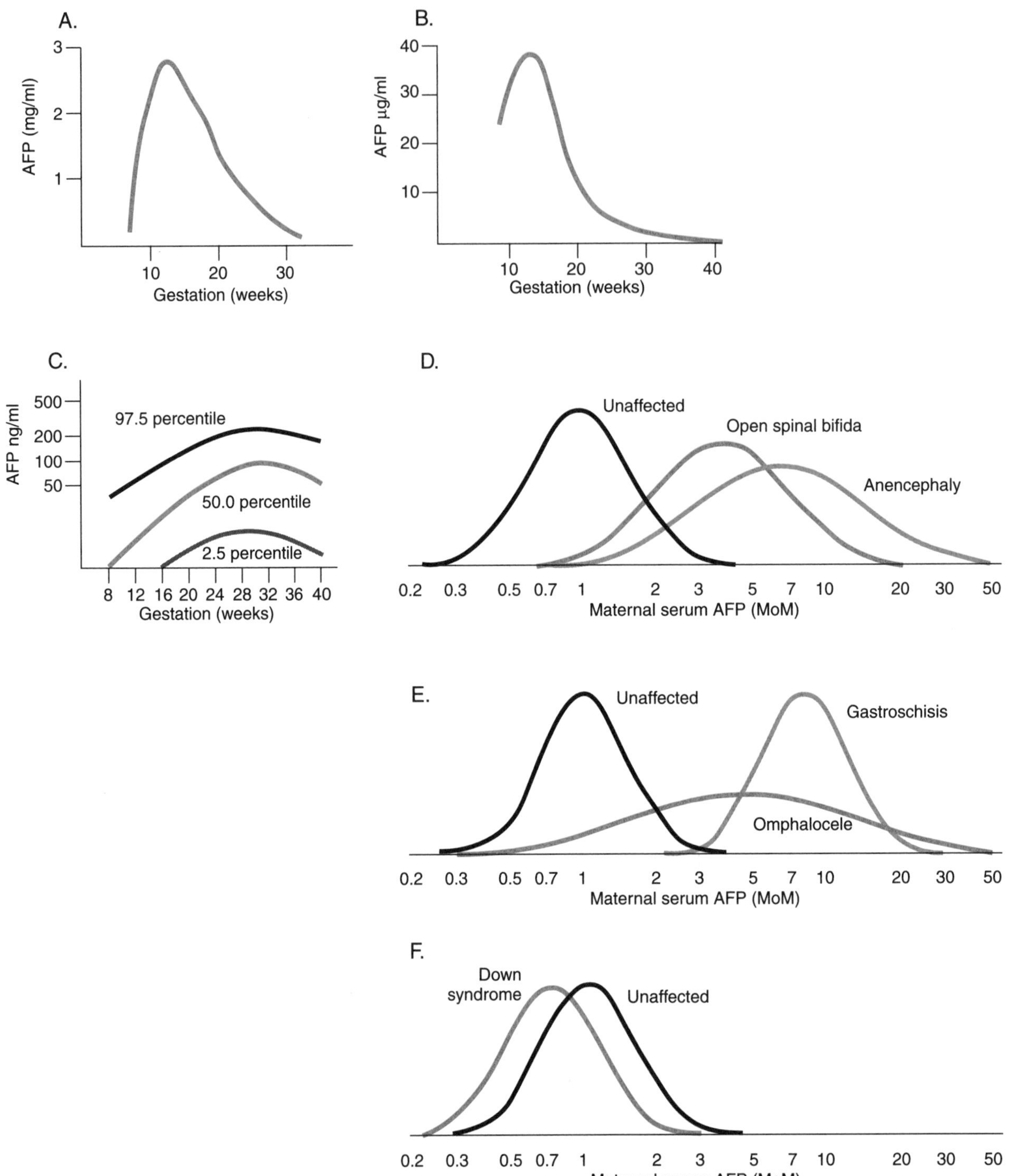

Figure 2.5 *Measuring levels of α-fetoprotein (AFP) can give useful information regarding the progress of fetal development. (A) AFP levels in serum of normal fetus during development. (B) AFP levels in amniotic fluid (AFAFP) during normal development. (C) Normal maternal serum AFP (MSAFP) levels during gestation. (D) and (E) MSAFP levels in four developmental defects, presented as multiples of the mean (MoM). (F) MSAFP levels often are low in Down syndrome pregnancies.*

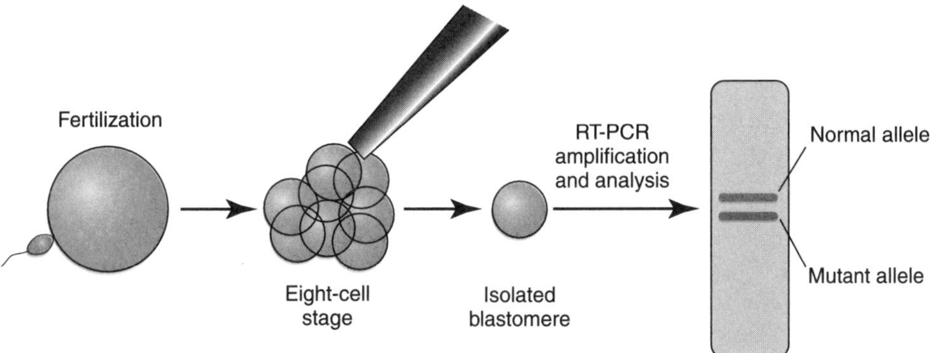

Figure 2.6 *Preimplantation genetic study of a blastomere isolated at the 8-cell stage of an IVF embryo. Amplification using PCR can identify discrete changes in single cells and still permit continued normal development and implantation of the seven remaining cells.*

have full developmental potential. Thus, prior to committed differentiation, an individual cell can be removed from a developing embryo for genetic study without compromising development. Prior to the availability of PCR, this was not a practical consideration, because the amount of material available in a single cell (about 6 pg) was too small to analyze. However, techniques including nested PCR and reverse transcriptase PCR (RTPCR; using PCR on products of RNA transcripts copied by the enzyme reverse transcriptase) make it possible to assay for the presence of specific genetic changes in a single cell from an early stage of an IVF embryo (see Figure 2.6).

One use of this approach has been for pregnancies at risk for an X-linked disorder. As discussed in detail in Chapter 6, males are at 50% risk for such disorders. Determining the sex at the 8-cell stage (for example, by using hybridization probes specific for the Y-chromosome) could permit selective growth and implantation of only female embryos. Alternatively, if the conceptus is at risk for a specific genetic change, testing can be directed toward establishing the presence or absence of that change. If the amplified cell DNA is found to be free of the change, the development of the remaining cells can continue and implantation can follow. If the change is detected, additional embryos can be tested to find an unaffected conceptus. An obvious advantage to this approach is that it eliminates the need for later invasive testing.

STUDY QUESTIONS

1 A couple has come to you to discuss a possible pregnancy. They tell you that they already have had a child who was said to have cystic fibrosis (OMIM #219700) and who died at age 2. They are reluctant to consider repeating this experience. They are of Northern European descent; there is no history of a similar problem in their family; and both are well and under 30 and are willing to spend any reasonable amount for a successful pregnancy. You tell them:

a You would like to see health records for their deceased child.
b Prenatal diagnosis is now possible based on mutation analysis.

c Prenatal diagnosis of cystic fibrosis by DNA is uncertain in their case, because the diagnosis and responsible mutation(s) are not established.
d The sweat chloride test cannot be performed on fetal cells.

2 A healthy 31-year-old woman comes to you concerned about having a child with Down syndrome. She and her husband are both healthy, and they have no history of the problem in their family. You tell her:

a Their risk for a pregnancy with Down syndrome is very low.
b A triple screening test of maternal blood could establish the diagnosis if the mutation is present.

The couple comes in a month later, still concerned. They indicate that the wife has a history of depression (she is currently doing well without treatment) and they cannot risk the anxiety during a possibly affected pregnancy. You tell them:

a The "gold standard" is amniocentesis, and that is what you recommend.
b CVS could be performed at 9–10 weeks.
c Triple screening is a good option.
d There is minimal likelihood that their offspring will have depressive illness.
e Perhaps they should consider adoption.

3 A couple is concerned about having another child with Q syndrome, a neurodegenerative disorder. Their first child is affected, and they have been told it is a recessive condition with a 25% recurrence risk. You examine the child (now 4) and confirm the diagnosis. Q syndrome has been mapped recently to a large gene of unknown function on chromosome 9; several mutations have been identified. You tell the parents that you need to establish which mutation(s) is(are) present in the affected child. An SSCP study on the child, using the Q syndrome gene probe, shows the pattern seen in Figure 2.7A. You tell the parents:

a Prenatal diagnosis should be possible.
b Prenatal diagnosis is not possible.

Despite your advice, the couple returns 3 months later. The woman is 8 weeks pregnant and they are anxious. They have been reading about prenatal diagnosis and urge you to make a diagnosis by CVS, which you agree to try to do. The result of the SSCP study is shown in Figure 2.7B. You tell them:

a The fetus is a carrier for Q syndrome.
b The fetus is affected with Q syndrome.
c You can't say what the fetal status is but can estimate that the likelihood that the fetus is affected is: (1) 1 in 10; (2) 1 in 5; (3) 1 in 3; (4) 1 in 2; (5) 1 in 3.

4 A 34-year-old woman is seen in the obstetric clinic. She is 15 weeks pregnant, and her MSAFP is three times the mean. What management do you recommend?

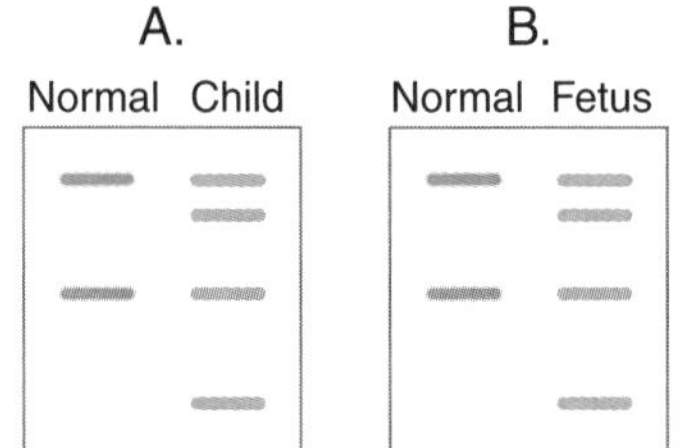

Figure 2.7 *SSCP patterns. (A) Patterns of a normal child and of an affected child. (B) Patterns of a normal fetus and of a fetus at risk.*

FURTHER READING

Boehm CA. Prenatal diagnosis and carrier detection by DNA analysis. *Prog Med Genet* 7:143, 1988.

Goldberg JD and Golbus MS. Chorionic villus sampling. *Adv Hum Genet* 17:1, 1988.

Harper PS. *Practical Genetic Counseling*, 4th ed. Wright, Bristol, United Kingdom. 1993.

Stetten G, Goodman BK, and Fox HE. New cytogenetic technology and its application in maternal-fetal medicine. *J Maternal-Fetal Invest* 7:155, 1997.

USEFUL WEB SITES

HELIX (INCLUDES GENLINE)

http://www.hslib.washington.edu/helix
Medical Genetics Knowledge Base. Electronic textbook under development relates testing to diagnosis, management, and counseling. Directory of laboratories providing testing for genetic disorders.

PROMOTING SAFE AND EFFECTIVE GENETIC TESTING IN THE U.S. (1997)

http://www.med.jhu.edu/tfgtelsi
Report of Task Force on Genetic Testing including discussions of safety, quality control, communication, and rare disease studies. Recommendations for providing gene tests in United States.

UNDERSTANDING GENE TESTING

http://www.gene.com/ae/AE/AEPC/NIH/index.html
Illustrated brochure from National Cancer Institute with useful information for lay public.

MARCH OF DIMES

http://modimes.org/
Resources for families and professionals about birth defects. Professional education section.

NATIONAL ASSOCIATION FOR RARE DISORDERS (NORD)

http://www.NORD-rdb.com/~orphan
Primary nongovernmental clearinghouse for information about rare disorders. Over 1100 disease reports. Broad resource guide. Includes Rare Disease Database, a wide range of information for families and professionals. Listings by symptoms, causes, affected populations, treatments. Listing of support groups. Includes orphan drug designation database.

CHAPTER 3

Chromosomes

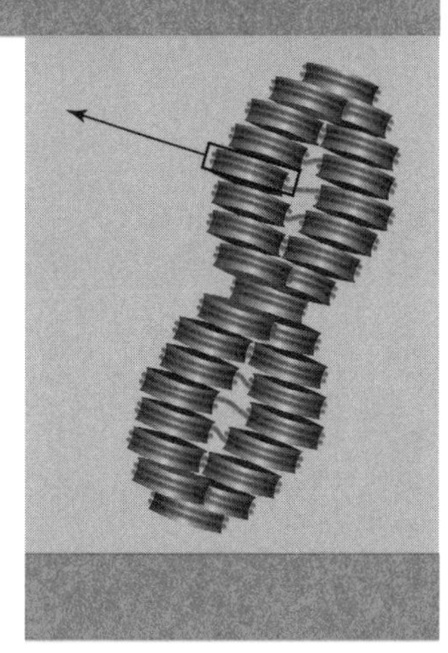

Basic aspects of chromosome structure and the important processes of meiosis and mitosis have been presented in Chapter 1. As the organizational units for nuclear DNA, chromosomes are responsible for transporting hereditary information. Thus, changes in chromosomes—in their number or structure—can have serious clinical consequences. It is important to realize that chromosomal changes may be present at conception, reflecting either inheritance or changes that occurred during meiosis. Chromosome changes also may be acquired. Acquired changes, particularly notable in cancer genetics, have important biological effects but do not necessarily have heritable consequences; they will be discussed in Chapter 4.

A. CHANGES IN CHROMOSOME NUMBER

Somatic cells have two copies of each of 22 autosomes and either one X- and one Y-chromosome or two X-chromosomes; they are said to be "diploid" (from the Greek for "double"). Any change in these numerical relationships (aneuploidy) has the potential for causing severe disease. The basis for disease under these circumstances is not always clear. This is especially confusing when it is obvious that even if one chromosome of a pair is missing, the other should contain the same genetic information. On the basis of clinical observations, however, it is clear that both a balance between the number of copies of a given gene and the relationships of that number to other genes on the same and different chromosomes are essential in regulating development. Also, as will be considered below, there may be differences in expression between genes on a maternally derived chromosome and those on a paternally derived chromosome ("imprinting").

The development of methods for early prenatal diagnosis (see Chapter 2), along with detailed study of fetal tissue from early spontaneous fetal losses, has led to important and surprising conclusions. The proportion of concept uses with abnormal chromosome numbers is considerably higher than that seen in liveborn infants. There is an underlying incidence of spontaneous abortions of about 15% of all recognized pregnancies, and 5–6% of all zygote loss reflects chromosome abnormalities. Most spontaneous fetal loss due to aberrant chromosomes occurs early, generally in the first 4 months of pregnancy. This early, spontaneous loss helps explain the incidence of aberrant chromosomes in liveborns of 0.5–0.6%.

As these numbers have been studied in more detail, it has become clear that several types of numerical chromosome changes are relatively prominent early after conception. Only a small number of fetuses with these changes survive to live birth. Figure 3.1 summarizes some of these observations. Aneuploidies with three copies (trisomy) of chromosomes 13 or 18 are relatively prominent in early pregnancies but are extremely rare in liveborn infants. When they do occur, they are associated with a broad spectrum of serious developmental anomalies that are usually lethal soon after birth. Fetuses trisomic for chromosome 16 do not survive to livebirth. The best known trisomy—trisomy 21, which causes Down syndrome—also becomes less frequent in fetuses of advanced gestational age, as shown in Figure 3.1. Although a significant number of concept uses with Down syndrome do survive, it is obvious that this is far less than the number conceived.

These important observations indicate that intrauterine develop-

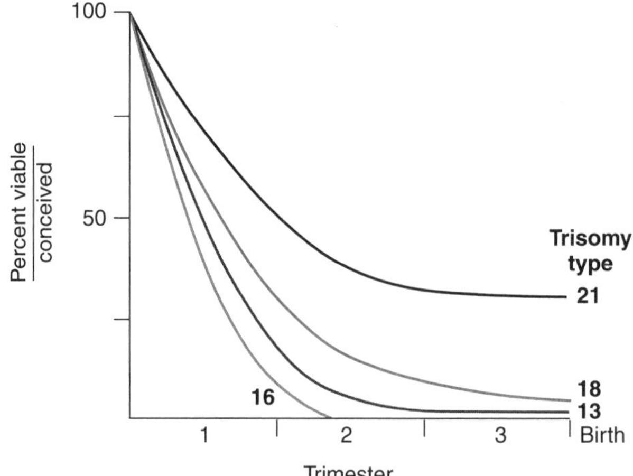

Figure 3.1 *Spontaneous fetal loss for autosomal trisomies. Note the poor survival of trisomies 13 and 18. There is no survival of trisomy 16 conceptions, although they are prominent early.*

ment itself places important constraints on fetal viability and many concept uses with aneuploidy are lost well before birth. It is likely, although the data are more difficult to obtain, that an additional substantial number of abnormal concept uses never even reach the stage of successful uterine implantation. Current estimates, gathered from several approaches to this important question, project that the proportion of viable concept uses may be surprisingly low—approximately one-third. This impressive fetal loss most likely is a protective device, minimizing the maturation, and ultimately the birth, of seriously genetically aberrant fetuses. Based on these data, chromosome studies made at early times after conception for prenatal diagnosis can be expected to show a larger proportion of abnormalities than might be expected in the population of liveborn infants. Even without intervention, it is likely that the majority of these abnormal concept uses will not survive the intrauterine environment (recall Figure 3.1).

B. DOWN SYNDROME

1. CLINICAL ASPECTS

By far the most readily recognizable individuals with abnormal autosomal chromosome numbers are those with Down syndrome. The individual with Down syndrome has an abnormality—usually trisomy—of chromosome 21. Down syndrome occurs throughout the world with an incidence of about 1 in 660 newborns. The developmental features characteristic of affected individuals can be recognized in all populations. Figure 3.2 shows photographs of several individuals with Down syndrome. Down syndrome should be recognized by practicing physicians in all specialties, both because of its relative frequency and because of the special medical considerations these individuals may present.

Individuals with Down syndrome have hypotonia, joint laxity, and soft skin. Prominent facial features include a flat facial profile; thin, straight hair; patulous lips, often with a protruding tongue; prominent inner canthal folds and slanted palpebral fissures; speckling of the irides (Brushfield's spots); small ears; and small nose. About 45% show a simian crease in the hands, and the middle phalanx of the fifth finger is often hypoplastic. They are generally short and remain so throughout their lives. Developmental abnormalities of the gastrointestinal tract also occur (see below). It should be possible to make the diagnosis of Down syndrome in the newborn. Even in the absence of chromosome studies, the presence in the newborn of 6 of the 10 criteria established by Hall should be diagnostic (see Table 3.1).

About 40% of individuals with Down syndrome also have congenital heart disease, particularly problems related to septation. These include (in order of decreasing frequency) A-V communis, ventricular septal defect, patent ductus arteriosus, atrial septal defect, and aberrant subclavian artery.

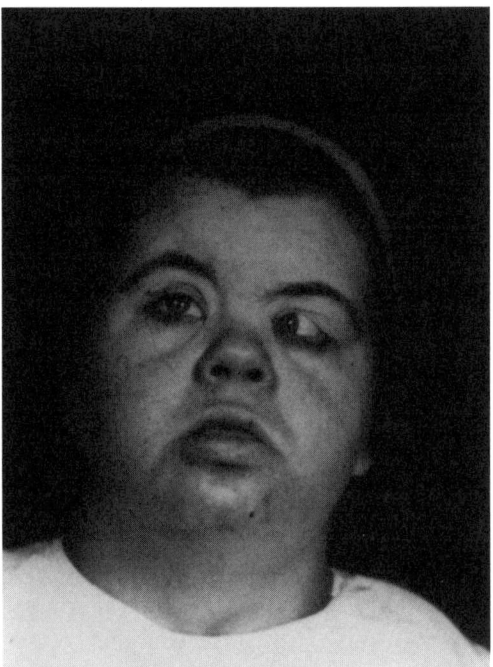

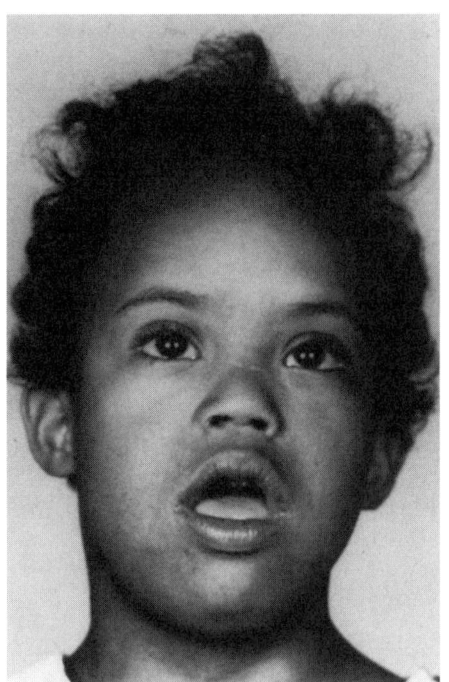

Figure 3.2 *Two individuals with Down syndrome. While the boy on the right has the expected karyotype of trisomy 21, the woman on the left has only 46 chromosomes. How can this be explained? See text.*

The most notable feature of individuals with Down syndrome, however, is mental retardation. Although the degree of retardation may range from moderate to severe, it is a universal finding. Despite this retardation, many individuals with Down syndrome can function with reasonable levels of independence. They can make considerable progress in school and learn basic skills. As they grow older, these people often find activities in sheltered workshops or other protected environments.

Individuals with Down syndrome frequently have a remarkably pleasant personality. Parents with more than one child often note the

Table 3-1 ▶ **Clinical criteria for the diagnosis of Down syndrome in the newborn**

Hypotonia
Decreased Moro reflex
Hyperflexibility
Flat face
Oblique palpebral fissures
Dysplastic (simple) ear
Redundant (loose) neck skin
Simian palmar crease
Clinodactyly (5th finger)
Dysplastic pelvis

(Hall, *Clin Pediatr* 5:4, 1966.)

remarkable docility of their Down syndrome child in comparison with the unaffected children. This friendliness often extends to adults, in which case openness and willingness to be approached by strangers can present social difficulties.

The successful surgical management of problems faced by individuals with Down syndrome in the newborn and infant period (repair of duodenal atresia and congenital heart defects) permits many to enter and progress through adult life. The health of adults with Down syndrome is an area of growing interest and concern. Unfortunately, many of these individuals are not good medical historians because of their mental retardation, so their care has to be administered in an environment of anticipation. Also, the clinician must be suspicious of problems that the individual may articulate poorly. For example, adults with Down syndrome may develop hearing loss and hypothyroidism that were not noted in their earlier years. Also, longer-term complications of congenital heart disease may develop as these people grow older. There is an increased incidence (nearly 1%) of acute myelocytic leukemia during childhood that crosses over into acute lymphocytic leukemia after late adolescence/early adulthood.

One of the most interesting and unfortunate complications for adults with Down syndrome is the frequent development of neurological and mental deterioration in a fashion resembling Alzheimer's disease. It is initially surprising to consider mental deterioration in individuals who already have a baseline level of mental retardation. Nevertheless, serial cognitive measurements often show a loss of skills. Pathological changes seen in the brain include neurofibrillary tangles and amyloid plaques that are typical of Alzheimer's disease, but they are seen at earlier ages. The observation that the gene for the parent protein (amyloid precursor protein, APP) of the relatively short polypeptide (AP) found in Alzheimer's amyloid plaque is located on human chromosome 21 has raised the possibility that aberrant control of this gene could contribute to mental deterioration.

2. GENETIC ASPECTS OF DOWN SYNDROME

Not all of the genetic changes that underlie the clinical findings in Down syndrome individuals are yet understood. Obviously, duplicating an entire chromosome involves thousands of genes. Are all of these genes necessary for producing the changes of Down syndrome? This important question has been addressed in several ways.

Some individuals who have the clinical findings of Down syndrome do not have a complete additional copy of chromosome 21. As will be discussed below, some of these individuals have translocations and other structural abnormalities that involve only a subsection of chromosome 21. Thus, it has been possible to establish that the short arm of chromosome 21 does not generally contribute genes to the final clinical picture of Down syndrome. On the long arm, the region 21q22→telomere (approximately the distal third of the long arm) is implicated in most of the phenotypic changes. It has been proposed that the genes in this region are sufficient to produce the Down syndrome phenotype. However, this relatively simple notion does not completely explain the situation. A small number of Down syndrome individuals with chromosomal translocations show duplication of more proximal regions of the long arm of chromosome 21. Thus, other genetic regions are important. One common feature has been the limitation of congenital heart disease to those individuals with Down syndrome who have duplications in and distal to region 21q22.

The implication of these observations is that Down syndrome reflects a complex set of interactions of multiple genes. It is still not clear, however, what the minimal region compatible with Down syndrome is. The number of genes present on chromosome 21 can be estimated. The length of the entire chromosome is approximately 1.5% of that of the total genome, and this implies that between 1000 and 2000 genes should be present on this chromosome. Although this is a relatively large number, it is not overwhelming. If we consider only the distal third of the long arm, it is possible that fewer than 500 genes might be in the region causing most of the phenotypic features. Unfortunately, this remains a potentially very complex interacting system, and considerably more studies will be necessary to determine the contributions of individual genes. In addition to the individual genes themselves, further consideration must be given to the potential interactions of these genes, particularly in regulating development.

Because it is such a prominent phenotype, the demographics of Down syndrome have been studied in considerable detail. The biggest risk factor for conceiving an individual with Down syndrome is advanced maternal age. This is shown most clearly in Figure 3.3. The likelihood of having a conceptus with Down syndrome begins to rise strikingly at maternal age 35, and with conceptions at age 45 or older the risk is substantial. There also is some increased risk for conceptions with increased paternal age, but it is not as striking. Of course, maternal and paternal ages frequently vary together.

Molecular genetic markers (such as $[CA]_n$ repeats) can be used to distinguish maternal and paternal chromosomes, and it has been possible to determine which parent provided the additional chromosome to trisomic individuals; overwhelmingly this turns out to be the mother. It is possible, although the mechanism is not understood, that the long latency period of oocytes in older mothers (meiosis I may have been stable for 45 years) makes reliable disjunction less likely. In the absence of standard disjunction, two chromosomes can pass more readily to a given germ cell. As described earlier, other trisomies also may occur, but these are more likely to prove lethal early in development and not appear in liveborns (recall Figure 3.1).

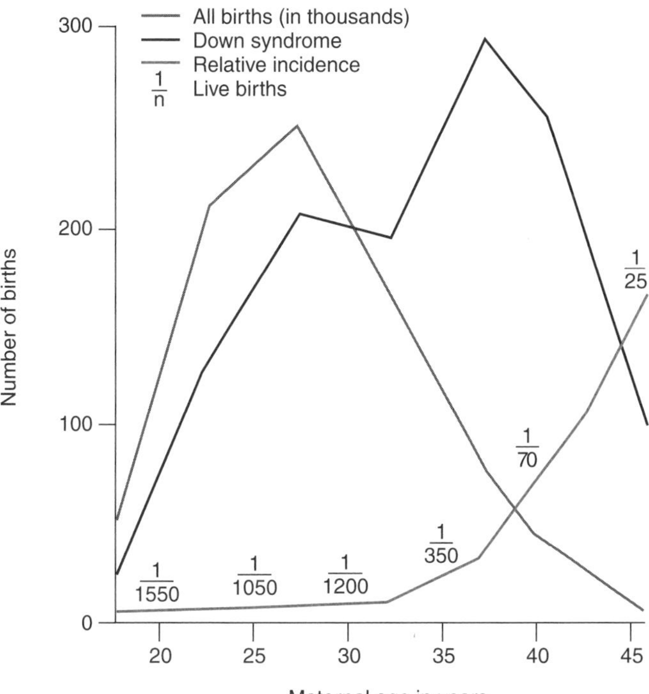

Figure 3.3 *Incidence of Down syndrome with maternal age. Note the prominent increase after age 35, the point at which prenatal diagnosis becomes particularly valuable.*

This important demographic observation has had many clinical implications. The first is that genetic counseling is very strongly directed toward older mothers in terms of the risks of Down syndrome. Currently, prenatal diagnostic laboratories perform fetal chromosome studies often, and the risk of Down syndrome due to advanced maternal age is the most frequent indication. The second implication, however, is that detecting Down syndrome in pregnancies of older mothers could ultimately eliminate the birth of individuals with Down syndrome; this has not occurred. The reason for this apparent paradox also is suggested by Figure 3.3. Although older mothers are at considerably greater risk for Down syndrome conceptions, they have fewer pregnancies. Thus, the large numbers of pregnancies in younger mothers end up producing more individuals with Down syndrome despite a reduced frequency of Down syndrome in these conceptions. Therefore, although the basic demographic aspects of Down syndrome incidence are well recognized, the number of affected pregnancies in older mothers (and thus those likely to be detected by prenatal studies) is relatively low.

Because of this ironic relationship between maternal age and Down syndrome, other approaches have been sought to search for additional, noninvasive markers of a Down syndrome pregnancy. A combination of ultrasound and the triple screen maternal blood laboratory changes (discussed in Chapter 2) is used as a possible indicator of a Down syndrome pregnancy. Although this testing indicates only an increased risk of a trisomy 21 pregnancy, it is useful for screening younger women during their pregnancies in order to identify those for whom chromosome analysis might be helpful. With anticipated refinement, this approach may help detect Down syndrome in the larger set of pregnancies in younger women.

3. LESS COMMON KARYOTYPE CHANGES IN DOWN SYNDROME

a. TRANSLOCATIONS

As noted above, not all individuals with Down syndrome have an independent additional chromosome (i.e., they are not "trisomic"). The woman on the left in Figure 3.2 has this situation. In such individuals, the problems arise due to a "translocation." In chromosomal translocations, a fragment or even an entire chromosome can be joined to another chromosome. As shown in Figure 3.4, this creates problems.

About 3% of Down syndrome individuals have chromosomal translocations. The best known of these is a translocation that involves chromosomes 14 and 21. This type of translocation is known as "Robertsonian" and involves fusion near the centromere with loss of most of the short arms of both chromosomes; it is designated t(14;21). For such a chromosome, a genetically balanced situation (i.e., diploid equivalency) can occur only in certain conceptions, as shown in Figure 3.4. (The other type of translocation—"reciprocal"—involves breakage of two nonhomologous chromosomes with joining of reciprocal segments; this will be discussed below and in Chapter 4.)

Remarkably, individuals who have one normal chromosome 21, one normal chromosome 14, and a t(14;21) translocation chromosome can be clinically normal. Such an individual has a diploid complement of the information on all chromosomes, because the information lost on the short arms of both chromosomes 14 and 21 encodes ribosomal RNA genes that are redundant and represented elsewhere in the genome. This person is called a "translocation carrier," and the translocation is said to be "balanced." Several sibs of the woman on the left in Figure 3.2 have this karyotype. Complications arise for conceptions from a balanced translocation carrier, however, as shown in Figure 3.4. Obviously, the products of meiosis will include germ cells containing the translocated chromosome, germ cells containing no information for chromosomes 14 or 21, and other germ cells that contain normal, independent chromosomes 14 and 21.

Fertilizing a germ cell containing no 14 or 21 information is presumably lethal. (Such a conceptus has only 45 chromosomes and is haploid for all genes on chromosomes 14 and 21; this situation has not been recognized clinically.) By contrast, fertilizing a germ cell containing a normal chromosome 21 as well as the t(14;21) translocation with a normal haploid germ cell will lead to a conceptus with 46 chromosomes

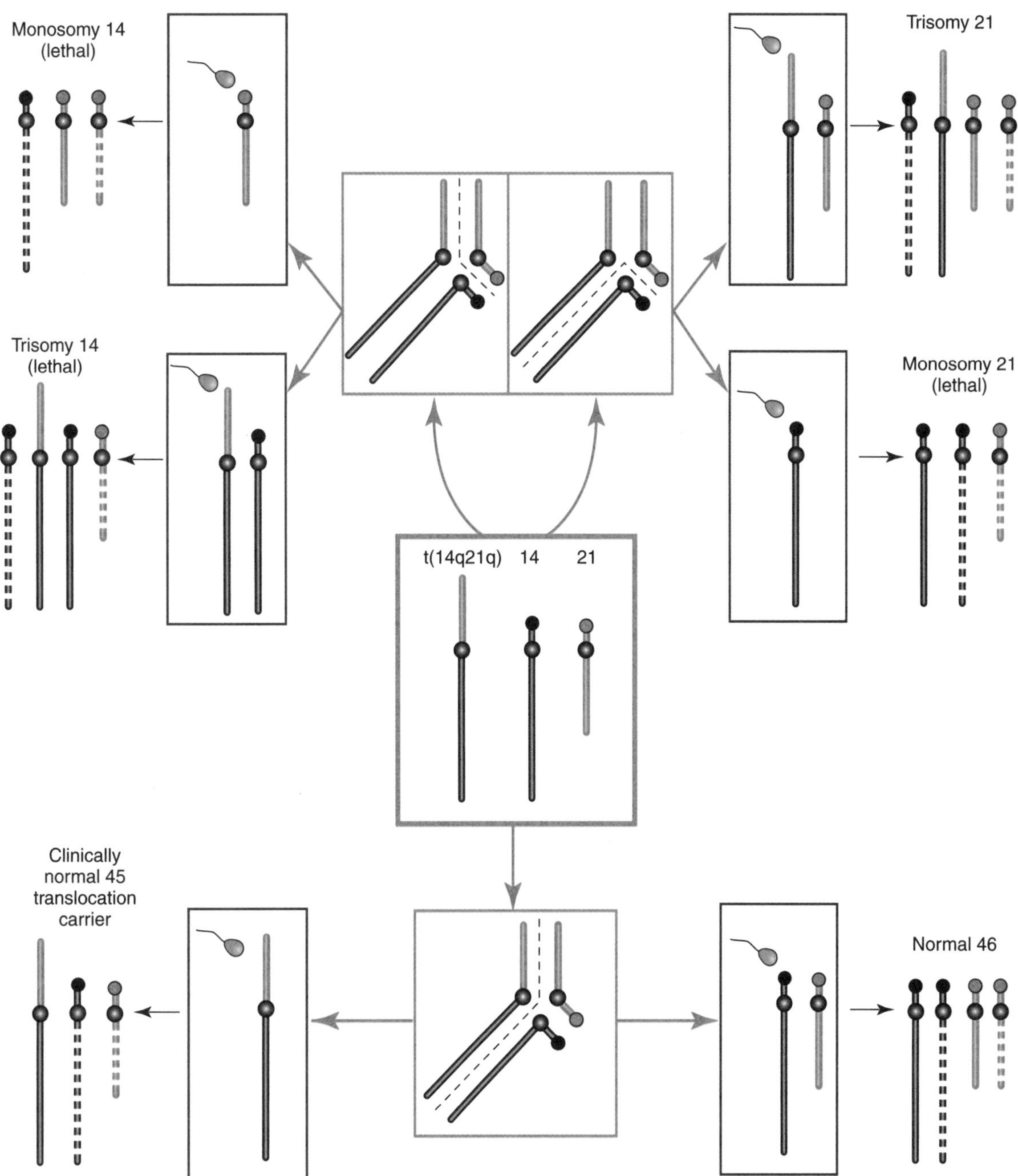

Figure 3.4 *Possible outcomes of fertilizing products of a Robertsonian translocation t(14;21) meiosis with a normal gamete. Note the lethal events as well as the likelihood of developing the balanced translocation carrier state and translocation Down syndrome.*

but 3 copies of chromosome 21 information (hence 21 "trisomy"). Such an individual will have Down syndrome. Obviously, for such an individual the genetic implications are substantially different from those for an isolated individual with trisomy. In the translocation situation, a family pattern can be expected to include both Down syndrome individuals and the possibility of multiple spontaneous abortions (presumably resulting from the haploid situation described).

b. MOSAICISM

A small number (1–2%) of individuals with Down syndrome have chromosomal mosaicism. The most frequent explanation is that a trisomic conceptus has lost one of the copies of chromosome 21 in only one or some of its cells during early development. Thus, there are two populations of cells—trisomic and normal. Depending on the proportions of these cells, affected individuals can vary in the severity of their manifestations; some may go undetected. An important point is that the germ cells of such a mosaic individual could be of two types—normal and 21+. Thus, this person could have an increased risk of having children with Down syndrome. The possibility of such an event is often considered when confronting a young mother with a Down syndrome child (recall Figure 3.3). If the mother has gonadal mosaicism, the recurrence risk could be high. The possibility of this situation often leads to recommending prenatal diagnosis for subsequent pregnancies.

The notion of germ-line mosaicism can be applied to chromosomal variation(s) as well as to single-gene mutations. As we will consider below and in later chapters, it is one at least theoretically important complication in counseling regarding possible recurrence of what are generally considered to be de novo events.

C. SEX CHROMOSOMES

As noted earlier, there are two copies of each autosomal chromosome in somatic cells. By contrast, the sex chromosomes have an entirely different relationship. Females have two X-chromosomes and males have one X- and one Y-chromosome. This sets up a paradox. How can the same amount of genetic information present on a single X-chromosome in males have the same level of effect as the genetic information present on two X-chromosomes in females? This is especially puzzling because, as noted above, the problems in Down syndrome develop because of an abnormal number of copies of the otherwise normal genes on chromosome 21. Although this question is not completely resolved at the molecular level, considerable progress has been made in understanding it; this will be discussed below. Another contrast between the sex chromosomes is the remarkable discrepancy in their sizes. While the X-chromosome is relatively large (approximately 6% of nuclear DNA), the Y-chromosome is quite small, and only a few genes have been assigned to it.

Obviously, germ cells in males must be of two types: one with 22 autosomes and an X-chromosome and the other with 22 autosomes and a Y-chromosome. By contrast, all end products of female meiosis must have 22 autosomes and one X-chromosome.

1. X-CHROMOSOME INACTIVATION

An understanding of the mechanism of compensation for the presence of two X-chromosomes in females was originally proposed by Dr.

Mary Lyon. In her now-famous "Lyon hypothesis," she proposed that the X-chromosomes in somatic cells of mature females are of two types. One is active and expresses its full complement of genetic information; the other is inactivated in some manner and does not serve as a source of genetic information. While some details of this proposal have been modified since it was made, it remains the underlying notion regarding X-chromosome control in females. From an operational perspective, it is recognized that soon after fertilization one of the X-chromosomes in female cells undergoes physical and molecular changes. This chromosome appears different on microscopic staining and also divides at a later time in mitosis. This so-called "late-replicating X" is nearly inert from the perspective of genetic information. An important aspect of Dr. Lyon's original notion is that this change in one of the X-chromosomes occurs randomly, so that in early embryonic life different cell lineages may have alternative X-chromosomes inactivated. The result of this in the maturing and finally in the adult individual is that different somatic cells will express the genetic information encoded on different X-chromosomes. Thus, the somatic tissues of females are said to be "mosaic" because they represent the contributions of genes from different X-chromosomes. As will be discussed in Chapter 7, this has important clinical implications.

The molecular basis for X-chromosome inactivation is not completely understood. The process begins with activation of a gene called "X-inactivation-specific transcript" (XIST) on the long arm of the X-chromosome (see Figure 3.5). XIST is expressed only on the inactive X-chromosome and produces an RNA molecule (not translated into protein) that transmits the inactivation signal throughout the chromosome. The process involves physical reorganization of the DNA within the chromatin and also the addition of methyl groups to the DNA bases of substantial regions of the inactivated X-chromosome. The changes of X-inactivation, including the persistent expression of the XIST gene, the visual appearance of the chromatin, the late replication of the inactive X-chromosome, and the lack of expression of other genes on the inactive X-chromosome, are preserved through subsequent somatic mitoses. In women, at least some somatic cells show a darkly staining mass on the edge of their nuclei. This is the inactive X-chromosome and has been called the "Barr body." Obviously, similar cells from a normal male should show no Barr body, because there is only a single X-chromosome.

The regulation of numbers and ratios of sex chromosomes is critical in development. Any changes in these ratios have important clinical consequences. These consequences, for obvious reasons, are manifested by changes in stature, sexual differentiation, and maturation. Because these developmental changes are so prominent, several have become well recognized from their clinical manifestations alone. These will be discussed below.

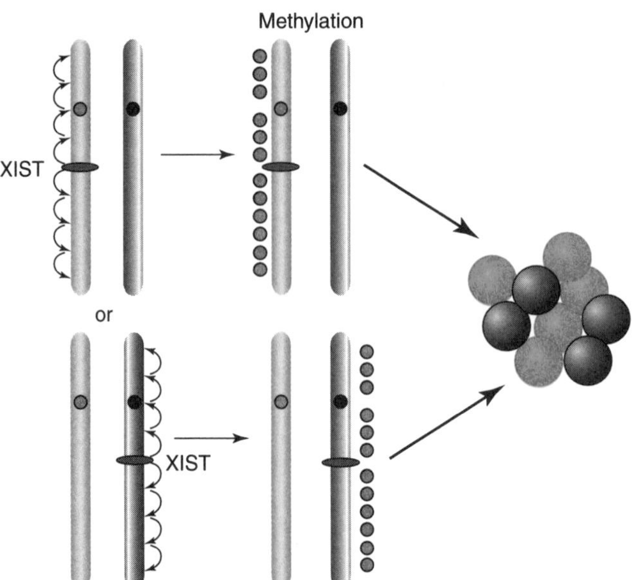

Figure 3.5 *Schematic view of X-chromosome inactivation. In cells of the early female embryo there is random inactivation of either the maternally or the paternally derived X-chromosome through events at the X-inactivation center that include expression of the XIST gene. RNA from this gene spreads the inactivation signal throughout the X-chromosome, leading to the addition of methyl groups (shown as dark circles) to the DNA. Once established, this inactivation pattern is then maintained through subsequent cell divisions, leading to a mosaic pattern in the progeny.*

2. XO FEMALES (TURNER SYNDROME)

Individuals with Turner syndrome are usually recognized readily, and the diagnosis should be considered for any phenotypic female who has short stature and/or delayed or absent menarche. These females are relatively short beginning at about 2 years of age. They have no adolescent growth spurt, and their mean final height is 143 cm (57 in). The limbs are affected somewhat more than the trunk. Some individuals have a series of physical changes, including, most frequently, a broad back of the neck (some times referred to as "webbing") and changes in the angle of the elbows, giving the lower portions of the arms an angular appearance (see Figure 3.6). Vascular anomalies include coarctation of the aorta and sometimes aberrant development of lymphatic vessels. Despite the appearances emphasized by Figure 3.6, many individuals with Turner syndrome appear normal except for short stature.

Although there is variation in the psychosocial features of individuals with Turner syndrome, mental retardation is not common, and mean verbal IQs are indistinguishable from those of age-matched controls. Mean performance IQs may be somewhat lower, however, but this is usually well tolerated. These individuals clearly identify themselves as females, an attitude that is reinforced when exogenous cyclic estrogens are used.

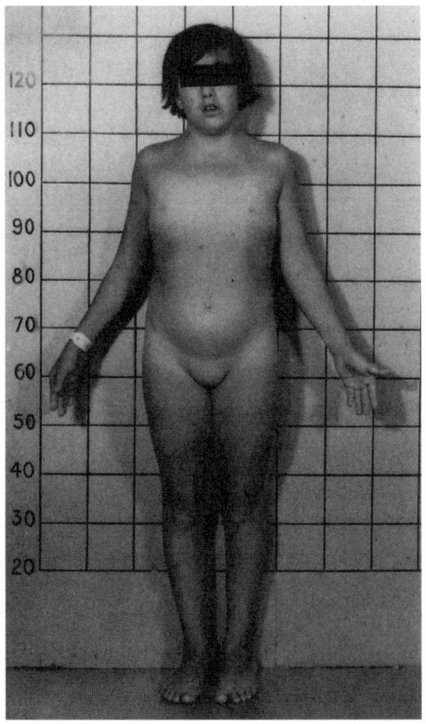

Figure 3.6 *Photograph of an individual with Turner syndrome. The findings are variable, but relatively short stature and infertility are the most common features. Note the prepubertal status and angulation of the elbows.*

One of the most prominent features of individuals with Turner syndrome is primary amenorrhea. This lack of normal menarche and adolescent development is frequently the basis for making the diagnosis, and, consequently, the diagnosis often is suspected after gynecological consultation. Significantly, these individuals respond appropriately to exogenous estrogens, confirming that appropriate receptors are present and that the uterus is intact.

The common genetic feature of individuals with Turner syndrome is the presence of some anomaly in one of the two X-chromosomes. The simplest anomaly is absence and, thus, the most common karyotype for an individual with Turner syndrome is "45 XO." However, Turner syndrome also can be associated with other changes of one of the X-chromosomes, including deletions, the formation of circular chromosomes, and chromosomal translocations. In fact, many individuals with Turner syndrome are themselves chromosomal mosaics, with 46-XX and 45-XO cells present in various cell lineages. As introduced above, this generally is the result of the loss of an X-chromosome from a female cell early in development. Thus, rarely, and based on the random chance that the germ-line cells have the 46-XX karyotype, an individual with somatic features of Turner syndrome may be fertile.

Despite infertility and the physical changes described above, most individuals with Turner syndrome make good adjustments in life. The use of exogenous hormone replacement in puberty can lead to appropriate adolescent maturation and remarkably normal development.

At first consideration, the presence of aberrant development in an individual with the XO karyotype appears to contradict the Lyon hypothesis. On the basis of that hypothesis (discussed above), normal development requires the activity of only a single X-chromosome, and thus an XO situation should provide all of the necessary genetic information. However, information from two X-chromosomes appears to be needed for normal early female embryonic development; female X-inactivation occurs only later. It is important to appreciate that individuals with the XO karyotype are recognizably female. Thus, there must be essential information on the Y-chromosome that contributes to the determination of the male developmental pathway. This observation is relevant to the group of conditions discussed below.

3. XXY MALES (KLINEFELTER SYNDROME)

The presence of two X-chromosomes and a Y-chromosome presents a very different clinical picture; known as "Klinefelter syndrome." The "XXY male" appears as a phenotypically normal male in early development, although gonadal development is reduced and the individual remains hypogonadal. During puberty, XXY males often have a prominent growth spurt, leading them to be tall with long arms and legs (Figure 3.7). These individuals have minimal testicular development and seminiferous tubule dysgenesis; they are characteristically

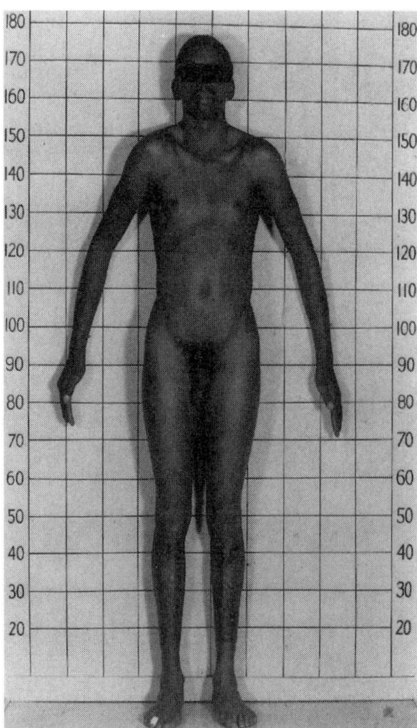

Figure 3.7 *Photograph of a male with Klinefelter syndrome. Note the very long arms and legs. This individual is infertile, but some secondary sexual characteristics are present. Gynecomastia was mild in this patient but is a variable feature.*

infertile. Many XXY individuals show some degree of gynecomastia. It is important to note that breast cancer, otherwise very rare in males, occurs with an increased frequency in these individuals.

This syndrome is relatively frequent, occurring once in 600–1000 liveborn males. Despite the physical changes, these individuals are raised as males. They may benefit from exogenous androgen administration during and after puberty. The possibility of breast cancer usually is seen only later in life. Despite their infertility, which is usually permanent except in the case of chromosomal mosaicism, the use of exogenous androgens during and after puberty can aid psychosexual development.

It is important to realize that the nondisjunction that leads to the XXY chromosomal constitution can lead to the presence of additional X-chromosomes as well. Thus, individuals with male habitus and XXXY, XXXXY, or even other combinations are recognized, although they are considerably less frequent. Unfortunately, individuals with higher numbers of X-chromosomes are more likely to show some degree of mental retardation. Because of its importance for both prognosis and diagnosis, it is valuable to perform a karyotype on these individuals. The majority (about 80%) of these individuals have an XXY karyotype; about 15% show XY/XXY mosaicism. The latter generally show milder changes and may be fertile.

D. CHROMOSOME STRUCTURE CHANGES

Any changes in the chromosomal complement of an individual imply aberrant amounts and/or control of critical genetic information. Thus, such changes must be considered seriously in any individual who presents with unusual features. The sections above emphasized the simplest chromosomal changes—numerical changes in either autosomes or sex chromosomes. Chromosomes are very complex macromolecular arrays of DNA and proteins, however. Thus, changes in the physical (and often visible) structure of individual chromosomes may be encountered as well. These changes may have implications not only for the individual who bears them but also for his or her meiosis and fertility.

1. DELETIONS

Relatively simple changes within individual chromosomes include the loss or reorganization of part of a chromosome. The loss, termed a deletion, is the physical absence of some region, notable by a change in the length of the chromosome as well as by the loss of banding patterns on light microscopy and of regional DNA sequences in FISH and other hybridization studies (recall Figures 1.16 and 2.3). Obviously, deletions may extend from the loss of a single DNA base pair (recall Figure 1.13) to very long sections. Only losses of large amounts of material will be visible by light microscopy. As will be discussed in Chapter 4, detailed characterization of such changes is of growing importance in cancer biology.

The loss of a fragment of a chromosome obviously implies the loss of the genes within that region. For autosomes, this means that the individual will be haploid for the genes in the region of loss (as opposed to having the normal diploid complement of information). In general, the remaining copy of the genetic information is intact on the other chromosome. However, important problems can arise when a gene or genes remaining on the chromosome unaffected by the deletion have changes (see below and Chapter 4).

In contrast to the autosomes, a deletion on the X-chromosome of a male will have important consequences, because there is no alternative source of genetic information. Such deletions frequently manifest as clinical disorders in males, whereas they are often not detected in females carrying the same deletion because of the presence of the other X-chromosome.

Identifying X-chromosome deletions in males has been essential in mapping and identifying genes for particular conditions on the X-chromosome. One of the most prominent examples of this is Duchenne muscular dystrophy (OMIM #310200). While most males with this condition do not have any visible chromosomal changes, several affected individuals have had microscopically visible deletions; these deletions permitted the assignment of the responsible gene to the short arm of the human X-chromosome, as shown in Figure 3.8. Applying

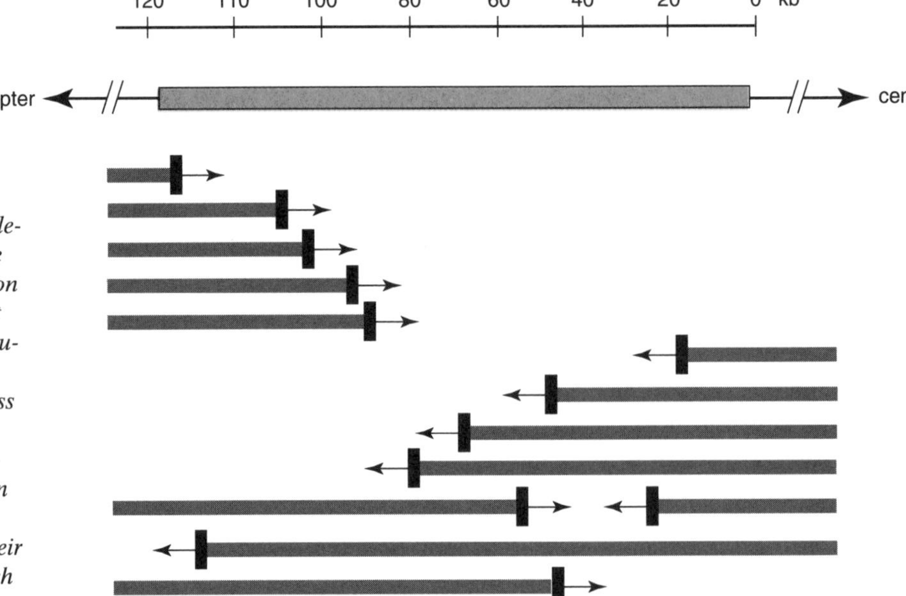

Figure 3.8 *Overlapping deletions on the short arm of the X-chromosome. Each deletion was isolated from a different patient with Duchenne muscular dystrophy (OMIM #310200) who had visible loss of chromosomal material in this region. Although the extents of the deletions differ in molecular terms, they implicate a common region by their overlap (see Chapter 7). Such localizations are valuable for subsequent gene isolation.*

molecular biological techniques to the DNA of such individuals permitted exploration of the region of the deletion and ultimately led to identification of the responsible gene. Other X-chromosomal conditions are discussed in Chapter 7.

The chromosomal extents of most visible deletions are far greater than the dimensions at which most single genes are organized. Thus, such changes often produce so-called contiguous gene defects. The manifestations of such defects may be complicated, because both the genetic information and the local organization and expression of more than one gene may be involved.

2. INVERSIONS, DUPLICATIONS, INSERTIONS

Another important chromosome anomaly is a change in the linear order or organization of sequences in a given chromosomal region; examples are shown in Figure 3.9. Information from a region may be inverted, so that the normal physical order is reversed. Although the actual genetic information may be intact, such a reversal (or "inversion") can change control regions and, at the points of breaking and rejoining, has the potential to interrupt important sequences (possibly even genes themselves). Of course, having an inversion disrupts pairing during meiosis.

Regions of chromosomes also can be duplicated (see Figure 3.9). Duplications may occur either in the same or in the opposite orientation as that found in the normal chromosome. Such duplications are recognizable in karyotypes by characteristic banding pattern changes. Moving genetic material from one chromosomal region into another leads to an insertion. An insertion puts genetic information completely out of its regional context and may even move it onto a different chromosome. Such an event changes the potential for control and also, as will be discussed in more detail below (and in Chapter 4), establishes the possibility of interrupting genes at the points of insertion.

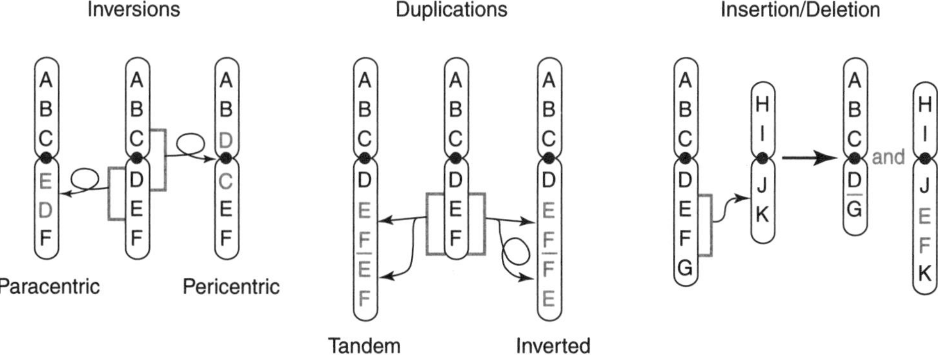

Figure 3.9 *Regional chromosome changes in outline form, including inversion, duplication, and insertion events. Note the consequences of these structural changes for meiotic pairing.*

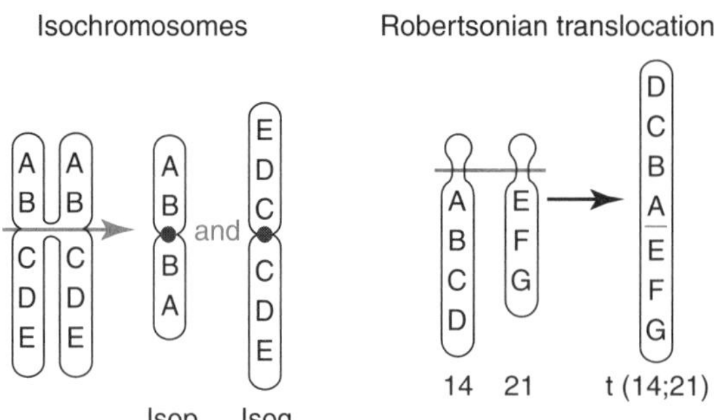

Figure 3.10 *Isochromosomes have a characteristic set of molecular marker and visible banding patterns extending from the centromere. In contrast, Robertsonian translocations contain the complete long arms of different acrocentric chromosomes.*

3. TRANSLOCATIONS

Several types of chromosome changes involve the reorganization of very large regions, sometimes entire arms, of chromosomes (see Figure 3.10). The simplest of these are isochromosomes and Robertsonian translocations. In isochromosomes, the short arms and/or the long arms of the same chromosome join at the centromere. The result of this joining—which obviously creates two copies of a relatively large genetic region—is usually very detrimental to development. Isochromosomes are readily recognized in karyotypes and show the characteristic mirror-image appearance of banding patterns extending in both directions from the centromere.

Robertsonian translocations may be considered special cases of joining at centromeres (see also Figure 3.4). As discussed in the section on Down syndrome (above), Robertsonian translocations occur between chromosomes that have tiny short arms (acrocentric chromosomes). Usually, as noted earlier, there is no obvious loss of genetic information, because the short arms contain repeated ribosomal RNA genes. Such an event usually forms a very recognizable and often long product.

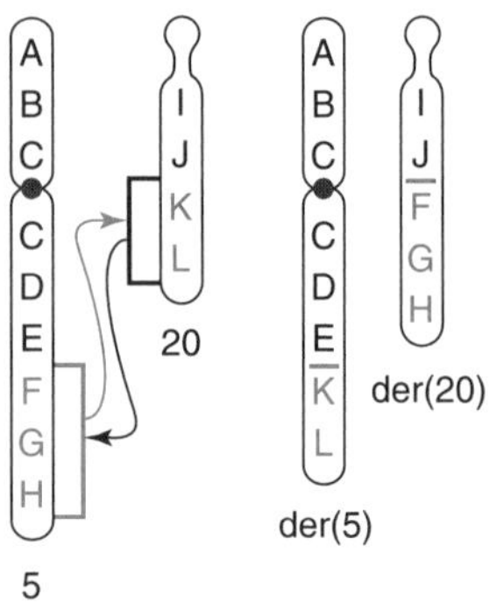

Figure 3.11 *Reciprocal chromosome translocations. These often are complex, with breakpoints that cannot be resolved by light microscopy. The points of joining can be used as regional chromosome markers and sometimes define important genes (see also Chapter 4).*

Finally, a group of translocations exists in which all or part of one chromosome is exchanged with another. This process also was introduced earlier in relation to Down syndrome, but it is particularly important because of its relationship to cancer cell biology (see Chapter 4). In these translocations, there is reciprocal exhange of sequences between one chromosome and another, generally leading to two anomalous chromosomes (see Figure 3.11) that are difficult to clarify by light microscopy. Molecular hybridization methods such as FISH and painting can help define the details of translocations. It should be obvious that such a change in physical structure has the potential to affect control and integrity of large regions of genetic information.

Chromosome anomalies present unusual problems during meiosis. Some of these problems are shown in Figure 3.12. It is, for example, possible for an individual with a balanced Robertsonian translocation (and who contains diploid copies of all genetic information) to appear clinically normal. However, the problems that develop in meiosis and

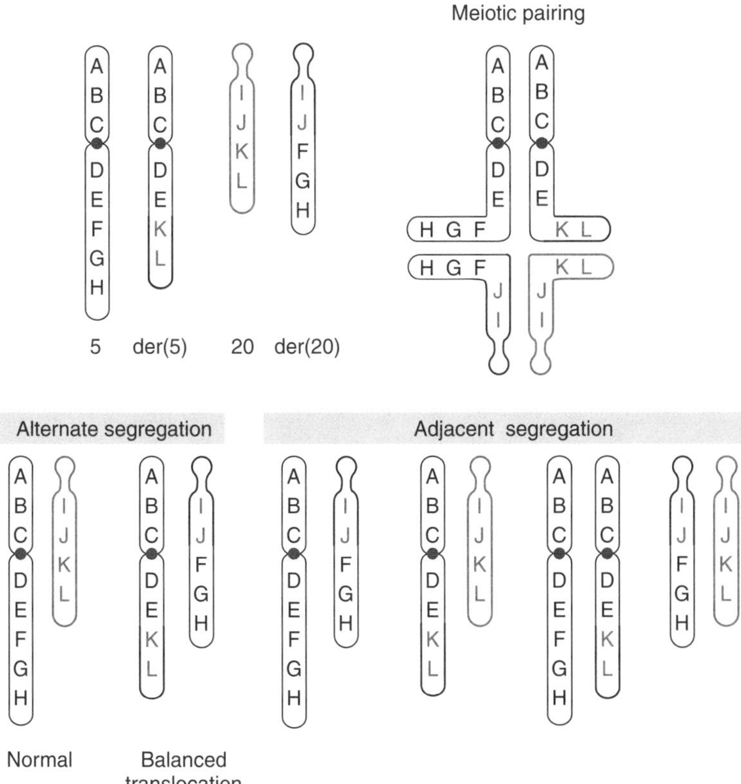

Figure 3.12 *Chromosomal anomalies cause difficulties during meiotic crossing-over. Note that many of the products of fertilization are not viable. Such situations often lead to recurrent fetal loss.*

in fertilization of the individual's meiotic products can be quite serious, and often lethal. As can be appreciated from Figure 3.12, few zygotes will contain a normal genetic complement after such matings. Thus, the presence of serious structural chromosome abnormalities—even if not detected—because of their important developmental consequences for the affected individual, often leads to a clinical picture of infertility and/or recurrent, spontaneous fetal loss. Sometimes chromosomal translocations and other anomalies are detected only after obstetric problems have been encountered, and this makes karyotype analysis an important part of infertility counseling. Approximately 4% of spontaneous first-trimester abortions lead to the diagnosis of unbalanced chromosomal abnormalities.

4. OTHER STRUCTURAL CHANGES

In addition to the situations considered above, several other types of abnormal chromosomes may be encountered. In general, these are considerably less frequent.

Ring chromosomes result when two broken ends of a chromosome are joined. If the ring thus formed does not contain a centromere, it will be lost in meiosis. Although ring chromosomes have been found for all human chromosomes, they are generally unstable, presenting particularly difficult problems during mitosis when the sister chromatids attempt to separate. Such problems can lead to changes in the size and content of the rings and often to their loss, leading both to additional anomalies and to a mosaic distribution in progeny cells. FISH and painting techniques have helped identify the constituent sequences in rings.

Marker chromosomes are rare. They are usually tiny and often contain little more than a short region around a centromere, making them a challenge to identify even with FISH and painting. When these appear in cell culture (for example, in prenatal diagnostic studies), they can be confusing, because their clinical implications often are unclear. A small subset of marker chromosomes contain functional genes, and some of these are associated with developmental problems.

Dicentric chromosomes, as their name implies, have two centromeres. These rare structures reflect breakage and rejoining of centromeric fragments from either the same or different parent chromosomes. Obviously, if both centromeres are functional, the dicentric will break during mitosis. Occasionally, mitotically stable structures are formed, usually because one of the centromeres is not used.

E. IMPRINTING

As described earlier, the fact that females have two copies of all genes on the X-chromosome implies that there must be some way of controlling the information expressed from these two sets of presumably identical genes. This is accomplished in the somatic cells of females by the process of X-chromosome inactivation. Historically, it has been the assumption that the expression of the genes on only one of a pair of chromosomes is limited to this situation of X-inactivation and that all autosomal genes operate with full expression from both maternally and paternally derived chromosomes. This is the basis of the assumptions underlying much of linkage analysis and the study of transmission patterns of inherited traits, which will be discussed in more detail in later chapters. However, recent evidence indicates that there may be differences in the expression, chromatin structure, and genetic implications of certain domains of autosomal chromosomes depending on their parent of origin. This implies that a gene(s) inherited from the mother can have expression pattern(s) different from those of the corresponding gene(s) inherited from the father. Although this is clearly not the case for most autosomal genes in humans, it nevertheless has been shown to be the case for several selected genetic regions. In fact, most current evidence supports the notion that multiple (possibly many) small chromosomal regions may be affected by this unusual transmission and expression process (imprinting).

Several examples of imprinting have been described in mouse genetics, but the most prominent situation in humans derives from a set of related phenomena all mapping to chromosome 15. Two remarkably contrasting clinical syndromes have been identified, both of which have been known for some time to involve chromosomal deletions in the region 15q11-13. The first of these, called "Prader-Willi syndrome" (PWS) (OMIM #176270), is associated with varying degrees of mental retardation, a large appetite with obesity, short stature, and characteristic neurological abnormalities. The other, called Angelman syndrome (AS) (OMIM #105830), has different features, including much more severe mental retardation, movement disorders, seizures, and a characteristically abnormal facial expression. The two remarkably different syndromes are easily distinguishable clinically (see Figure 3.13).

Although these conditions have been recognized as distinct from one another for over 30 years, only with the application of molecular dissection and detailed chromosome studies has it become apparent that they involve identical regions of chromosome 15. More recent studies have shown a pecularity of the genetic changes, however. The majority of the deletions in PWS involve loss of *paternal* chromosome sequences, while deletions in AS reflect loss of *maternal* chromosome sequences in the same region. This means that PWS patients are able to express only genes contributed by the mother to that region and AS patients can express only genes contributed by the father to the same region.

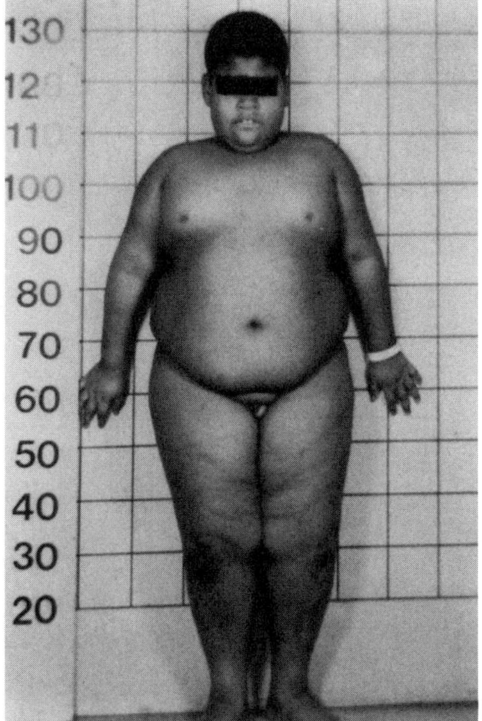

A

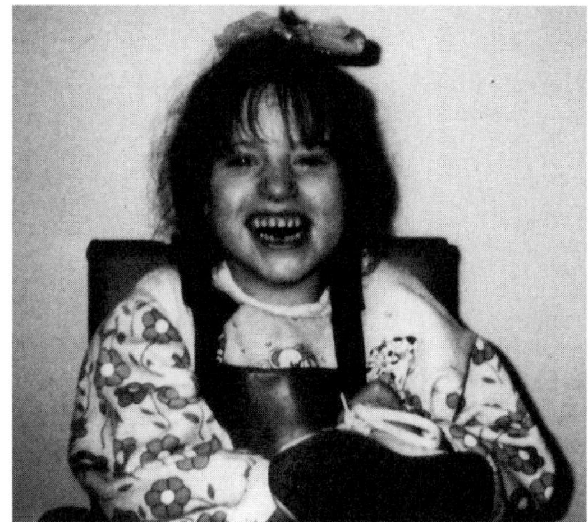

B

Figure 3.13 *Photographs of children with Prader-Willi (A) and Angelman (B) syndromes are readily distinguished. There is no confusion between these two phenotypes clinically. (Photographs courtesy of Dr. George Thomas.)*

An additional phenomenon—uniparental disomy—has been identified that accounts for at least some cases of both PWS and AS and may affect expression of other imprinted phenotypes as well. While all nucleated somatic cells are disomic for autosomes (meaning that two copies are present), one chromosome copy comes from each parent. However, using molecular markers that distinguish between maternal and paternal chromosomes has shown that, rarely, both come from the same parent. This is likely to be a very unusual occurrence, but it must be considered when anomalous transmission or gene expression patterns are found. Marker studies have identified situations in which the two chromosomes from the same parent are the same (isodisomy) and situations in which they are different (heterodisomy). Obviously, finding uniparental disomy in PWS patients means that two maternally derived copies of chromosome 15 are present; conversely, two paternally derived copies of chromosome 15 may rarely be found in AS.

The picture of a different pattern of gene expression and different clinical consequences (AS vs PWS) based on the parent of origin of the chromosomal region involved is unusual and has been the center of considerable study. Many of the deletions identified in these patients are relatively large, certainly capable of including several genes.

The search for molecular counterparts of the differential gene expression seen in imprinting has emphasized alternative patterns of DNA methylation. As described in Chapter 1, some restriction enzymes are sensitive to the presence of methyl groups added to DNA bases within their recognition sites. Thus, it is possible to take advantage of pairs of restriction enzymes in which cleavage by one is blocked by the presence of methyl groups while the other can continue to recognize and cleave DNA at the same site despite the methylation. The relative sensitivity of DNA to cleavage by these two enzymes thus can provide an estimate of the amount of methylation at the susceptible restriction enzyme site(s) and, by implication, in a larger, contiguous chromosomal region (see Figure 3.14).

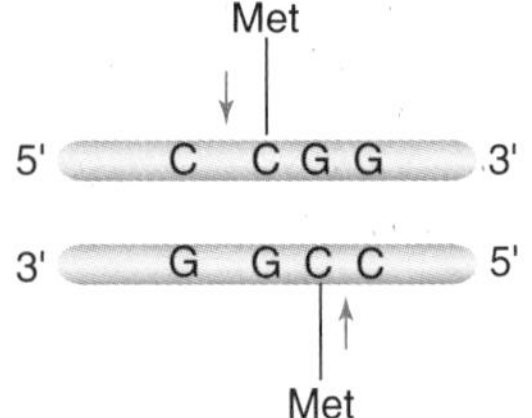

Figure 3.14 *Detecting site-specific DNA methylation by restriction enzyme cleavage patterns. Comparing the pattern of DNA cleavage by enzymes sensitive or resistant to the presence of cytosine methylation at the enzyme recognition site demonstrates the presence or absence of local methylation. MspI can cleave at the site as shown; HpaII cannot do so.*

Although all genes within the AS-PWS region have not been fully characterized, at least one has been studied in detail. This is a gene for a small ribonuclear protein (called SNRPN). Studies of sequences around the site where transcription is initiated for the SNRPN gene have shown that the region is extensively methylated on the maternally derived chromosome in AS (i.e., only the methylation-resistant enzyme can cleave the DNA) and unmethylated on the paternally derived chromosome (i.e., cleavage occurs with both test enzymes). The opposite situation exists in PWS.

These observations provide a molecular explanation for control of at least one of the genes in this small region. Differential methylation of either the maternal or the paternal copy of the gene is a potential initial regulatory mechanism.

DNA base methylation is associated with repression of transcription in other systems as well. For example, fetal globin genes are generally methylated in adult cells, where they are not expressed. However, the unusual feature about imprinting is that it involves only those genes derived from one parent. It also is important to realize that methylation is an example of the sort of DNA modification that can be transmitted genetically. Once double-stranded DNA that contains methyl groups is replicated, one of each of the new double strands will retain its methylated bases. Enzymes can then recognize these so-called "hemi-methylated" regions and add the missing methyl groups to the newly replicated complementary strand. In this way it is possible to transmit a maternal or paternal imprint pattern through mitosis and cell division.

The full extent of imprinting in genetic control is unknown. The AS-PWS dichotomy has been studied in considerable detail because the syndromes are readily recognized and, although they are rare, sufficient numbers of individuals have been studied to justify useful conclusions. There are undoubtedly additional imprinted regions in human chromosomes, however. An important reason for their not having been studied in detail has been that they have not been recognized on the basis of prominent, distinguishable clinical phenotypes. Imprinting is most likely an important contributor to developmental processes. PWS and AS both involve extensive developmental changes, particularly in the nervous system, where the expression of one or the other parent's genes appears to be critical.

The consequences of imprinting are likely to extend over a surprisingly large genetic region. Studies of other genes within the AS-PWS region have shown that methylation patterns can be carried through loci that are as far apart as 1.5 megabases (Mb; 1 Mb = 10^3 kb = 10^6 bp). This observation implies that there must be regional control over susceptibility to imprinting and, possibly, over the maintenance of the imprinting itself. This is reminiscent of the X-inactivation control regions discussed earlier.

The term *imprinting control element* (ICE) has been suggested for the identity of a chromosomal segment that is responsible for imprinting genes in its region. The notion of involving differences in larger chromosomal regions rather than only single genes also is consistent with the observation that imprinted chromosomal regions replicate later in mitosis. The paternal allele usually replicates ahead of the maternal allele. (Recall the late replication of the inactivated X-chromosome in females.) The methods by which the ICE is activated and imprinting-specific regions are maintained are still not understood fully, but they obviously have important implications in terms of gene expression, development, DNA replication, chromatin structure, and clinical phenotypes. There is additional evidence that imprinting also is involved with neoplastic changes (see Chapter 4).

CASE STUDY

A 36-year-old woman with Down syndrome lives at home and is able to go to a sheltered workshop daily. There she is active with packing small bags of electrical equipment. She is remarkably affable and pleasant to everyone. She has mild hypertension for which a simple diuretic provides good control. She also is hypothyroid, having taken supplemental hormone for over 15 years. Two years ago her mother indicated that the patient was quieter than usual and did not always respond to directions or questions. After some simple hearing tests in the clinic were equivocal, she had a formal audiogram, showing bilateral hearing deficiency in high frequencies. Although this problem was not sufficiently severe to require amplification, awareness of it has permitted helpful home and work adjustments and she is now interacting well again.

COMMENT

This woman points out several important features of Down syndrome:

- Many reach adulthood.
- Sheltered workshops can offer useful social and activity options.
- Hypothyroidism is more common in these people. Because of the patient's habitus and retardation, some of the cardinal presenting features of their condition may be missed; Down syndrome should be suspected during patients' routine care.
- Hearing loss also is an important feature. Its recognition can permit useful social and environmental adjustments to assist social interactions.

STUDY QUESTIONS

1 Two sisters have brought their much younger brother to the clinic. They are immigrants and speak little English. The small family group has kept everyone at home, and their brother does reasonably well there. They are concerned because he has become breathless over the last several months. On examination, you find a pleasant, rather short 19-year-old who is slightly pale. He speaks very little. He has prominent epicanthal folds and a unilateral simian crease. He has a moderately impressive (Grade 3/6) systolic murmur at the left sternal border, in the 4th intercostal space. There is a bruise on his left thigh. You suggest:

a He may have Down syndrome and a karyotype must be obtained.
b Bring in other family members to see if any look like the patient.
c Refer him to your hearing and speech center to evaluate possible dysarthria.
d Arrange for x-rays to determine bone age.
e Send him to the cardiology department for a stress test.

2 You get a telephone call from cardiology to explain that they decided not to do a stress test. Instead they have recommended an echocardiographic study. It shows changes compatible with an atrial septal defect. You reply:

a Fortunately, this defect has a low recurrence risk in family members, (see Chapter 9), but you will need to examine them all.

b The patient needs catheterization.
c The patient needs an ACE inhibitor.
d The patient needs digitalis.

3 Fortunately, a colleague who has been covering for you has obtained basic laboratory studies. You are surprised to find a leukocyte count of 56,000/mm^3 and a hemoglobin of 9.2 g/dl; other values are normal. You conclude:

a The patient has an upper respiratory infection.
b The patient has endocarditis.
c The patient has infectious mononucleosis.

4 Based on the possible diagnostic array from question 3, you suggest:

a A throat culture
b Multiple blood cultures
c A monospot test

5 All of your studies are negative. Where do you go from here?

6 A 24-year-old woman comes into your office with her 14-month-old daughter, who has Down syndrome. She is concerned because she and her husband are healthy and this is their first child. Although they have found their daughter a pleasant baby, they are not eager to have more children with Down syndrome. You tell her:

a Such events are not unknown, although they are rare. Her likelihood of having another child with Down syndrome is very low.
b She should be certain to have "triple screening" for subsequent pregnancies.

7 The woman returns 20 months later. Her daughter is growing but with slightly delayed milestones. The couple has repeatedly tried, unsuccessfully, to conceive. She has had some menstrual cycle abnormalties, with three episodes of delay and heavy flow. Her husband has accompanied her today. He recalls a second cousin who died with heart disease at 2 years of age in a hospital overseas, but no other details have ever been available. You suggest:

a Ultrasound should help evaluate subsequent possible pregnancies.
b The mother needs a gynecological referral for menstrual cycle abnormalities.
c The recurrence risk of Down syndrome is still low, and "triple screening" will be helpful.
d The father needs to have a karyotype done.
e Perhaps using oral contraceptives for several months will reestablish the mother's cycle.

8 A 38-year-old man has come to you with a breast lump. Biopsy shows that it is an early infiltrating carcinoma. After it has been removed surgically, he returns to you and you suggest:

a Mammography for his younger sister
b Prophylactic contralateral mastectomy
c Testicular biopsy
d A karyotype

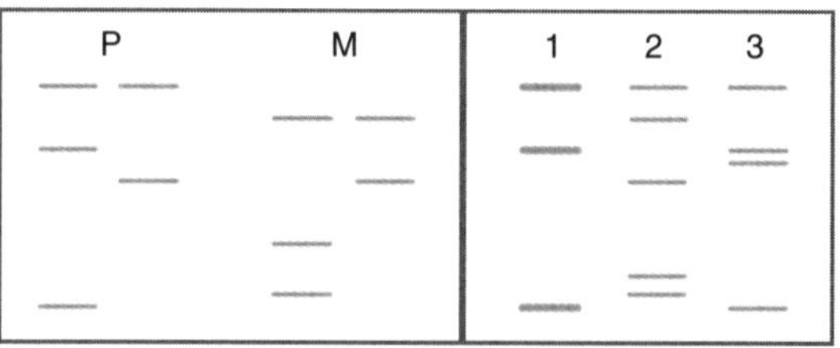

Figure 3.15 *Macho disease is present in three unrelated children; their parents are unaffected. BMOC probe hybridization to Southern blots shows that all three sets of parents have the same pattern, as shown, but that the children differ.*

9 Parents of a 15-year-old girl have come to you for advice. She is pleasant, attractive, and of average intelligence, but her family physician gave her oral contraceptives because of amenorrhea. They were still worried about her and insisted on chromosome studies; they have brought you the report, and it shows 46 XX (38%), 46 XO (62%). They want to have the study repeated. What do you tell them? Should karyotypes be obtained for the parents as well?

10 You are working in a chromosome laboratory and have been assembling a report on the results. The data from CVS studies have shown two ring chromosome, one isochromosome 2, and two trisomy 16 reports. These numbers seem high to you and have not been found in your review of amniocentesis studies. How can you explain the differences?

11 Macho disease is immediately recognizable because affected infants have dense muscles and a round face; they also are mentally retarded. The condition has been mapped to the long arm of chromosome 2, and a highly polymorphic DNA probe (BMOC) is available that probably is within the gene region. Southern blot patterns of restriction digests for parental DNA and three affected children hybridized to the BMOC probe show the patterns in Figure 3.15. Although the three children are unrelated, each had parents with the same pattern. What can you conclude about the Macho phenotype?

FURTHER READING

Daniel A. *The Cytogenetics of Mammalian Autosomal Rearrangements.* New York, AR Liss, 1988.

Epstein CJ. Down syndrome (trisomy 21), in Scriver CR, Beaudet AL, Sly WS, and Valle D (eds.). *The Metabolic and Molecular Bases of Inherited Disease, 7th ed.* New York, McGraw-Hill, 1995, Chap. 18, pp. 749–794.

Gardner RJM and Sutherland GR. *Chromosome Abnormalities and Genetic Counseling.* New York, Oxford University Press, 1996.

Ledbetter DH and Ballabio A. Molecular cytogenetics of contiguous gene syndromes: Mechanisms and consequences of gene dosage imbalance, in Scriver CR, Beaudet AL, Sly WS, and Valle D (eds.) *The Metabolic and Molecular Bases of Inherited Disease*, 7th ed. New York, McGraw-Hill, 1995, Chap. 20, pp. 811–839.

Lyon MF. X-chromosome inactivation and the location and expression of X-linked genes. *Am J Hum Genet* 42:8, 1988.

Saitoh Y and Laemmli UK. Bands arise from a differential folding path of the highly rich AT scaffold. *Cell* 76:609, 1994.

Therman E and Susman M. *Human Chromosomes: Structure, Behavior, and Effects.* New York, Springer-Verlag, 1993.

Willard HF. The sex chromosomes and X-chromosomal inactivation, in Scriver CR, Beaudet AL, Sly WS, and Valle D (eds.). *The Metabolic and Molecular Bases of Inherited Disease*, 7th ed. New York, McGraw-Hill, 1995, Chap. 16, pp. 719–737.

USEFUL WEB SITES

CYTOGENETICS GALLERY

http://www.pathology.washington.edu:80/Cytogallery
Includes acquired and constitutional chromosome anomalies and karyotypes. Useful comparison of banding patterns. FISH examples in color. Access to idiograms for generating custom figures.

CYTOGENETIC RESOURCES

http://www.kumc.edu/gec/geneinfo.html
A particularly valuable entrance to multiple sites including Cytogenetics Images Index, Cytogenetic images and animations, Gene Map of human Chromosomes, Human Cytogenetics Database, Chromosome Databases, Chromosome empiric risk calculations, and Karyotypes of normal and abnormal chromosomes. Excellent cross-references to other sites.

MOLECULAR CYTOGENETICS AND POSITIONAL CLONING CENTER (BERLIN)

http://www.mpimg-berlin-dahlem.mpg.de/~cytogen/
Describes a family of yeast artificial chromosomes (YACs) cytogenetically and genetically anchored, spread evenly over the entire human genome. A resource for FISH mapping. There also is a collection of YAC probes for all human chromosomal ends, averaging several cM from actual ends. This is a reference panel for studies of telomere rearrangements.

RESOURCES FOR HUMAN MOLECULAR CYTOGENETICS

http://bioserver.uniba.it/fish/Cytogenetics/welcome.html
Libraries of partial chromosome paints, recognizing a definite region of a chromosome. >900 fragments characterized from both normal and radiation-induced somatic cell hybrids.

CYTOGENETIC IMAGES AND ANIMATIONS

http://www.tokyo~med.ac.jp
Animations of meiotic interchanges with normal and anomalous chromosome structures.

MARCH OF DIMES

http://modimes.org/
Resources for families and professionals about birth defects. Professional education section.

CHAPTER 4

Genetics and Cancer

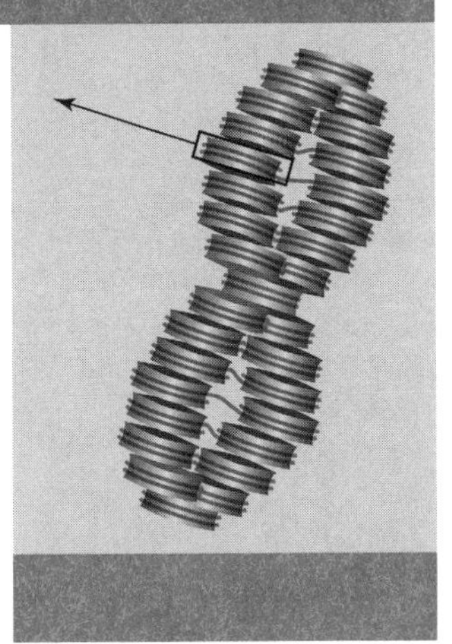

A useful view of the development of cancer is that some basic alteration in a cell gives it aggressive growth properties that are then transmitted to the lineage of that cell. This implies that cancer must be intimately involved with genetic changes, because the growth pattern then becomes a heritable feature of the progeny of the initial abnormal cell. This notion has undergone considerable study, and several important lines of supporting genetic evidence can be considered in cancer biology.

A. CHROMOSOME CHANGES

The idea that chromosome changes (and, by implication, gene changes) could be consistent and conspicuous features of cancer cells is not new. Some chromosome changes are virtually diagnostic of specific cancer categories. The most prominent of these is the so-called Philadelphia chromosome (see Figure 4.1), a characteristic, easily recognized chromosome reflecting breaking and rejoining events that produce a reciprocal translocation. The abnormal chromosome (written as "Ph^1") can be detected in leukemic cells of individuals with chronic myelogenous leukemia (CML). The Philadelphia chromosome is a marker for the presence of CML. Successful treatment of CML leads to disappearance of the Philadelphia chromosome from white blood cells. The Philadelphia chromosome is only one of many aberrant but predictable chromosome changes that are recognized as characteris-

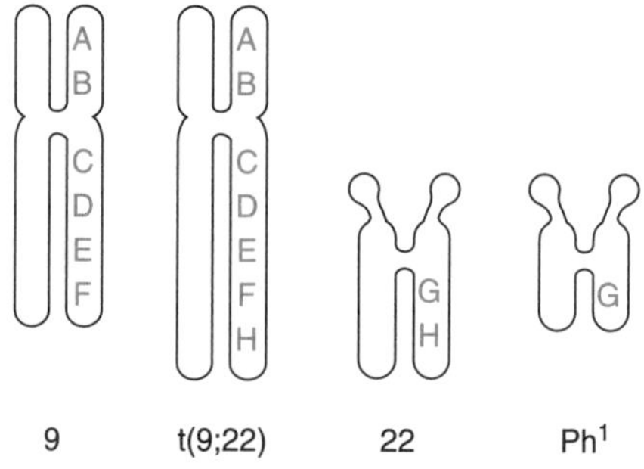

Figure 4.1 *Diagram of the formation of a Philadelphia chromosome. This chromosome results from a translocation between the distal long arm of chromosome 9 and the distal long arm of chromosome 22. It is characteristic of most patients with chronic myelogenous leukemia. What sort of translocation is this?*

tic for different types of tumors. These changes can be diagnostic when present, and they also can serve as useful monitors of the spread and aggressiveness of the malignancy. Obviously, eliminating the malignancy (curing the cancer) is associated with disappearance of cells containing the chromosomal abnormality or abnormalities.

Molecular biological studies have now been brought to bear on many of these aberrant chromosomes. It has been possible to dissect the genes involved at the points of breakage and rejoining. Several of these genes have been identified, and the fact that the structural changes occur at predictable positions in and around the genes has implicated at least some of them in the biology of the cancer cells. An entire group of genes has been defined in this way. A list of some of these is shown in Table 4.1. The genes found at these positions are frequently

Table 4.1 ▶ **Genes at translocation breakpoints associated with neoplasia**

Gene	Location	Translocation	Disease
SRC family (tyrosine protein kinases)			
ABL	9q34	t(9;22) Ph1	CML/ALL
LCK	1p34	t(1;7)	T-ALL
ALK	5q35	t(2;5)	NHL
Serine protein kinase			
BCR	22q11	t(9;22)	CML/ALL
Cell surface receptor			
TAN1	9q34	t(7;9)	T-ALL
Growth factors			
IL-2	4q26	t(4;16)	T-NHL
IL-3	5q31	t(5;14)	PreB-ALL
Mitochondrial membrane protein			
BCL2	18q21	t(14;18)	NHL
Cell-cycle regulator			
CCND1 (BCL1-PRAD1)	11q13	t(11;14)	CLL/NHL
Myosin family			
MYH11	16p13	inv(16),t(16;16)	AML-M4Eo
Ribosomal protein			
EAP (L22)	3q26	t(3;21)	t-AML/CML BC
Unknown			
DEK	6p23	t(6;9)	AML-M2/M4

(From Rowley, *Am J Hum Genet* 54:403, 1994.)

involved with controlling cell growth, protein synthesis, or DNA replication or recombination. The remarkable consistency of these chromosomal translocation events is still not completely explained in terms of the underlying mechanism(s), but it fits in with the notion that a change in the control or structure of the gene or genes in a critical region is essential for tumor development.

Another type of chromosome change that has been recognized in cells of various tumors involves the loss of chromosomal material. Historically, these losses have been characterized on the basis of deletions visible by conventional light microscopy. Visible deletions, however, necessarily involve very large regions of chromosomal DNA and therefore must involve many genes. Obviously, the loss of a large number of genes could create any number of complex biological results. Therefore, the study of large chromosome deletions has not been particularly helpful in terms of understanding the detailed biology of genetic changes in cancer.

More useful information has arisen from studying small deletions affecting, in at least some cases, only single genes, or at least very small numbers of genes. Here an important notion has arisen that must be considered in cancer biology, as well as in the biology of other genetic problems that develop after birth or even during adult life. The presence of one abnormal autosomal gene may have no consequences for the individual, because a normal copy exists on the other chromosome. This, obviously, is the situation in heterozygotes. It is clear, on the basis of many population studies, some of which will be considered in later chapters, that the frequency of heterozygotes for many conditions is collectively high (see Chapter 6). Most of these are clinically undetectable. If the remaining normal gene were to be lost or inactivated (as a rare random event in a somatic cell due to an acquired mutation such as a replication error), however, the only source of information for that locus would necessarily be the mutant gene. Thus, loss of the normal gene in a heterozygous situation either unmasks the expression of the mutant allele or means that no gene product can be made at all. After such an event, the progeny cells would have no copies of the normal gene, because one was defective by inheritance and the other acquired a mutation during growth. Such change would occur well after the initial fertilization event, possibly much later in life (recall the discussion of mosaicism in Chapter 3). One would expect that such an event would be relatively rare; however, it could have important medical consequences.

The loss of a normal allele in a heterozygous situation could be particularly important if there were a way of detecting the resulting changes in the physiology of the affected cell. Because uncontrolled cell growth, recognized clinically as cancer, would be a dramatic marker for such an event, allele losses or specific gene changes associated with cancer have become the focus of considerable study.

B. GATEKEEPER GENES

An important class of genes, collectively referred to as "gatekeepers," has been recognized. In the past, these also have been considered "tumor suppressor genes" and have been defined operationally as genes whose normal function is important for controlling cell growth and guaranteeing that aberrant cell division does not occur (usually by inhibiting cell growth or promoting cell death). Obviously, loss or aberrant function of such a gene could lead to uncontrolled cell growth. Such gatekeepers have generally been found to be specific for certain cell or tissue types, and their individual failure would be expected to lead to specific distributions of tumors. Table 4.2 lists several "gatekeeper" genes and tumors associated with them.

An individual heterozygous for a gatekeeper mutation gene might never have a problem in his or her entire life, because a normal copy of the gene always was present and capable of controlling cellular metabolism. However, if there were any change in the function or effectiveness of the normal allele, tumor suppression would be weakened and a detectable neoplastic event could occur. In such a situation, the likelihood of losing the protection of the normal allele would be related to the likelihood of having some sort of mutational event in that allele. Thus, in a cell heterozygous for a tumor suppressor gene mutation, the frequency of mutation(s) at the normal allele would be related to the frequency of tumor development. Given the normal rate of mutation in human cells (approximately 1 in 10^8 per generation per locus), there is a finite (and rather low) chance that a mutagenic event would occur in the normal allele. Obviously, if we consider a cell with rapid turnover (more mitosis) throughout the life of the individual (for example, a blood cell precursor, an intestinal epithelial cell, or a skin cell), there would be an increased number of chances for a mutagenic event to occur.

C. THE KNUDSON HYPOTHESIS

The likelihood of such an event was recognized by Knudson. He proposed that the incidence of unmasking a mutant allele would be related to the spontaneous mutation rate of the remaining normal allele. According to this hypothesis, the frequency of

Table 4.2 ▸ **Syndromes with gatekeeper gene mutations**

Syndrome	OMIM #	Tumor type
Neurofibromatosis I	162200	Schwann cells
Retinoblastoma	180200	Retina
von Hippel-Lindau	193300	Kidney
Adenomatous polyposis coli	175100	Colon

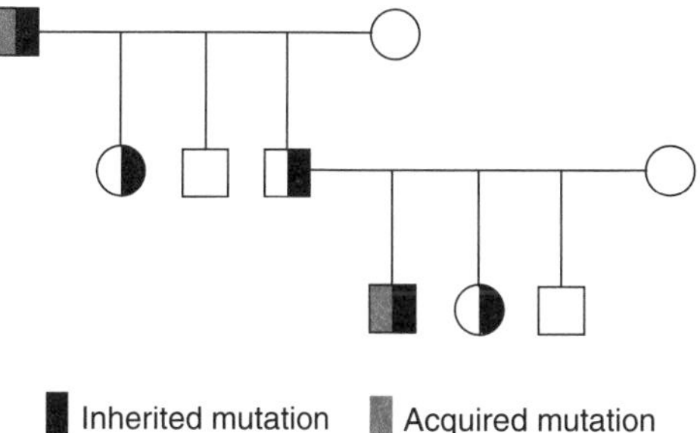

Figure 4.2 *The Knudson hypothesis: two "hits" (mutations) are needed for tumorigenesis. When the first hit in one of a pair of autosomal genes is inherited, only a single hit in the other allele is needed to lead to tumor growth. Obviously, since one of a pair of genes is already mutated from the time of conception, the frequency of visible tumors is far higher than if independent mutations were needed in both. (Recall that the likelihood of having both genes mutated is the product of the individual mutation frequencies.)*

tumor development (and probably the number of tumors) due to failure of or change in a gatekeeper gene's function in an individual who is heterozygous for an inherited mutation in that gene would be considerably higher than the rate in an individual who inherited a normal copy of the gene on both chromosomes. Homozygous normal individuals would require a mutation in both alleles (two "hits"); this occurs at a frequency representing the product of the frequencies of mutation at each site (see Figure 4.2).

Obviously, the "Knudson hypothesis" is that individuals requiring a mutation in both alleles of a tumor suppressor gene would develop the resulting tumor far less frequently than individuals in whom one aberrant allele is inherited. As we will discuss in later chapters, this is an important model for understanding many types of genetic phenotypes and is essential in understanding many aspects of inherited predispositions to cancer and other diseases.

D. LOSS OF HETEROZYGOSITY

The idea of losing normal alleles during cell growth and repeated mitoses has become testable. It is possible to search, using Southern blotting and restriction enzyme analysis, for evidence of loss or change(s) of specific genes in tumors and normal cells. A particularly instructive set of studies has been based on investigations of colon cancer. Neoplastic changes occur often in the cells lining the colon. These cells are constantly undergoing mitosis and replacement; mutations are inevitable. A large number of these acquired mutations lead to benign polyps, and many are of no clinical consequence. On the other hand, some of the changes progress to become initially local and finally aggressively invasive colon cancer. With the advent of colonoscopy and other methods for retrieving tissue in early and later stages of disease, and with PCR now permitting DNA studies on small numbers of cells, it has become possible to assess the presence of genetic changes as cells develop from benign growths to malignant tumors; here the emphasis has been on searching for deletions implying loss of specific genes or chromosomal regions.

The analysis is performed when maternal and paternal copies of a specific region can be distinguished, usually on the basis of distinctive restriction enzyme patterns or some trivial polymorphism [such as $(CA)_n$ variation] unrelated to their biology. Study of the two alleles (maternal and paternal) in increasingly aberrant tissues can disclose the loss of one or both (see Figure 4.3). The loss of one allele immediately establishes a situation in which the remaining allele has no counterpart. The phenomenon is known as "loss of heterozygosity" (LOH). By examining hundreds of biopsy samples of colon polyps and tumors, Vogelstein, Kinzler, and coworkers were able to establish a relative sequence for the loss of alleles in the clinical progression of colon tumors. They found that there were progressive stages of genetic change, detectable as LOH, that were consistent throughout the biology of these tumors. Of course, as the tumors became more aggressive with less reliable growth control and metastasis, there often were multiple genetic changes and the metastatic colon tumors showed multiple sites of LOH.

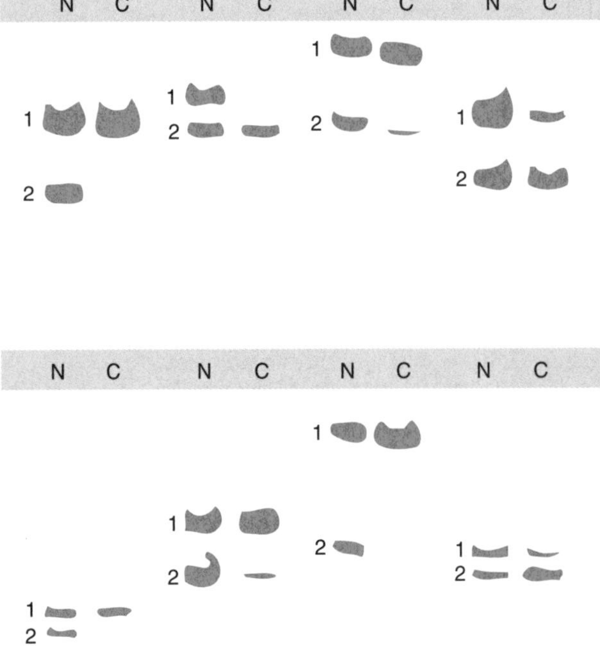

Figure 4.3 *Loss of heterozygosity (LOH) studies in colon cancers. Note the different patterns in cells from different individuals. Such studies detect only changes that lead to altered mobility or disappearance of the specific DNA band; many point mutations (which also might be important in terms of tumor formation) will not be found in this way. N, normal tissue; C, cancer. (Modified from Fearon et al. Science 238:193, 1987.)*

E. CARETAKER GENES

An interesting possibility is that losing a gene that helps control or provides proofreading for genetic errors could be a particularly damaging event, because subsequent mutational events or replication errors could no longer be corrected. This also is an important notion in terms of understanding the progressive alterations of cancer cell biology. As LOH studies at different points in colon tumor development have been analyzed, it has become clear that there is a relentless progression with a relatively predictable sequence of allele losses. This has led to the detection and characterization of several genes that are essential for colonic integrity, including the gene for the hereditary nonpolyposis colorectal cancer (HNPCC, OMIM #120435). This gene encodes a protein that is important for repair of DNA base pair mismatches that develop during normal DNA replication as well as in response to external mutagens. Such a gene can be considered a "caretaker gene."

Not surprisingly, caretaker gene mutations have been identified in several genetic syndromes associated with malignancy; Table 4.3 lists several of these. It is important to note that these mutations are probably not directly responsible for the aberrant cell growth characteristic of cancer. Rather, they lead to what has been called "genetic instability," a situation in which there is absent or inefficient repair of genetic damage during the lives of cells. Such damage could be due to mutagens, such as drugs or radiation, or even to the low rate of spontaneous mutation described in Chapter 1. Because maintaining genetic integrity is essential for all cells, one could anticipate that multiple types of cells would be affected and multiple mutations could accumulate in these disorders. Mutation of gatekeeper genes also could occur, at which point a full neoplastic growth picture would develop.

For the full neoplastic picture to develop in an individual heterozygous for a mutation in a caretaker gene, the process is different from that for the gatekeeper mutations described above. Instead of a single mutation, three mutations would be required. The first would inactivate the normal allele of the caretaker gene. This would lead to a broad increase in mutations, which in turn ultimately would include mutations in both alleles of a gatekeeper gene. At this point, the full picture of a cancer cell would ensue.

Table 4.3 Syndromes with caretaker gene mutations

Syndrome	OMIM #	Defect
Xeroderma pigmentosum	278700	Excision repair
Hereditary nonpolyposis colorectal cancer	120435	Mismatch repair
Ataxia–telangiectasia	208900	Cell cycle control
BRCA1/BRCA2	113705/600185	DNA break repair

Because of the additional mutations needed, one would not expect the frequency of tumors in an individual heterozygous for a caretaker gene mutation to resemble the predictions of the Knudson hypothesis. In fact, the predisposition to cancer might be only moderately increased (probably under two orders of magnitude). Nevertheless, recognizing aberrant behavior of a caretaker gene could be helpful in predicting cancer development and even in designing effective treatment.

Several studies have now been directed toward analyzing many sorts of tumors. This is particularly useful in situations in which biopsy material is available at different stages of tumor growth and development. Thus, chromosome changes are now being characterized in neoplasms of the bladder, prostate, thyroid gland, lung, and other important organs. It is likely that continued efforts in this regard will unravel more of the events that collectively result in uncontrolled neoplastic cell growth. Unfortunately, it also is likely that the events will be complicated. The evidence to date indicates that there are progressive losses of reliable control of mutation (DNA damage) repair, mitosis, cell-cell interactions, and growth that may not be easily interpretable because so many changes have occurred. Nevertheless, the fundamental genetic notion underlying tumor development is becoming clear. This is that neoplasia involves the accumulation of a series of individual genetic changes that ultimately affect basic cell biology.

CASE STUDY

A 28-month-old boy had normal growth and development but showed poor attention to objects over 2 months. He was often seen turning his head to the right when looking at objects. Basic examination showed an apparently healthy child but the expected red reflex was absent from his right eye (see Figure 4.4, *left*); in its place a white spot was seen. Ophthalmoscopy showed a large mass in the right eye, largely obscuring the retina; it was a retinoblastoma. Because of the size of the mass it was treated by enucleation. The boy did well for another 16 months and was seen for a routine follow-up visit. Ophthalmoscopy showed several small white tumor masses in the left retina (see Figure 4.4, *right*).

COMMENT

This child has recurrent retinoblastoma. Several features confirm aspects of the Knudson hypothesis:

- Both eyes are involved.
- There are multiple tumors in the left eye.

Retinoblastoma accounts for 5% of childhood blindness and 1% of childhood cancer deaths. Approximately 30% of cases are hereditary with a spectrum of lesion locations and sizes as predicted from the Knudson hypothesis. Isolated retinoblastomas are usually solitary and virtually always unilateral.

Some individuals with retinoblastoma have deletions involving the (RB) gene on chromosome 13q14. This gene encodes a large protein that is essential in maintaining cell division control and also is a protein that interacts with other proteins to promote or limit cell division. Obviously, an inherited RB gene mutation renders the individual highly susceptible to any change (loss, mutation, inactivation) of the remaining RB gene. The frequency of such events is manifest by multiple, recurrent tumors.

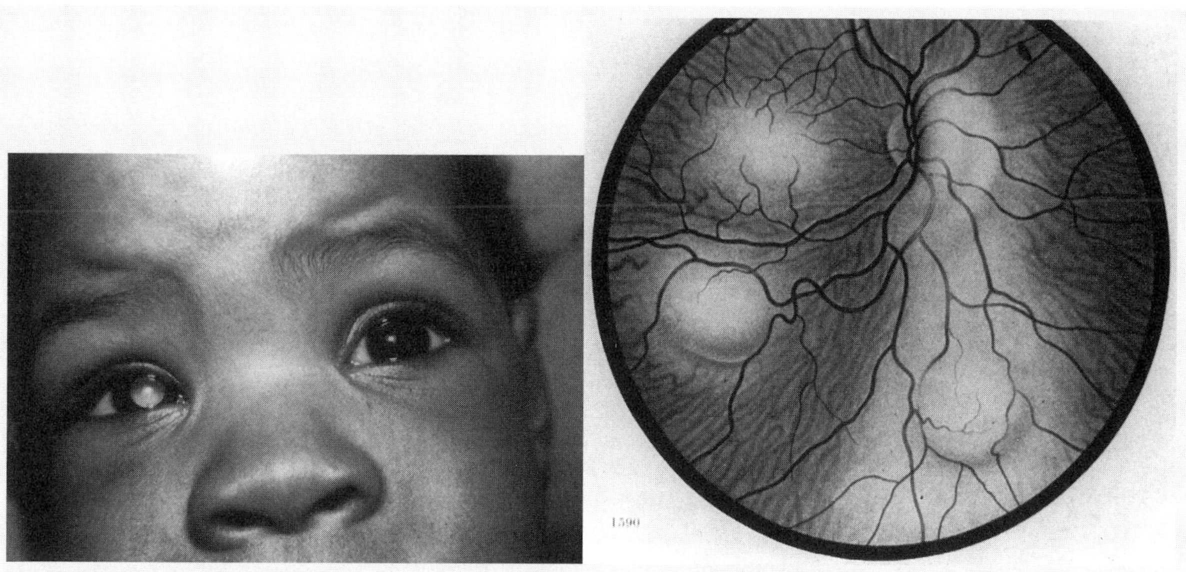

Figure 4.4 *Initial presentation of the patient showed the right eye change as shown to the left. A drawing of the later findings in the patient's left eye is shown to the right. (Photographs courtesy of Dr. I. H. Maumenee.)*

STUDY QUESTIONS

1 Cockayne syndrome (OMIM #216400) and Fanconi anemia (OMIM #227650) are rare inherited disorders that may be complicated by malignancy. Although their mutations differ, both are associated with defective repair of DNA damage. Can these statements be related?

2 A 28-year-old woman comes to you for a general health checkup. You find that she has a striking family history for colon cancer in her father's family (his brother, sister, and mother died with the disease, but he had two unaffected sibs), although her father died in a car accident at age 30 and had no known problems. She has three younger sibs. You tell her:

a She should have a colectomy.
b She has no basis for concern because her father and two of his sibs were unaffected.
c She should increase her intake of fiber.
d She should have colonoscopy.
e Her younger sibs should have colonoscopy.
f) She should have stool specimens tested for occult blood twice a year.
g Genetic testing may be available for her.

3 Wilms tumor (OMIM #194070) is a nephroblastoma that usually develops in children and can be encountered in sibships. Two important clinical observations are: (1) sibs with Wilms tumor often have multiple tumors and (2) sporadic cases of Wilms tumor (which may appear pathologically indistinguishable from the others) are generally unilateral. Consider the

implications of these observations. How do they fit in with what you know about tumor biology? Would LOH studies be valuable?

4 You are the primary care physician for a 70-year-old man with CML. The problem was stable until last year, when he required an intensive period of chemotherapy. Following this he did quite well and is now back in your office for a routine visit. You review the oncologist's report, in which she states that prior to his chemotherapy Ph^1 was seen in 72% of peripheral blood cells but that now the level is between 3% and 5%. The family is interested in your opinion about how he is doing. What will you tell them?

5 Cytogenetic studies of solid tumors:

a are used to predict metastasis patterns
b can assist with diagnosis
c can be difficult to perform and interpret because the tissues grow aberrantly
d are less valuable than FISH and painting
e employ many cytogeneticists
f define sites for gene therapy

FURTHER READING

Cahill DP et al. Mutations of mitotic checkpoint genes in human cancers. *Nature* 392:300, 1998.

Kinzler KW and Vogelstein B. Gatekeepers and caretakers. *Nature* 386:761, 1997.

Knudson AG. Genetics of human cancer. *Ann Rev Genet* 20:231, 1986.

Rowley JD. The Philadelphia chromosome translocation: A paradigm for understanding leukemia. *Cancer* 65:2178, 1990.

Vogelstein B and Kinzler KW. *The Genetics of Cancer.* New York, McGraw-Hill, 1997.

USEFUL WEB SITES

DICTIONARY OF CELL BIOLOGY
http://www.mblab.gla.ac.uk/~julian/Dict.html
Internet edition of reference book of same name. Permits searching of 5450 entries with links to other, related entries.

INDIANA UNIVERSITY BIOTECHNOLOGY
http://biotech.chem.indiana.edu
Basic introduction to principles of genetics and molecular biology.

HOWARD HUGHES MEDICAL INSTITUTE
http://www.hhmi.org
Summaries of sponsored research in genetics. Timely new updates. Useful section on structural biology.

CYTOGENETICS GALLERY

http://www.pathology.washington.edu:80/Cytogallery
Includes acquired and constitutional chromosome anomalies and karyotypes. Useful comparison of banding patterns. FISH examples in color. Access to idiograms for generating custom figures.

CYTOGENETIC RESOURCES

http://www.kumc.edu/gec/geneinfo.html
A particularly valuable entrance to multiple sites including: Cytogenetics Images Index, Cytogenetic images and animations, Gene Map of Human Chromosomes, Human Cytogenetics Database, Chromosome Databases, Chromosome empiric risk calculations and Karyotypes of normal and abnormal chromosomes. Excellent cross-references to other sites.

CYTOGENETICS

http://www.waisman.wisc.edu/cytogenetics
Useful review of chromosome changes in tumors.

CHAPTER 5

Autosomal Dominant Conditions

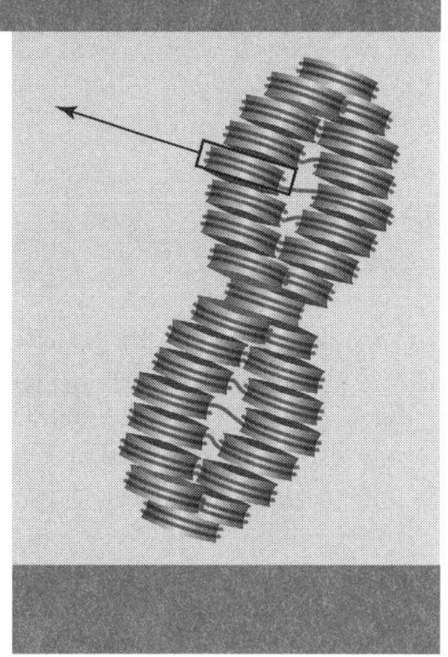

A. OVERVIEW

A dominantly inherited condition is detectable in an individual who has only one altered gene of a pair. Such a person is called a "heterozygote." Because having only a single aberrant copy of the gene is sufficient to develop a recognizable clinical phenotype, the transmission pattern of autosomal dominant conditions is very characteristic. The chance that any individual germ cell will contain the chromosome with the abnormal gene is obviously 50%, because there is a 50% chance of receiving either of the chromosomes of any autosomal pair. This means that the chance for an affected individual to transmit the abnormal gene is 50% with each conception. In general, the aberrant gene is detectable clinically in all individuals carrying it, and thus it is expected that an average of one-half of the offspring of an affected individual also will be affected.

Several autosomal dominant pedigrees are shown in Figure 5.1. Certain conspicuous features can be derived from these patterns; other aspects are important as well:

- Careful questioning may reveal multigenerational family histories of problems, establishing what has been termed the "vertical transmission pattern."
- The sexes are involved equally, because there is no sex limitation

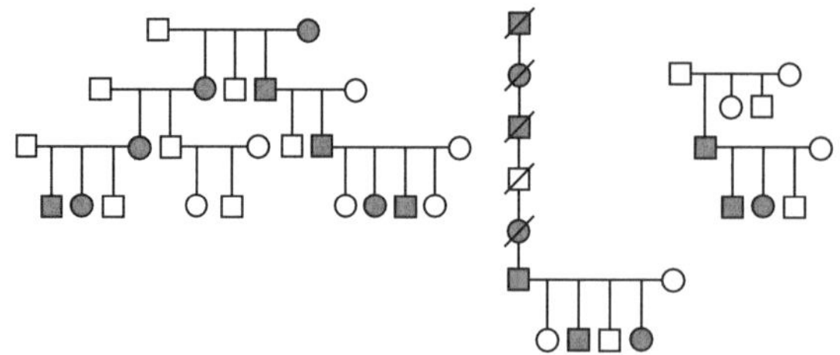

Figure 5.1 *Kindreds with autosomal dominant inheritance. Note that even partial ascertainment (i.e., not all siblings have been identified or some generational data are missing) can still establish the basic inheritance pattern. Such longitudinal information can help in interpreting clinical presentations and also may be helpful for determining prognosis.*

for manifesting the aberrant gene (although some manifestations in males may differ from those in females).

- Some individuals present with recognizable clinical pictures that reflect new mutations. Although there is no antecedent family history in these individuals, subsequent transmission of the trait will follow the predictable pattern, with a 50% chance of transmitting the trait with each conception. Often this idea of having new mutational events establish the potential for subsequent, predictable Mendelian transmission is difficult for families and individuals to understand.
- Because dominant conditions are detectable in the presence of the normal allele of the responsible gene (on the unaffected chromosome), their physiologic bases often, but not always, are related to aberrant structural or developmental problems. Only a relatively small number of dominant conditions (acute intermittent porphyria being an excellent example—see Chapter 10) reflect enzymatic or metabolic blockades.
- Dominant transmission includes the mechanism of so-called triplet repeat disorders. This relatively newly recognized phenomenon has added particularly valuable insight into certain neurologic disorders and what have been considered anomalous aspects of their transmission.
- Dominant disorders often show "pleiotropy." This means that multiple, overtly unconnected biologic and clinical changes may develop from a single mutation. Pleiotropic effects often involve multiple organ systems. With improved biochemical and physiologic understanding, such effects are found to be consistent with the biology of change.
- The severity or prominence of a particular aspect of a dominant disorder may be unpredictable, despite the obvious presence of the mutation. This has been termed "variable expressivity."
- So-called dominant tumor syndromes provide clinical support for the Knudson hypothesis, presented in Chapter 4, regarding neoplastic transformation.

Consideration of several dominantly inherited conditions will point out the operation of these individual principles.

B. NEUROFIBROMATOSIS (VRNF)

This remarkable condition, also associated with the eponym "von Recklinghausen disease" (VRNF; OMIM #162200) is found in all ethnic groups and is relatively common (on the average, one in 10,000). The name derives from the soft tumors that develop (see Figure 5.2). The tumors are derived from supporting tissues of the nervous system; they do not arise from the nerves themselves. This means that they are generally outgrowths of Schwann cells and have a very characteristic, usually rather bland and monotonous, appearance on biopsy. Neurofibromas can develop wherever there are nerves. When superficial, they can be obvious, pedunculated, and freely movable on stalks or detectable only as lumps below the skin surface. Careful palpation often reveals deeper neurofibromas that are not even visible. Neurofibromas continue to develop over the lifetime of the affected individual. Careful and continued observation can document their development. A particularly characteristic location for their development is in the ciliary body of the iris in the eye, where they are

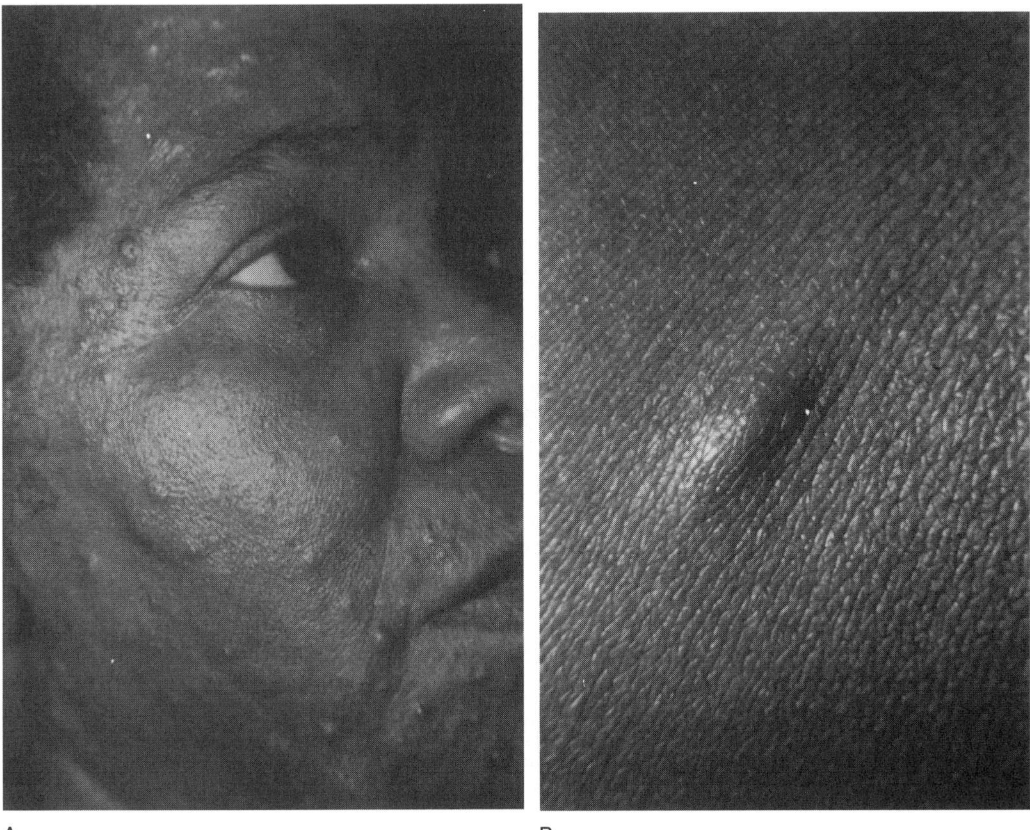

Figure 5.2 *Neurofibromas. These may be frequent and very prominent, as in the individual in A. They also may be more subtle and isolated, as in B.*

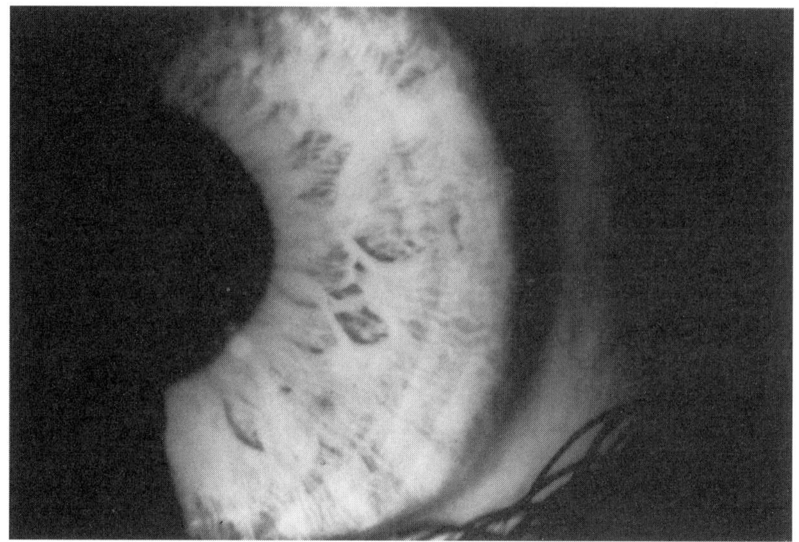

Figure 5.3 *Lisch nodules. These growths around ciliary nerves can be useful in recognizing and helpful in confirming the diagnosis of VRNF. Some can be seen without magnification; others require a hand-held ophthalmoscope or slit lamp for inspection. See also Plate 1.*

known as Lisch nodules (see Figure 5.3). The presence of Lisch nodules can be one of the most useful confirmatory observations in cases where the diagnosis is uncertain.

In addition to neurofibromas, another characteristic feature of this condition is the presence of skin pigmentary changes. The most frequent is the pigmented macule (see Figure 5.4). These macules are generally the color of coffee with cream and thus are known as "café-au-lait" spots. Café-au-lait spots have a characteristic hue in most population groups. They have smooth borders and can appear anywhere on the body. These spots are not unique to VRNF and can be seen occasionally in otherwise healthy individuals, as well as in those with a few less common conditions. Nevertheless, seeing large numbers is highly correlated with the presence of the condition. An important population study showed that the presence of six or more café-au-lait spots is usually diagnostic of VRNF (see Figure 5.5). Although these macules may have considerable cosmetic prominence, they do not, in themselves, cause problems.

Complications develop in individuals with VRNF, but they are generally difficult to predict. This is characteristic of the variable expressivity recognized in autosomal dominant conditions. Most of the complications of VRNF are related to the neurofibromas themselves. As described, they represent benign, soft, freely movable tumors, usually of relatively small size. They may grow to a large size, however, particularly in areas such as the abdomen, where there is no constraint on their enlargement and where they may not be noticed. The large size in itself may not be a problem; some abdominal tumors of remarkable size have been well tolerated. If such a tumor develops within an area of the nervous system that has bony confines, however, the situation

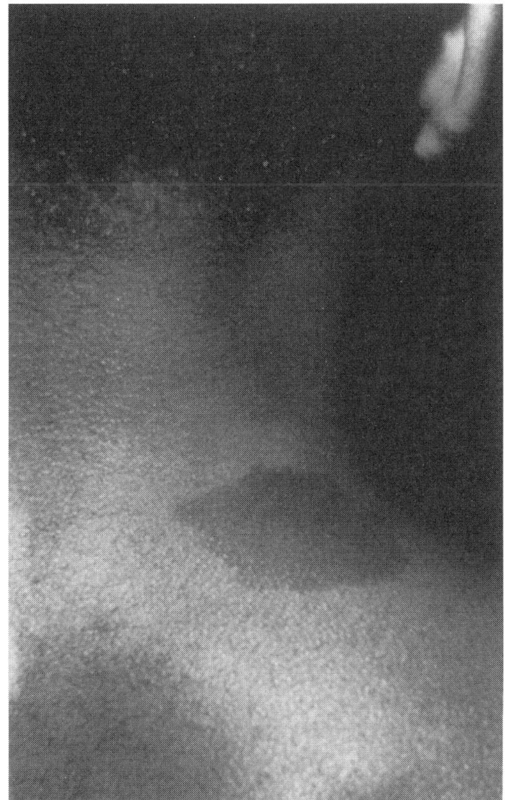

Figure 5.4 *Café-au-lait spot. These pigmented macules can be of various sizes but have smooth borders and are recognizable in individuals of all races. See also Plate 2.*

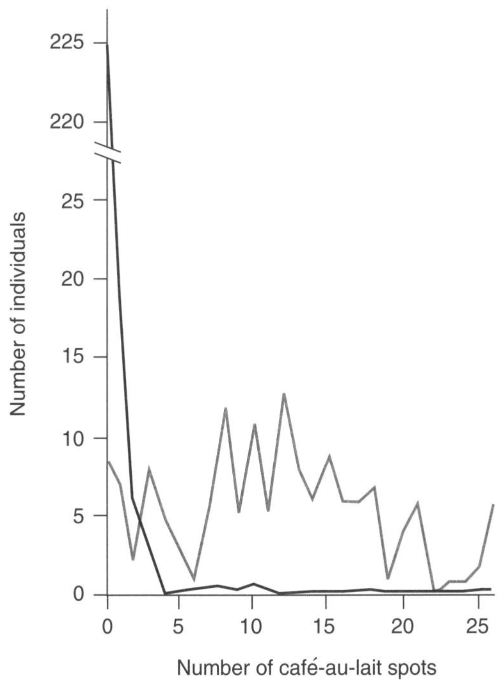

Figure 5.5 *Graph emphasizing the fact that small numbers of café-au-lait spots are relatively frequent in the population and infrequently indicative of VRNF. As the number of spots rises, however, the diagnosis is progressively more likely. The black line represents unaffected individuals; the colored line represents patients with VRNF. (Adapted from Crowe et al, Springfield, Charles C Thomas, 1956.)*

can be considerably more serious. Neurofibromas that develop in the spinal column can compress the spinal cord and cause characteristic regional neurologic changes (see Figure 5.6). Similarly, their development within the cranium can cause hydrocephalus and damage to local brain structures. Because of the possibility of these changes, long-term observation of individuals with VRNF is essential. Imaging techniques have greatly simplified the assessment of neurofibromas and can provide useful benchmarks for clinical comparisons throughout life.

VRNF is a result of changes in the gene for the protein "neurofibromin," located at chromosome position 17q11.2. This gene spans over 350 kb of genomic DNA and encodes a messenger RNA of 11-13 kb, containing at least 59 exons. Many types of mutations have been found in this gene, including deletions, point mutations, and frameshifts; 82% of fully characterized mutations lead to production of a truncated protein.

Molecular studies of neurofibromin show that it is related to so-called GTPase-activating proteins (GAPs). GAPs interact with proteins such as the RAS gene product in gatekeeper roles in the cell (see Chapter 4). The control of cell division and other metabolic processes is mediated by the association of RAS and other proteins with GTP. Hydrolysis of this triphosphate changes the regulation of these processes. Thus, any change in neurofibromin's role as a GTPase activator can substantially affect cellular metabolism and growth.

Obviously, altered control of this gatekeeper system could cause aberrant cell proliferation. This raises the presumption that neurofibromas and café-au-lait spots arise from unique events in individual cells; in other words, these are clonal lesions. This has been found to

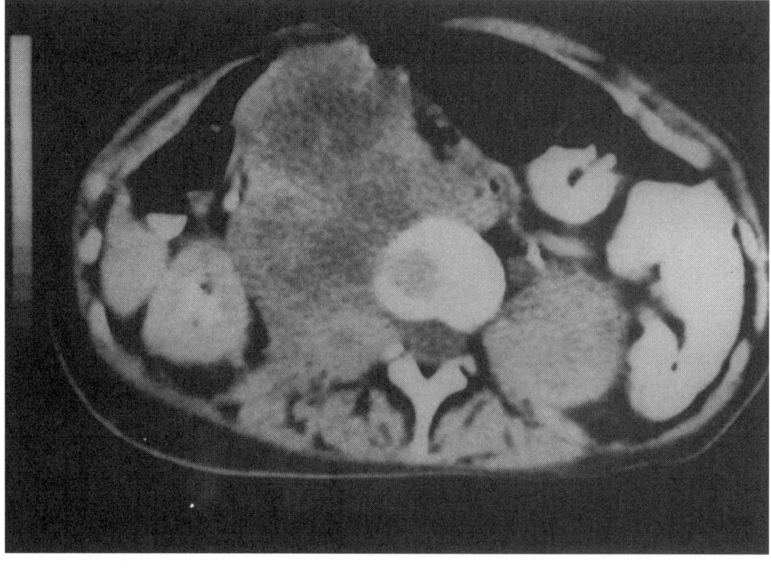

Figure 5.6 *Computed axial tomogram showing a histologically benign neurofibroma that began within the spinal canal and extended through a neural foramen to create a large abdominal mass. This individual was fortunate to have few symptoms, but signs of spinal cord compression were present. Surgical removal was difficult.*

be the case for VRNF on the basis of studies of polymorphic DNA markers and/or enzymes in tumors. Each skin spot represents the clonal progeny of a single melanocyte, a cell of neural crest origin.

Consistent with the Knudson hypothesis (see Chapter 4), individuals with VRNF are peculiarly susceptible to second mutagenic events in the neurofibromin gene. In fact, one of the reasons that the condition may be relatively frequent is that the very large size of the neurofibromin gene and its complexity present a large target for different mutations. Although mutations presumably occur in all genes at a finite frequency, most genes are not so pivotally placed as gatekeepers for cellular metabolism, and hence alterations in their behavior would not be so readily detectable. Another perspective derived from the same observation is that a gene must have sufficient complexity to be able to tolerate mutations that do not lead to immediate lethality. If the mutation killed the cell right away, it would never be propagated to the point where it reached clinical detectability.

Because mutations can occur during gestation as well as after birth, some consequences of VRNF can include developmental variations. This is particularly likely when tumors develop in or around bones and joints, preventing their subsequent normal development. Although rare, such conditions as pseudoarthroses and bone cysts are readily recognizable and can be a part of the broad, pleiotropic spectrum of this condition.

Because of the broad range of complications that may develop over what may very well be a normal lifespan, individuals with VRNF need careful longitudinal observation, preferably by one group of practitioners. In this way, the potential for sequential changes can be appreciated and managed expectantly, increasing the chance of detecting complications early. Obviously, benign tumors are more frequent in VRNF—these are the neurofibromas that develop in Schwann cells of the nervous system. In addition, longitudinal studies of multiple kindreds indicate an increase in the incidence of general neoplastic problems, particularly in younger individuals. Presumably this reflects a broader effect of aberrant gatekeeper gene function.

The diagnosis of VRNF should be straightforward based on clinical features. The pleiotropic combination of café-au-lait spots, neurofibromas, axillary freckling, and Lisch nodules in the eyes is usually diagnostic, but, as noted above, the variable expressivity of these may be confusing. VRNF presents an excellent example of the importance of emphasizing all clinical data in arriving at a diagnosis. Although individuals with this disease generally do well, early recognition of the underlying nature of the condition permits more thoughtful long-term follow-up and can minimize potential complications.

There is, unfortunately, no specific treatment for the aberrant cellular metabolism in VRNF. Current treatment is largely surgical, with symptomatic management of tumors and developmental changes as they occur. Longer-term prospects for controlling the gatekeeper function of the GTPase activating protein are exciting and undoubtedly will be the focus of considerable study. The situation is complicated, however, by the broad range of gene mutations and protein variations recognized and by the broad spectrum of consequences of individual variations. Thus, it is not possible to predict the consequence(s) of a specific mutation in the large neurofibromin gene; some abnormal proteins retain residual function but may lack appropriate responsiveness to control mechanisms.

C. ACHONDROPLASIA

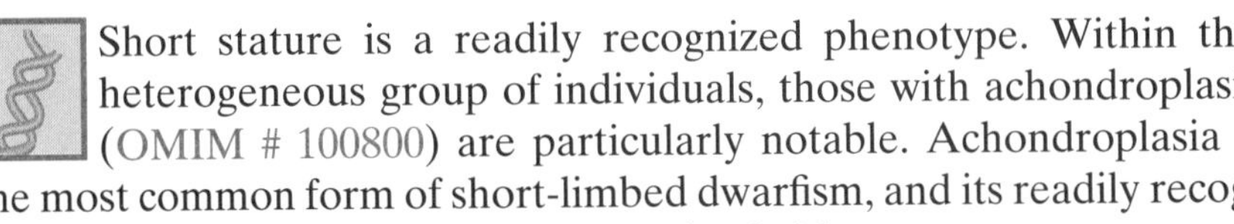

Short stature is a readily recognized phenotype. Within this heterogeneous group of individuals, those with achondroplasia (OMIM # 100800) are particularly notable. Achondroplasia is the most common form of short-limbed dwarfism, and its readily recognizable features make it virtually unmistakable.

Individuals with achondroplasia (Figure 5.7) reach predictable adult heights on the basis of established growth curves. The characteristic shortening of the long bones gives affected individuals relatively short limbs in comparison with their trunk length. The aberrant formation

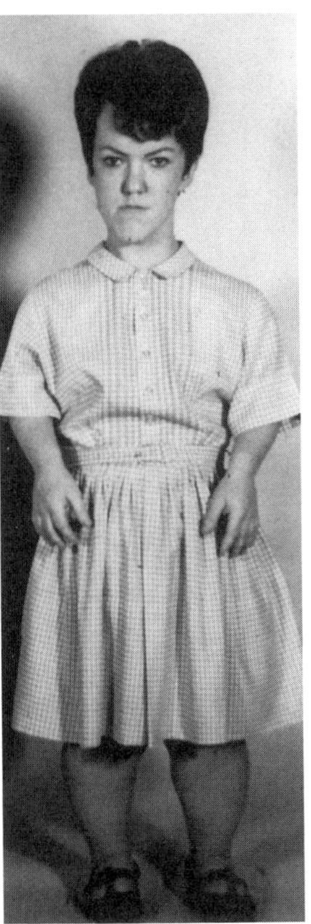

Figure 5.7 *Adult woman with achondroplasia. Note the relatively short arms (reflecting poor growth of long bones) and prominent frontal region of the skull.*

of enchondral bone also gives a characteristic appearance to the skull where the bones of the face, or chondrocranium, are relatively small, while those of the calvaria, or neurocranium, are of at least average size (Figure 5.8). Individuals with achondroplasia are small from birth and often have diagnostic radiologic and clinical abnormalities when first encountered.

Normal neurologic function in achondroplasia is important in assisting adjustment to skeletal difficulties and in helping affected individuals live productive lives. Nevertheless, there remains the possibility of developing skeletal problems that can affect the nervous system. In the first place, relative stenosis of the foramen magnum can lead to hydrocephalus; the possibility of this complication must be considered often throughout development because of the prominence of the neurocranium. Fortunately, neurologic and imaging studies can evaluate the development of this complication.

A later complication of neurologic function in achondroplasia also is related to skeletal changes. Although the vertebral bodies are relatively normal in structure, their posterior elements may be relatively small, and they do not widen caudally, as shown in Figure 5.8. This can lead to stenosis at various levels of the spinal cord. Usually the problems generated by spinal stenosis develop in late adolescence through middle age and present as typical findings of bladder spasticity and weakness developing after exercise ("spinal claudication"). Recognition of the possibility of this complication, combined with the effectiveness of laminectomy for reducing spinal cord pressure, has been important in

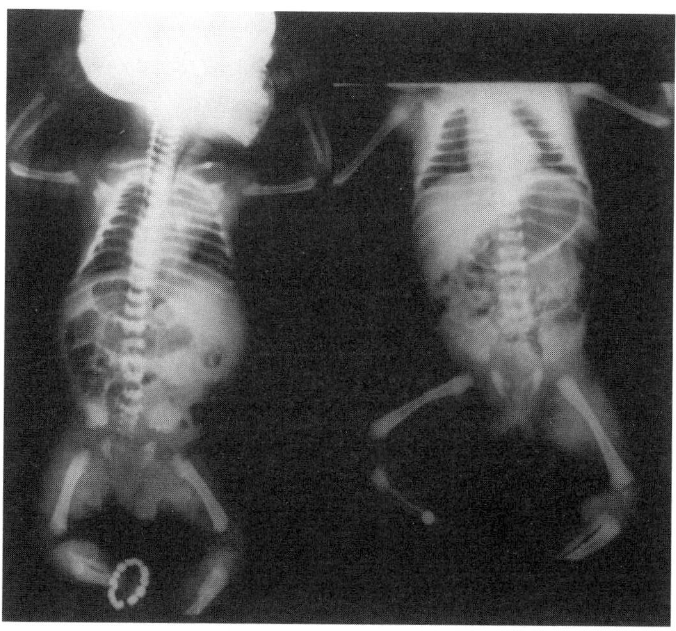

Figure 5.8 *Skeletal changes in achondroplasia in infancy (left side) include shortening of the long bones and narrowing of the spinal spaces in the lumbosacral regions. Compare these with an unaffected individual (right side). These changes remain prominent throughout life.*

maintaining the quality of life of affected individuals. Again, imaging studies have proved useful in establishing the locations and severity of changes and in guiding surgical intervention.

The relatively small size of the chondrocranium predisposes individuals to inadequate drainage of the middle ear. This is particularly noticeable in infancy and childhood, when chronic middle ear infections are common. Aggressive management is important to eliminate permanent hearing damage; external drainage is frequently necessary and helpful. Another complication that is related directly to the unusual bone formation is inappropriate alignment of the lower limbs. These changes result in combinations of varus and valgus deformities particularly affecting femoral alignment. These are of particular importance because they lead to abnormal pressure on the articulating surfaces in the knees, ankles, and hips and cause degeneration. Abnormal angulations can be managed effectively with wedge osteotomies, bringing the limbs into improved alignment. Performed during childhood, such osteotomies not only can produce a good cosmetic effect but also can minimize joint damage from later weight bearing.

Achondroplasia differs from many autosomal dominant disorders in the relative consistency of its clinical presentation (variable expressivity is not prominent). It also is said to have complete "penetrance," meaning that any individual carrying the genetic change will manifest the condition. Although these statements appear to go counter to features of dominant conditions as summarized earlier, the likely explanation for this apparent contradiction is the finding of the identical mutation in all affected individuals. Unlike many dominant conditions, some of which are discussed in this chapter, all individuals with achondroplasia studied to date have a glycine-to-arginine substitution at codon #380 of the gene for fibroblast growth factor receptor 3 (FGFR3) on the long arm of chromosome 4. Interestingly, most of these mutations represent a G→A transition, but a few individuals have a G→C transversion, making this guanosine nucleotide appear particularly susceptible to undergoing mutation (recall Table 1.2). This change from a neutral to a charged amino acid occurs in the transmembrane domain of this receptor. The receptor is expressed in a diffuse pattern in the adult brain, in cartilage during early bone development, and in resting cartilage in sites of endochondral ossification. Thus, the areas where changes occur are consistent with the observed effects on bone formation.

Although achondroplasia is readily transmitted as an autosomal dominant trait, population studies indicate that the majority (possibly as high as 90%) of individuals with this phenotype have new mutations, with a mutation rate of 1×10^{-5}. This may reflect mating biases. It also is important to remember that women with achondroplasia cannot have normal deliveries because of pelvic structure abnormalities, necessitating cesarean sections. An important epidemiologic finding is that new mutations for achondroplasia are associated with increased paternal age. This feature is recognized for other dominant conditions as well and may reflect multiple cell divisions that have occurred in the male germ line and their potential for the cumulative risks of mutation (see Chapter 1). There is no clear explanation for the uniformity of genetic changes in achondroplasia, however. It is possible that mutations elsewhere in the same receptor gene may be lethal during development; alternatively, they may be clinically silent. Either or both of these possibilities could explain the situation, although few mutations elsewhere in this gene have been detected in population studies of individuals of average stature.

Because the clinical picture of achondroplasia is relatively consistent, prognostic and management decisions can be made with reasonable confidence. Thus, affected individuals can be followed carefully to detect and minimize the consequences of their skeletal changes. With such management, and interventions as necessary, individuals with achondroplasia can lead remarkably productive and adjusted lives.

D. MARFAN SYNDROME

Individuals with Marfan syndrome (OMIM # 154700) show many of the characteristic features of dominant inheritance discussed earlier. Pleiotropism is notable, with multiple system involvement. Variable expressivity is prominent, with different degrees of clinical change in different individuals. The remarkable variation of

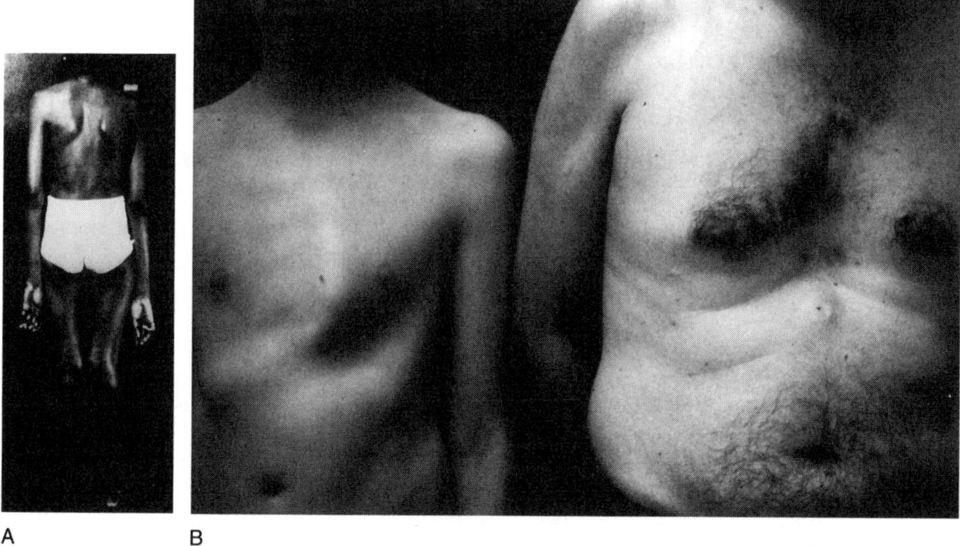

A B

Figure 5.9 *Marfan syndrome. (A) Note long, slender limbs and arachnodactyly. Mild scoliosis also is present. (B) Contrasting anterior chest wall structure of father and son, both with Marfan syndrome. Father (right) has mild pectus excavatum; son (left) has asymmetric pectus carinatum. (Photographs courtesy of Dr. Hal Dietz.)*

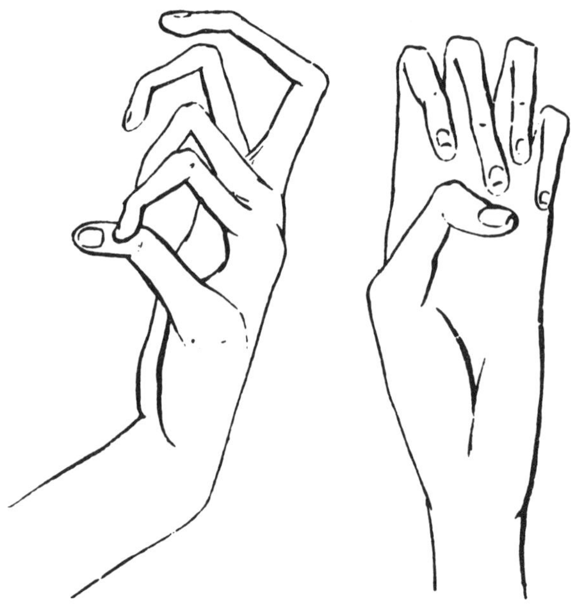

Figure 5.10 *Drawings of arachnodactyly from original description (Marfan, Bull Mém Soc Med Hosp Paris 13:220, 1896).*

features in individuals with Marfan syndrome was very difficult to rationalize until molecular studies provided an understanding of the underlying genetic changes.

Marfan syndrome is characterized by changes in growth of bones and supporting tissues. Excessive height is prominent (see Figure 5.9). The growth of long bones and the long, slender, gracile fingers and toes lead to arachnodactyly (see Figure 5.10). The long fingers permit overlapping of the thumb and fifth finger around the circumference of the opposite wrist, the so-called wrist sign, or Walker-Murdoch sign (see Figure 5.11). The length of the thumb also often permits it to protrude from the clenched hand, the so-called thumb sign, also called the "Steinberg thumb sign" (see Figure 5.11). Bone growth also can be associated with torsion in the spine and the development of scoliosis. Usually detected during childhood and worsening during adolescent growth, scoliosis is an important complication that needs prospective management. In addition to increased bone length, laxity of joints

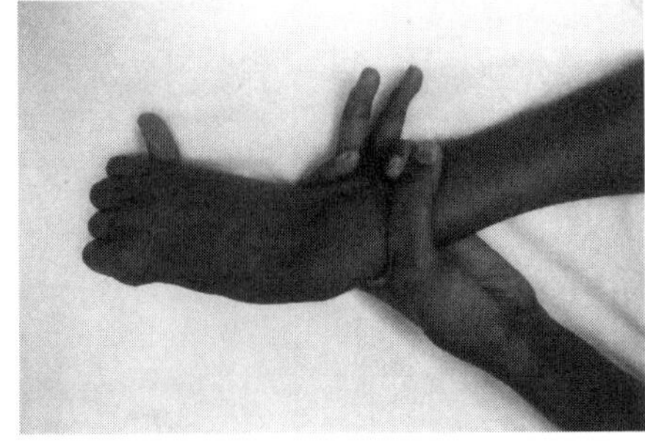

Figure 5.11 *Marfan syndrome. Note thumb and wrist signs in the same individual. (Photograph courtesy of Dr. Hal Dietz.)*

and supporting tissues can be prominent, leading to joint instability, dislocations, and weakness, particularly in joints involved with weight bearing, such as the ankle.

Another area of complications in Marfan syndrome is the eye, where laxity of the zonular fibers supporting the lens leads to spontaneous lens dislocation, or "ectopia lentis." (Figure 5.12). The fiber laxity and consequent lens mobility lead to iridodonesis, a waving motion of the iris upon movement of the eye that often can be appreciated with the hand-held ophthalmoscope. The ocular dimensions also can change, leading to a longer globe measured from front to back. This puts excessive tension on the retina and establishes the potential for retinal detachment.

The most life-threatening features for many individuals with Marfan syndrome are related to blood vessels. Here, the laxity of supporting tissues leads to progressive dilation of the aorta. This is most prominent at the proximal aorta just above the aortic valve and produces a characteristic series of changes progressing from valvular regurgitation through aneurysm formation to dissection. Other valves also may be involved in Marfan syndrome. Mitral valve prolapse is common. Aortic aneurysms also develop beyond the proximal aorta and may even appear in the abdomen.

On first glimpse, these three areas of change—bones and supporting tissues, eyes, and large blood vessels and heart valves—appear to have relatively little relation to one another. Yet they clearly are different manifestations of the same genetic problem, characteristic of pleiotropy. In addition, variable expressivity may lead to different degrees of prominence in different individuals. This perplexing situation is sometimes manifested within extended kindreds. For instance, a father may have blindness but no aortic difficulties at all, while a son has serious scoliosis and a daughter has aortic regurgitation, although both children have relatively normal vision. While the identical gene mutation is present in all three individuals, the expressivity differs considerably.

Another reason for differences in manifestations between kindreds comes from the broad spectrum of genetic abnormalities that underlie

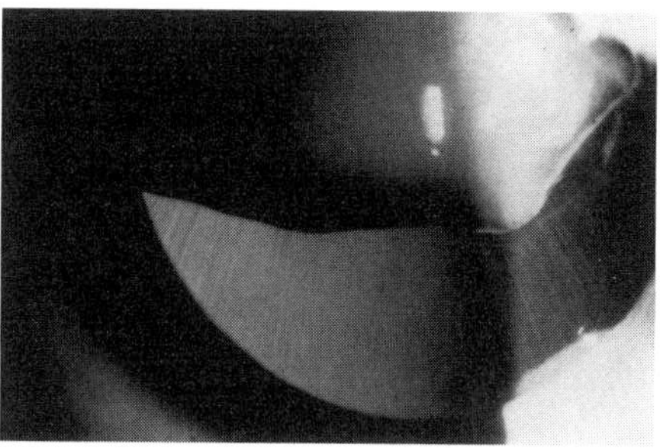

Figure 5.12 *Ectopia lentis in Marfan syndrome. Note zonular fibers visible beneath the displaced lens. (Photograph courtesy of Dr. Hal Dietz.)*

the syndrome. These genetic changes occur in the gene for the protein fibrillin, which is found on chromosome 15. The spectrum of mutations is remarkable, ranging from various point mutations to deletions and truncations of the encoded protein. The changes in fibrillin are important to consider from the pathophysiologic perspective. Clearly, one allele of the gene remains intact and is presumably responsible for producing normal fibrillin protein. Fibrillin, a 350-kDa glycoprotein, is a structural component of microfibrillar fibers, threadlike filaments that serve as scaffolding during elastin deposition. Because a fibrillin molecule interacts with other copies of the same molecule during elastogenesis, the notion has developed that the presence of both normal and abnormal proteins causes changes in the integrity, length, and/or physical properties of elastin in supporting tissues. There is no consistent way to predict what the consequences of any of these changes will be, however, and this obviously contributes to at least some of the clinical variation. What is consistent is that fibrillin is involved in forming elastin-containing tissues in the eye, in the aortic wall and heart valves, and in tendons and joints. All of these areas are at risk for change in individuals with Marfan syndrome.

In the clinic, individuals with Marfan syndrome must be treated prospectively, and many management techniques have been developed for individual complications. Joint laxity may be approached both with strengthening exercises and with bracing. Obviously, scoliosis can be managed through bracing, which is best begun early to minimize the need for surgery (while recognizing that it still may be required in certain cases). Ectopia lentis is helped by removing the ectopic lens and making appropriate optical corrections. Early ophthalmologic evaluation also is useful to detect and manage retinal detachment.

New approaches to managing cardiovascular complications in Marfan syndrome patients have been developed. Formerly, there was an important risk of lethal aortic dissection, but this has been reduced by preventive strategies and new surgical techniques. The echocardiogram provides noninvasive measurements of aortic valve function and aortic root dimensions. Additional and more detailed information can be obtained through magnetic resonance imaging. Combined with clinical follow-up observations, these noninvasive approaches permit detection of aortic enlargement as it occurs and before it threatens dissection.

While measuring aortic and valvular changes is important, efforts to prevent or minimize these changes also have been studied. The most effective of these is the use of β-blocking drugs. The rationale of this approach is that reducing the contractile force of the left ventricle minimizes the stress on the aortic valve and proximal aorta. β-Blockade can slow the progression of aortic root enlargement and delay the need for surgical repair. This treatment is well tolerated and provides a physiologically rational approach to minimizing the consequences of the underlying tissue defect.

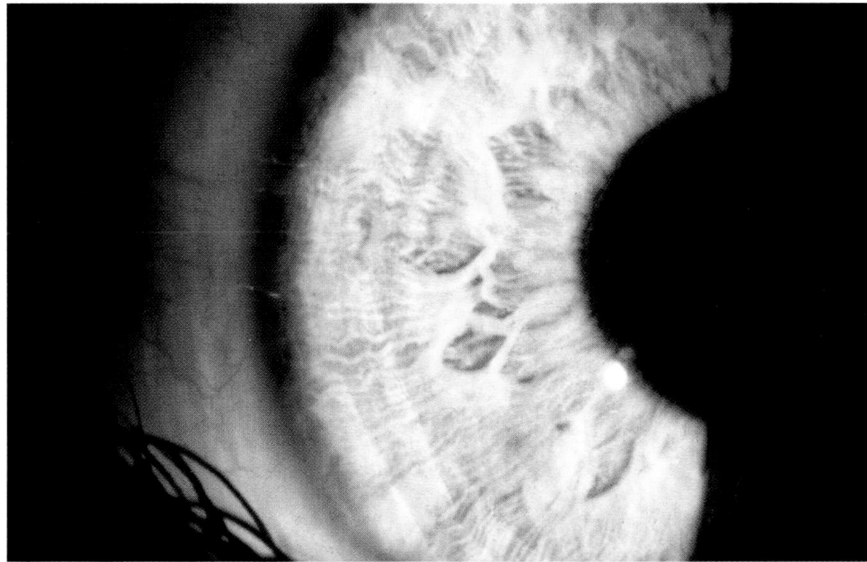

Plate 1 *Lisch nodules. These growths around ciliary nerves can be useful in recognizing and helpful in confirming the diagnosis of VRNF. Some can be seen without magnification; others require a hand-held ophthalmoscope or slit lamp for inspection.*

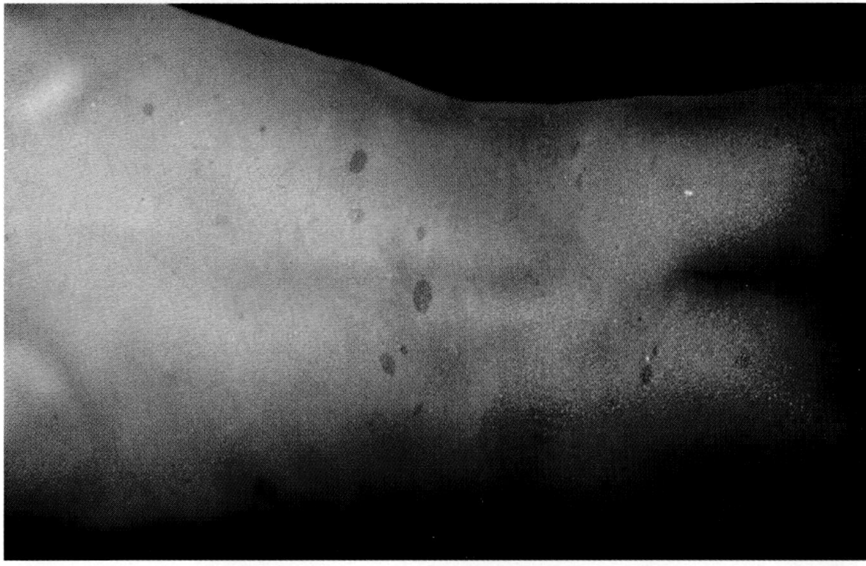

Plate 2 *Café-au-lait spots on the back of a patient with VRNF. Note that the spots are easily distinguishable despite the background skin pigmentation.*

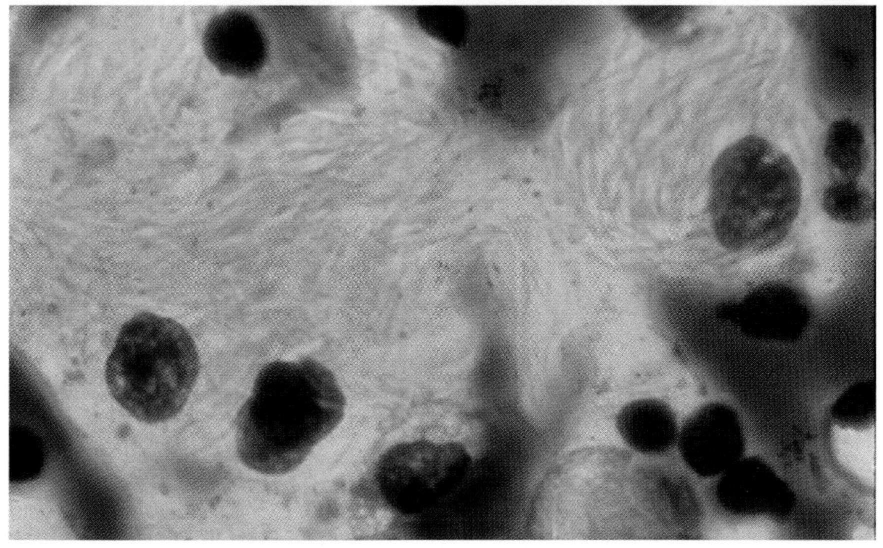

Plate 3 *Gaucher cells in bone marrow aspirate. These cells have pale purple cytoplasm on conventional staining and contain multilayered membranous inclusions. The cells can be very large—some are 100 μm in diameter. They are derived from macrophages.*

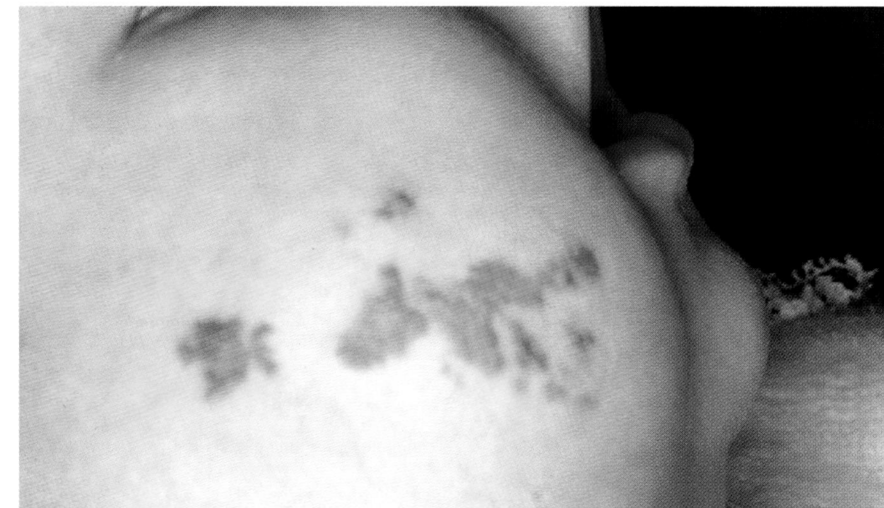

Plate 4 *Hemangiomas. These lesions can range from superficial "port-wine stains" to complex, deep changes with high degrees of vascularity. (Photograph courtesy of Dr. Grant Anhalt.)*

Plate 5 *In situ hybridization of a chromosome-21-specific DNA probe to an uncultured amniocyte. Three sites of hybridization are present, indicating the presence of trisomy 21. (TriGen assay, courtesy of Vysis, Inc.)*

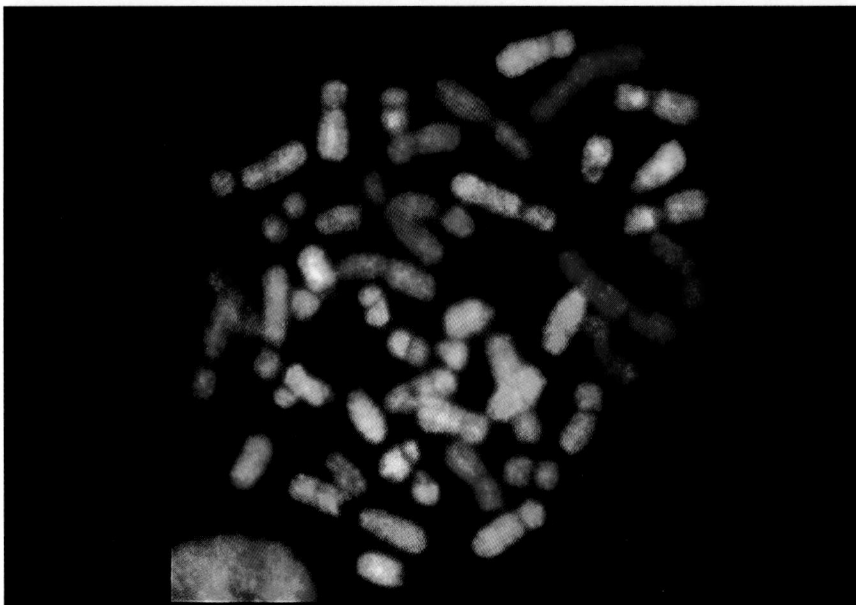

Plate 6 *Color merge photograph of multiprobe FISH showing a translocation between chromosomes 6 and 7 and identifying all chromosomes by color. (Courtesy of Dr. Christa Lese and Dr. David Ledbetter, University of Chicago, and Steve Lundell, Vysis, Inc.)*

Reliable surgical intervention to replace the aortic valve and proximal aorta has been developed. The approach recognizes the abnormal properties of the tissue and thus differs from similar repairs performed in individuals without Marfan syndrome. Because of the connective tissue laxity, it is important to minimize a situation in which the result depends on a single suture line connecting the prosthesis to the aortic tissue of the host. Figure 5.13 shows the insertion of a prosthesis combining both valve and proximal aorta into the existing aorta of the individual. This approach maintains the linear integrity (such as it is) of the patient's own aortic wall, permitting longitudinal strengthening of the wall, preventing further dilation, and solving the problem of valvular regurgitation. Aortic surgery for individuals with Marfan syndrome can be offered prophylactically—before aortic enlargement reaches dangerous dimensions. This, combined with the use of β-blockade, has been an important advance in the long-term care of these individuals. Because the mitral valve also is involved frequently, Marfan syndrome patients may need multiple valve replacements.

Women with Marfan syndrome present important considerations during pregnancy and delivery. The increased blood volume that normally develops during pregnancy places the mother at risk for aortic

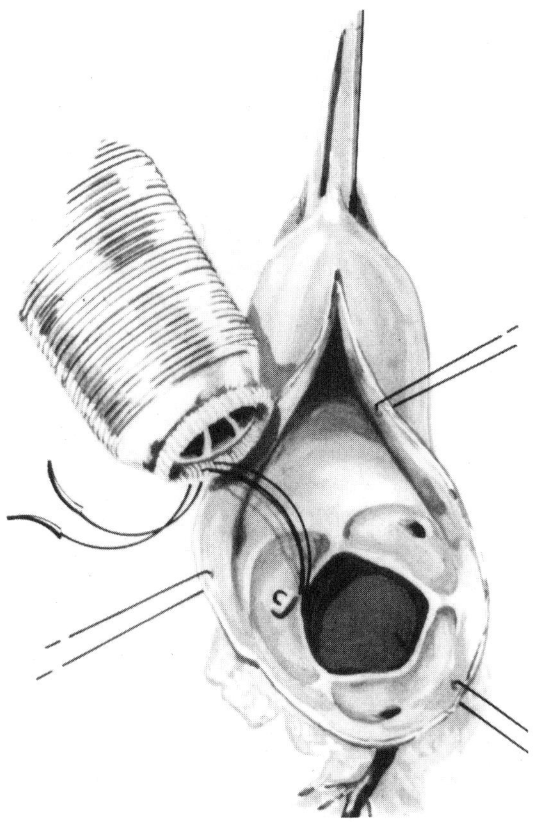

Figure 5.13 *Repair of ascending aortic aneurysm in Marfan syndrome. The artificial valve is attached to the base of the Teflon graft, and the entire assembly is sutured into the existing aortic valve ring. The coronary arteries are reattached, and the aorta is closed around the prosthesis. The continuity of the existing aortic wall is preserved.*

dissection, particularly in the immediate postpartum period. Fortunately, the use of β-blockade and noninvasive measurement of aortic dimensions and valve function have minimized maternal complications.

The pleiotropic manifestations of fibrillin defects make individuals with Marfan syndrome remarkably complex in their presentation. Managing the various clinical problems requires coordinating different medical specialties. In addition, there currently is no way to predict all potential manifestations reliably. Also, it is not currently possible to relate specific fibrillin gene changes to specific clinical manifestations. However, even within kindreds, where the mutation is necessarily the same, both the pleiotropy and the variable expressivity are puzzling and challenging.

Individuals with Marfan syndrome thus present a remarkable contrast to those with achondroplasia in regard to both clinical and underlying gene changes. They also present a contrast to individuals with VRNF, in whom, although the gene changes vary widely, the problems generally remain confined to supporting tissues of the nervous system. Thus, the clinical consequences of individual mutations cannot be separated from the underlying physiologic function(s) of the genes and gene products involved.

E. MYOTONIC DYSTROPHY

Myotonic dystrophy (OMIM # 160900) is the most common form of adult muscular dystrophy, having a prevalence of ~1 in 8000. It is characterized by muscle weakness and particularly by the phenomenon of myotonia. Myotonia implies a state of chronic spasm of the muscle, resembling tightness, and although this is progressive, it usually is noticeable relatively early in life. Additional clinical features help distinguish myotonic dystrophy from other forms of muscle disease. The presence of cataracts is notable, along with early hair loss in the frontal regions, electrocardiographic changes, and relative hypogonadism in some individuals. Muscle wasting often occurs in the neck and temporal areas. The combination of muscle weakness and myotonia with the other clinical features usually makes the diagnosis apparent. Its transmission in dominant pedigrees also is clear.

Unlike other dominant conditions, such as achondroplasia, in which the clinical phenotype makes the diagnosis obvious, individuals with myotonic dystrophy may differ remarkably in the rate of progression of their symptoms. Myotonic dystrophy is an example of a problem that has been explained in molecular terms only recently, in contrast to its long history of clinical observations.

It is an old observation that individuals with myotonic dystrophy may present earlier and with more severe manifestations in later generations; this phenomenon has been referred to as "anticipation." For example, in one kindred the grandfather was diagnosed as having myotonic dystrophy in his sixth decade, while his daughter began to show muscle problems in her thirties and her son showed cataracts, electrocardiographic changes, and muscle weakness in adolescence. However, because many of the clinical observations were made relatively early in the study of the disease, it was assumed that anticipation reflected improved awareness of features of the disease and hence an increased likelihood that even very subtle changes would be sought and detected early in individuals at risk. For example, it is now becoming more clear that cataracts may be the earliest clinical manifestation, sometimes present at birth or developing in early childhood. Cataracts may occur many years before the muscle changes appear.

This paradoxic situation has now been explained by an appreciation of the underlying genetic mechanism. This mechanism is not unique to myotonic dystrophy, but its consequences are particularly apparent in this disorder. The biologic basis is the repetition (expansion) of a triplet nucleotide sequence (CTG) in the DNA of the responsible gene, 19q13.3, on chromosome19 (see Figure 5.14). The repeat is transcribed and is found in the 3′ untranslated region of the mRNA. Unaffected individuals have from 5 to 35 tandem copies of the repeat, but the repeat can become longer. After a critical length of repeats has been reached, symptoms of the disease develop. The lengthened repeat group itself becomes progressively more unstable, so that additional repeats can be added more readily. These repeats can become as long as 5 kb, and there is an approximate correlation between the length of the repeat and the severity of the disease and an inverse correlation with the age of onset (or clinical detection). The repeated sequence continues to lengthen in affected individuals in later generations, presumably due to its intrinsic instability (and liability to expand) during meiosis. There also is evidence that the repeat may continue to lengthen in somatic tissues during the lifetime of the host; this could explain the delayed onset of some clinical problems.

The gene for myotonic dystrophy produces a protein called "myotonin protein kinase"; this protein is a member of the family of protein kinases that phosphorylate other proteins, but its precise function is not known. It is likely that the repeat amplification in myotonic dystrophy is a good example of a "threshold phenomenon." This means that a critical repeat length ("premutation") must be achieved before the

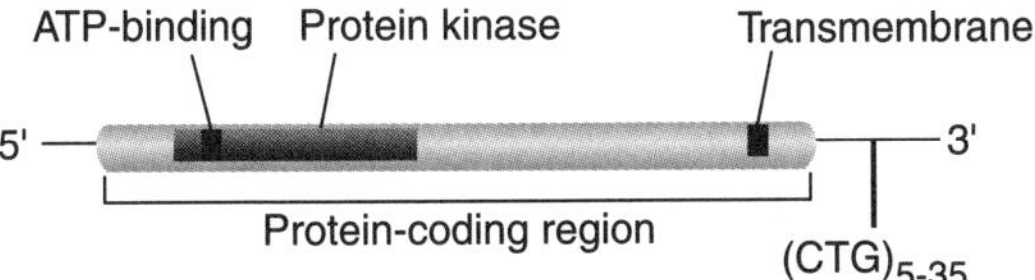

Figure 5.14 *The trinucleotide unit that is expanded in myotonic dystrophy. Note that this is within the transcribed region of the gene and becomes part of the 3' untranslated region of the mRNA transcript. See text for details. (Modified from Caskey et al, Science 256:784, 1992.)*

sequence becomes genetically unstable. Also, it is likely that the number of repeats imparting instability to the gene is approximately the number at which symptoms begin to be manifest. The relatively low prominence of symptoms (frequently of later onset) shown by individuals with low numbers of repeats may explain how a dominant inheritance pattern can be traced to apparently asymptomatic individuals in earlier generations. Such individuals may have had modest increases in triplet repeat number and very few (if any) symptoms. They certainly would not have had substantial muscle disease or hypogonadism, although they may have had mild cataracts and frontal balding. Such relatively asymptomatic individuals could have passed their lengthened, and hence potentially unstable, repeat segments to subsequent generations, permitting further lengthening and more obvious clinical changes to develop. Figure 5.15 presents data relating the length of the triplet repeat segment to the onset of detectable symptoms.

Congenital myotonic dystrophy is a severe neonatal form involving high numbers of CTG repeats in the same gene. Affected individuals receive their amplified gene from an affected mother. The molecular basis for this is not clear. One explanation is imprinting, as discussed in Chapter 3. Another possibility is that affected males may produce sperm in which the amplified region is unstable, leading to reduced sperm viability. Still another possibility is that expansion of the repeat may be more likely during female meiosis (see also the section on Fragile X syndrome in Chapter 7).

It is important to realize that this triplet repeat problem is not unique to myotonic dystrophy. It also has been found in several other diseases characterized by trinucleotide repeats, including fragile X syndrome (OMIM # 309550) and Huntington disease (OMIM #143100). An inter-

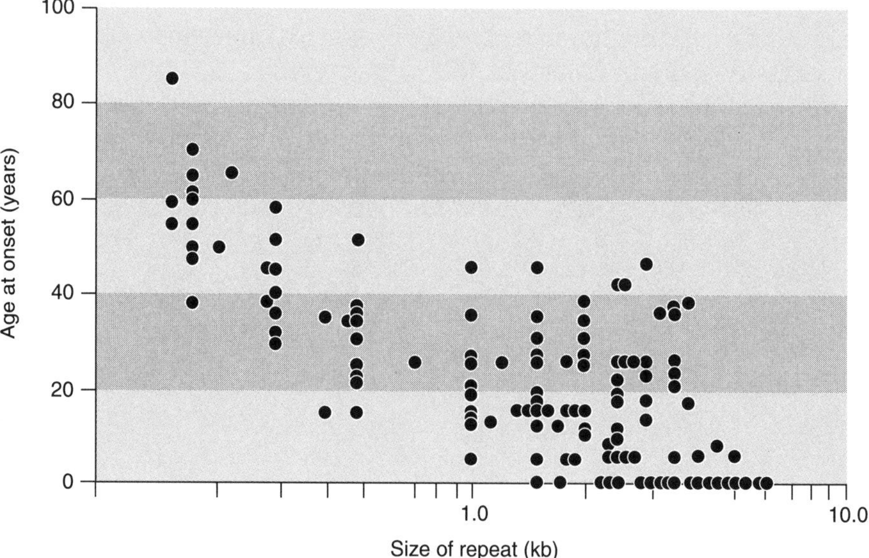

Figure 5.15 *Age of onset of clinical problems versus trinucleotide repeat length for multiple kindreds of myotonic dystrophy. Note that there is a group of short repeat lengths that are apparently not responsible for clinical findings. Such lower repeat lengths may predispose to further lengthening and the development of symptoms in later generations.*

esting feature of most of these conditions is that they involve the nervous system, but the precise molecular details of all of the conditions in this category are not yet understood. When contained within the coding region of the affected genes, trinucleotide repeats can lead to very abnormal proteins due to long stretches of single amino acids (this occurs in Huntington disease, in which repetition of CAG leads to a long polyglutamine sequence). Such proteins might have no function or behave unpredictably. In contrast, a similar repeated segment upstream (5′) or downstream (3′, as in myotonic dystrophy) of the coding region could suppress transcription, essentially behaving as a null mutation. Alternatively, such a change could affect the survival of the mRNA or cause a gain of function of the gene by, for example, changing the binding of the mRNA to a control element. Because of the dominant transmission pattern, the mutant (expanded) allele must exert its effect in the presence of a normal allele.

F. RECURRENCES AND RISKS

The most obvious feature of the transmission of autosomal dominant conditions is the 50% likelihood of transmitting the trait in each conception. As noted earlier, this derives simply from the likelihood of transmitting a single affected chromosome in each meiosis. Dominant conditions do, however, present the important opportunity to observe new mutations. Different conditions vary in the frequency with which individuals represent new mutations. For instance, as noted above, the majority of individuals with achondroplasia represent new mutational events. In contrast, many kindreds with Marfan syndrome or VRNF show dominant transmission pedigrees covering multiple generations. Certainly, dominant transmission with triplet repeat expansion is a prominent feature in myotonic dystrophy and Huntington disease. At least some differences in the proportion of new mutational events may reflect the likelihood of mating for affected individuals and the difficulties that may develop in pregnancy or childbirth for individuals affected with achondroplasia or Marfan syndrome.

In several of the conditions described in this chapter and in a large number of other dominant phenotypes, both pleiotropy and variable expressivity are important. This makes counseling and clinical discussions more challenging because it often is difficult to predict the spectrum and severity of changes in an individual even when the same gene mutation is involved. Obviously, between kindreds that have different mutations in the same gene (Marfan syndrome and VRNF are particularly striking examples), some of the variation may reflect different molecular changes. In other conditions, for example, achondroplasia, the uniform finding of a single amino acid change is likely to help keep the clinical picture more consistent. This cannot be the full explanation, however, because within individual kindreds affected with VRNF or Marfan syndrome, the same mutation is necessarily present. In these

CASE STUDY

William is 17 and has VRNF. His mother and brother also are affected and have generally done well except for the removal of three benign, pedunculated tumors over his mother's shoulder and two around his brother's waist. His mother's brother is retarded because of hydrocephalus that developed at age 3 secondary to a neurofibroma blocking ventricular drainage.

William himself has grown well and done well in school. He has always been active and agile. He has many superficial tumors but none has required removal. He has over 20 café-au-lait spots of various sizes. On a routine clinic visit he had a slight limp that he attributed to a fall several weeks before while playing football; he felt well otherwise. Examination showed the multiple tumors and spots, largely unchanged. Neurologic examination showed absent ankle jerks and a reduced knee jerk on the right but no other changes. He had good strength. Toes were downgoing. A CT scan showed a dense mass compressing the spinal cord and cauda equina from L1-L4. The mass extended through the neural foramina on the right side at L2-L4.

A laminectomy and dissection was performed showing a histologically benign neurofibroma whose precise origin could not be determined. Most of the mass could be removed, and decompression across the lumbar spine relieved the spinal cord pressure. Postoperatively, he did well. His knee jerks became symmetrical and normal and the left ankle jerk returned fully; the right ankle jerk remained reduced, although present. He resumed full activity.

COMMENT

This is a characteristic picture in the follow-up of an individual with VRNF. He had a subtle set of findings related to the slow development of spinal cord pressure due to a neurofibroma. The benign mass arose from the meningeal cells and could not be completely dissected away from the nerves without damaging them further; so the laminectomy, combined with the removal of the bulk of the tumor mass, reduced the cord compression. The family history is also interesting—his uncle developed brain damage due to hydrocephalus, again because of pressure from a benign lesion (before CT scans were available). Any of the multiple mutations in the neurofibromin gene might be responsible for this clinical picture with dominant transmission and the development of problems relatively late in life. Noninvasive techniques including CT and MRI are valuable when combined with clinical suspicion.

situations, trying to anticipate the multiple and frequently unpredictable effects and degrees of clinical severity developing from the same genetic change is particularly challenging. Recognition of triplet nucleotide repeat disorders has added an important new dimension to the appreciation of dominant phenotypes. These disorders present the possibility of increasing their severity with increasing numbers of repeats and hence justify the term "anticipation," which was applied to them in the past (for other reasons).

Rarely, individuals are born who are homozygous for dominant conditions. Such individuals are often severely affected. Many of these conditions may be lethal in utero, and the remainder present severe problems after birth. A particularly prominent example occurs with achondroplasia, the homozygous form of which is usually lethal, with severe skeletal manifestations, thoracic compression, and neurologic problems reflecting abnormal skeletal development. Homozygosity for dominant traits also is noted for other conditions, including those affecting the nervous system; in such individuals severe and prominent manifestations are seen early. Such individuals may be found as a result of mating circumstances in which both parents are affected. They present significant clinical and management challenges.

STUDY QUESTIONS

1 Asymmetric septal hypertrophy is a recognized dominant condition affecting cardiac muscle (OMIM #192600). A 21-year-old college student comes to you because he wants to play football and he has been told that both his father and his aunt are affected. He is asymptomatic. You tell him:

- a It is difficult to establish his prognosis because he is young.
- b Catheterization should clear up the situation.
- c You would like more information about his father's health.
- d Having both first- and second-degree relatives affected increases his risk.
- e Has he considered chess?
- f You cannot tell him much without a linkage study.

2 Familial hypercholesterolemia (FH; OMIM #143890) is an autosomal dominant disorder associated with high levels of serum cholesterol and premature atherosclerosis. A 23-year-old woman is in your office for a routine examination and remarks that she thinks her father may have the problem because he had a myocardial infarction at age 38 and his father died at 49 with "heart trouble." She has been married for a year and is thinking about starting a family. You tell her:

- a There is less concern because no females in her family have been affected with FH.
- b She is too young for meaningful laboratory studies.
- c You need to examine her husband.
- d A fasting cholesterol level will establish the diagnosis.
- e You need more information about her father's health.
- f Prenatal diagnosis looking for FH may be a good idea.

3 A 27-year-old patient with Marfan syndrome who has been your patient for 10 years calls you over the weekend. He has had brace treatment for scoliosis, but that is stable. ECHO studies have been normal in the past, and he does not wear glasses. He has had indigestion and back pain for 3 days while he was traveling, which showed some response to antacids; unfortunately, it has become more severe. No fever or nausea has been present, but he has had some diarrhea. You tell him:

- a He should have a colonoscopy.
- b He should use a prescription H2-blocker, which you can prescribe.
- c He needs Lomotil.

d He needs an abdominal ultrasound study.
e He needs a urinalysis.

4 The 37-year-old sister of a 43-year-old man with Huntington disease and a strong family history has come to you distraught. She is pregnant and very worried. She has never had any neurologic problems. You tell her:

a She should have neuropsychiatric testing.
b She should go to the prenatal diagnosis center for Huntington disease testing.
c She is at greater risk for a neural tube defect.
d She needs prenatal testing for Down syndrome.
e She needs to have a triplet repeat test for Huntington disease for herself.

5 You are pursuing a research elective in a laboratory studying myotonic dystrophy and are measuring triplet repeats in colleagues to improve the baseline. Your colleagues have agreed to participate with informed consent. Mary has a triplet repeat of 42; what do you tell her?

a She should have a neurologic examination.
b Because her family history is negative, the likelihood of her having problems is very low.
c Her offspring could have full-blown disease.
d She has a premutation.
e She will develop symptoms within the next decade.

6 You are working in a student health clinic doing routine examinations. While examining two students, you find that the first has two prominent café-au-lait spots on his trunk and one on his left leg. The other has three in a dermatomal distribution on her abdomen. What do you tell them?

FURTHER READING

Brook JD et al. Molecular basis of myotonic dystrophy: Expansion of a trinucleotide (CTG) repeat at the 3' end of a transcript encoding a protein kinase family member. *Cell* 68:799, 1992.

Caskey CT et al. Triplet repeat mutations in human disease. *Science* 256: 784, 1992.

Crowe FW et al. *Multiple Neurofibromatosis.* Springfield, Charles C Thomas, 1956.

Harper PS et al. Review Article: Anticipation in myotonic dystrophy: New light on an old problem. *Am J Hum Genet* 51:10, 1992.

Ptacek LJ et al. Genetics and physiology of the myotonic muscle disorders. *N Engl J Med* 328:482, 1993.

Pyeritz RE. Marfan syndrome and other disorders of fibrillin, in Rimoin DL, Connor JM, Pyeritz RE (eds), *Emery and Rimoin's Principles and Practice of Medical Genetics*. New York, Churchill Livingstone, 1996, p 1027.

Riccardi, V. Genotype, malleotype, phenotype and randomness: Lessons from neurofibromatosis-1 (NF-1). *Am J Hum Genet* 53:301, 1993.

Shen MH et al. Molecular genetics of neurofibromatosis type 1 (NF1). *J Med Genet* 33:2, 1996.

Skuse GR et al. The neurofibroma in von Recklinghausen neurofibromatosis has a unicellular origin. *Am J Hum Genet* 49:600, 1991.
Sørensen SA et al. Long-term follow-up of von Recklinghausen neurofibromatosis. *N Engl J Med* 314:1010, 1986.
Tsipouras P et al. Genetic linkage of the Marfan syndrome, ectopia lentis, and congenital contractural arachnodactyly to the fibrillin genes on chromosomes 15 and 5. *N Engl J Med* 326:905, 1992.
Wong L-JC et al. Somatic heterogeneity of the CTG repeat in myotonic dystrophy is age and size dependent. *Am J Hum Genet* 56:114, 1995.

USEFUL WEB SITES

HELIC (INCLUDES GENLINE)

http://www.hslib.washington.edu/helix
Medical Genetics Knowledge Base. Electronic textbook under development relates testing to diagnosis, management, and counseling. Directory of laboratories providing testing for genetic disorders.

OMIM (ONLINE MENDELIAN INHERITANCE IN MAN)

http://www.ncbi.nlm.nih.gov/omim
Basic catalogue listing mendelian genes and traits. Clinical descriptions, references, and links to other databases.

HEALTHLINKS

http://healthlinks.washington.edu/
Broad access to many related areas of basic and clinical genetics. Includes "Genome Machine II," database of genes and diseases mapped to a particular chromosome location. Also chromosome information.

MARCH OF DIMES

http://modimes.org/
Resource for families and professionals about birth defects. Professional education section.

NATIONAL ASSOCIATION FOR RARE DISORDERS (NORD)

http://www.NORD~rdb.com/~orphan
Primary nongovernmental clearinghouse for information about rare disorders. Over 1100 disease reports. Broad resource guide. Includes Rare Disease Database, a wide range of information for families and professionals. Listings by symptoms, causes, affected populations, treatments. Listing of support groups. Includes orphan drug designation database.

OFFICE OF RARE DISEASES (NIH)

http://cancernet.nci.nih.gov/ord/index
Broad-based NIH index including support groups, clinical research database, investigators, and extensive glossary.

CHAPTER 6

Autosomal Recessive Disorders

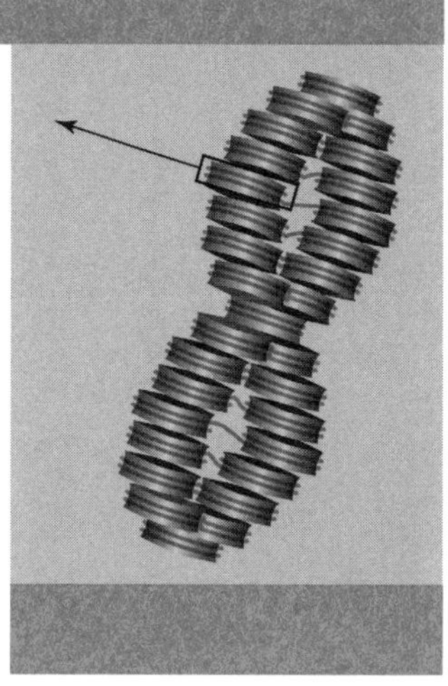

A. OVERVIEW

Autosomal recessive conditions appear when there is some abnormality in the copy of the gene contributed by each parent. The unaffected parents, who carry only a single copy of the abnormal gene, are said to be "carriers," or "heterozygotes." An individual who inherits the abnormal copy of the gene from each parent is said to be a "homozygote." The important genetic consideration is that the homozygote cannot synthesize the normal product from the relevant genetic locus because he or she has received an abnormal copy of the gene from each parent. Because each parent has only one abnormal gene copy, there is a one-in-four chance that any given mating will result in a conceptus with two abnormal copies of the gene. This 25% likelihood is an important number for counseling considerations. It also can be useful in terms of providing information about recurrence risks.

Figure 6.1 shows pedigrees with autosomal recessive inheritance. Several important aspects of recessive inheritance should be considered:

- Heterozygotes are generally unaffected clinically. (This is an important contrast to dominant disorders.)
- An individual manifesting a recessive disorder usually has heterozygous parents. Thus, the appearance of an affected person is often

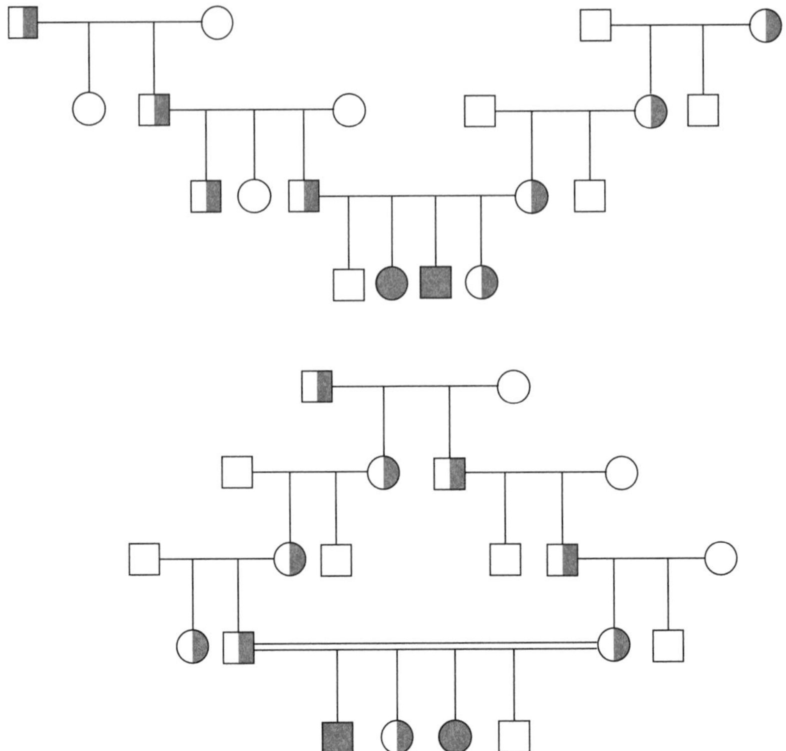

Figure 6.1 *Kindreds with individuals expressing autosomal recessive phenotypes. Note the consanguinity in the lower kindred, reflecting connections between the families to a brother and sister four generations earlier.*

unprecedented in the family. (In comparison, the appearance of a newly affected individual with a dominant disorder usually means that a new mutation has occurred.)

- Once a homozygote is identified, the recurrence risk for other matings of the same parents is 25%. This can lead to a "horizontal" pedigree pattern, with multiple affected persons in the same generation and none in earlier generations.
- Two-thirds of the unaffected siblings of an individual with an autosomal recessive disorder are likely to be heterozygotes.
- The likelihood that any two individuals selected randomly will be heterozygous for the same mutant allele is low, explaining the relative rarity of homozygotes for recessive phenotypes.
- Consanguinity can lead to an increased likelihood of matings between heterozygotes (see Figure 6.1). Because of inbreeding, consanguineous populations demonstrate a "founder effect," meaning that they are enriched for whatever genetic traits were present in and propagated from the founders. Such groups may be relatively small and concentrated, such as the Parsis in India and the Old Order Amish. On the other hand, they may be less concentrated but still tend toward marriages within the community, either overtly or because of geographic limitations. Examples include religious communities and islanders. It is important at least to consider the

possibility of consanguinity when counseling is offered or when a new homozygous individual is encountered.

- A homozygote must contribute one abnormal gene copy to any offspring, because he or she has no normal copies of the gene at the responsible locus. This often is of no clinical consequence for the offspring, however, because the risk is relatively low that the mate will be a carrier. Thus, the pattern of seeing affected individuals in only a single generation is the most common.
- If a homozygote mates within a community where the frequency of heterozygotes is high (e.g., a geographically isolated community), a pattern resembling dominant inheritance may occur. (Recall that the homozygote must contribute an abnormal gene and that there is a high frequency of asymptomatic carriers.) The pattern is called "pseudodominance" (see Figure 6.2) Although this pattern is relatively uncommon, it is important to recall when dealing with an unusual transmission pattern for what should be a recessive condition. Interestingly, because of purposely selective breeding practices, such patterns are recognized more frequently in animal breeding.
- Metabolic abnormalities are common in recessive disorders. Many of these can be considered "Inborn errors of metabolism"—a phrase introduced by Archibald Garrod. With no normal copies of the responsible gene, recessive metabolic diseases may be severe; in fact, they may be lethal. Homozygotes for many recessive conditions often die early—sometimes in infancy—and others may not be able to reproduce.
- Despite severe (possibly lethal) manifestations in homozygotes, heterozygous individuals can disseminate the abnormal allele, preserving the chance for homozygotes to appear again.
- Many recessive disorders have been studied in detail from both biochemical and molecular genetic perspectives; these often are detectable prenatally and/or presymptomatically.
- The metabolic barriers created by recessive mutations have received considerable study, and some effective treatment strategies have been devised.

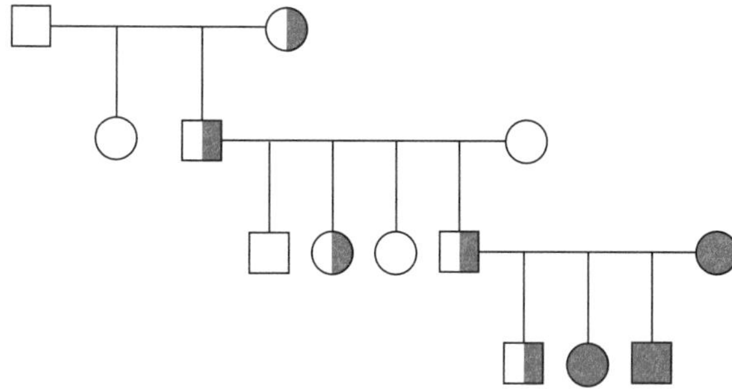

Figure 6.2 *The pattern of pseudodominance has occurred in this kindred because of the high frequency of asymptomatic heterozygotes. This raises the risk of homozygous conceptions to 50% in the mating of the homozygous woman shown. (Why is this the case?)*

Discussion of several autosomal recessive disorders in detail will show these features:

B. SICKLE CELL DISEASE

Sickle cell anemia (OMIM # 141900.0243) has been the basis for much teaching and understanding of recessive genetic disease. This condition results from a specific point mutation at codon 6 in the β-globin gene, causing a substitution of glutamic acid for valine at that position. (What triplet codon was involved? Recall Table 1.2.) This mutation was recognized early in the study of proteins, and specific study of hemoglobin led to a remarkably accurate understanding of the molecular basis of this disorder. This single amino acid change permits the hemoglobin molecule to become unstable in the presence of relatively low oxygen tension. Thus, in any situation where the erythrocyte becomes challenged by increased oxygen demand, tissue anoxia, or other problems, the entire three-dimensional structure of the hemogloblin tetramer undergoes a conformational change. This change from the conventional globin tetramer to the sickle cell globin tetramer changes the shape of the affected erythrocyte. Erythrocytes whose hemoglobin has the sickle conformation (hemoglobin S, or HbS) cannot spontaneously return to their normal shape (see Figure 6.3). Thus, they become trapped in constricted areas, including, very prominently, the spleen. The sickled cells are destroyed after sickling and sequestration and a chronic hemolytic anemia develops.

Individuals with sickle cell anemia are prone to recurrent attacks of pain. These pains, referred to as "sickle cell crises," represent acute episodes of vessel occlusion and hemolysis due to sickling. Such recurrent episodes can lead to kidney damage, muscle disease, and progressive infarction of the spleen. Of course, being without a spleen increases the individual's susceptibility to infections. Recurrent sickle cell crises are traumatic for affected individuals and present considerable challenges to medical management. Prophylactic antibiotics may be useful

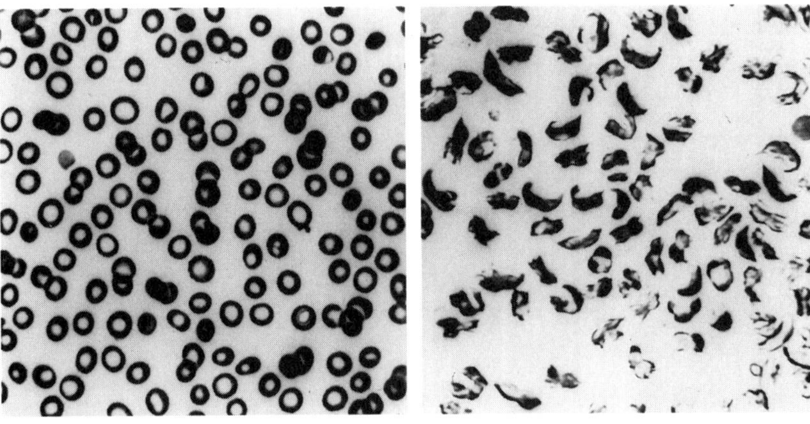

Figure 6.3 *Comparison of normal erythrocytes (left) and those containing sickled hemoglobin (right). The cells on the right cannot spontaneously reassume their normal biconcave shape and undergo entrapment and hemolysis.*

in certain individuals in whom splenic infarction has occurred; these patients also benefit from pneumococcal vaccine.

The mutation in sickle cell disease is common in African-Americans and other persons of African descent; approximately 8% in the United States are heterozygous for the allele. In certain areas of the Mediterranean basin and Western Africa, the carrier frequency can reach 25%. This high carrier frequency increases the likelihood of matings between carriers. For this reason, sickle cell anemia appears once in about 600 births in such populations.

Despite the uniform genetic change underlying sickle cell disease, there is an impressive range of clinical expressivity. For some homozygotes, the frequency of crises and medical difficulties is so severe that the condition becomes lethal in childhood. Others have relatively mild clinical courses. All the reasons for this variation in clinical severity of what obviously is the identical molecular change are not known. One very important variable that affects the course of the disorder relates to the concentration of sickle hemoglobin in erythrocytes. The gene that produces the mature β-globin is an adult globin gene, but other genes within the β-globin gene cluster (see Figure 1.5) are expressed earlier in development. Small amounts of "fetal" hemoglobin (also called "hemoglobin F" [HbF]) can be found in erythrocytes of all adults. Fetal hemoglobin is derived from a β-chain gene different from that carrying the sickle cell mutation. Thus, cells containing HbF cannot have the sickle cell mutation, because their β-globin protein is derived from another gene. Individuals differ in the amount of HbF present in their circulation, however. This may range from a trivially low level to a substantial proportion of so-called "F cells." Although its oxygen affinity is slightly different from that of adult hemoglobin (hemoglobin A, or HbA), HbF cannot undergo the sickling reaction. Furthermore, the presence of HbF can act to increase the oxygen-carrying capacity of the blood, reducing the likelihood of a sickle cell crisis. This represents an important instance of having genetic variation at one locus (i.e., the locus controlling the fetal β-globin/adult β-globin switch) affect the manifestations of a totally unrelated ("unlinked") genetic change. In this case, the unlinked mutation operationally dilutes the concentration of HbS in the blood. This situation, in which a change at a distant genetic site affects the phenotype of another gene, is called "epistasis" and has general applicability to other genetic disorders, as well as common disorders (see Chapter 12).

The diagnosis of sickle cell disease is made with simple laboratory tests of the structure of the hemoglobin protein. Because the genetic change is always the same, it also can be identified easily through DNA studies. Figure 6.4 presents several examples of DNA-based diagnostic tests for sickle cell disease.

Treatment of individuals with sickle cell disease has improved. In the past, the main approach was to attempt to minimize symptoms

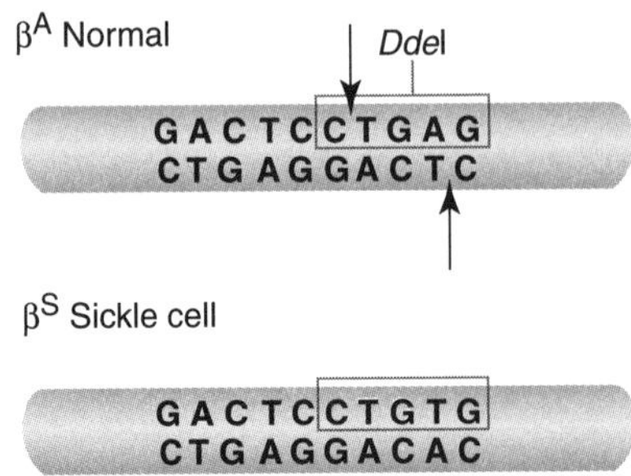

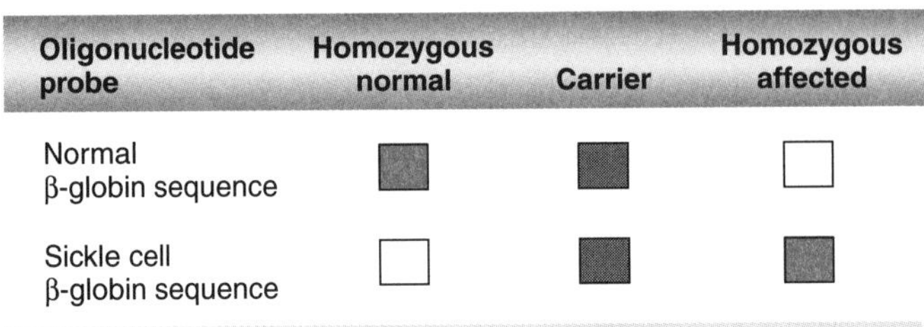

Figure 6.4 *Two DNA-based methods for detecting the sickle cell gene mutation. (Top) The A→T transversion at codon 6 destroys the recognition and cleavage sequence for the restriction enzyme DdeI, thereby changing the DNA fragment sizes following DdeI digestion and distinguishing between the alleles. The arrow indicates the sites of phosphodiester bond cleavage by DdeI when the correct sequence is present. (Bottom) The alleles can be distinguished by hybridization using two synthetic oligonucleotides, one corresponding to the normal β-globin sequence and the other to the sickle cell gene sequence. The hybridization stringencies are adjusted so that each oligonucleotide binds with maximum efficiency to the DNA corresponding to its sequence. Each spot is a sample of DNA from the test or from a reference individual and is bound to a filter. Because heterozygous carriers contain one copy of each sequence, the hybridization intensity is at an intermediate level. This approach does not require electrophoresis or restriction enzyme digestion.*

during the hemolytic crisis by using hydration, analgesics, and transfusions. Although these measures are still valuable during crises, they offer no assistance in preventing future problems. More recently, studies have emphasized efforts to increase the amount of fetal hemoglobin in erythrocytes. This has included a large trial with hydroxyurea. This drug reduces the frequency of hemolytic crises in test populations and holds promise for use in larger populations. Butyrate has been used for the same purpose, but experience with it is not as widespread. Another approach is to perform frequent transfusions to maintain a relatively normal total number of erythrocytes (of course, the transfused cells do not contain HbS). Repeated transfusions have their own complications, however, including iron overload. Another approach that, in theory, would be successful is to perform bone marrow transplantation. Obviously, replacing the erythrocyte precursors with cells containing the normal β-globin gene must, of necessity, eliminate the production of HbS. Unfortunately, the complications of bone marrow transplantation have currently reduced the appeal of this approach, particularly since improved management has considerably improved the life expectancy and quality of life for affected individuals.

As noted above, sickle cell disease is one of the best-studied genetic conditions in the world. Because the prevalence of the heterozygous state is so great and the severity of the homozygous state is considerable, many workers have attempted to explain the persistence of a high frequency of heterozygosity in certain populations. The best explanation for this situation relates to the relative resistance of heterozygous individuals to falciparum malaria. The malaria parasite colonizes sickle erythrocytes less efficiently. Obviously, such relative resistance to a devastating infectious disease, especially in the era before malaria treatment was available, would increase the likelihood of survival of carriers. The perception thus arose that the occasional appearance of homozygotes, with their attendant morbidity—and frequently mortality—was a relatively small price to pay for the population advantage provided by the broad distribution of the carrier genotype throughout regions where malaria is endemic. Through over 40 years of study, this perspective has persisted and gained broad acceptance. It is important to note that other hematologic disorders have a similar geographic distribution, including the thalassemias and glucose-6-phosphate dehydrogenase deficiency (OMIM #305900, see Chapters 7 and 10).

C. GAUCHER DISEASE

Gaucher disease reflects malfunction in a specific enzymatic pathway. In this case, the enzyme affected is β-glucosidase. Heterozygous carriers of mutant alleles for this gene appear normal. However, homozygotes for mutations in β-glucosidase develop a series of problems. Gaucher disease stands in impressive contrast to sickle cell disease, and this contrast is important for a general perspective on recessive conditions. Sickle cell disease is due to a unique mutation, occurring consistently with the same molecular change in all affected individuals. By contrast, many mutant alleles can lead to Gaucher phenotypes. The different mutations, occurring at different places within the responsible gene (see Figure 6.5), can cause different problems. For example, some mutations produce no functional protein at all. Others, in contrast, produce a protein that has enzymatic function but altered activity. Obviously, there is the potential for considerable difference in phenotypes between individuals who make no protein and those who make a partly functional molecule. Not all of the mutations underlying Gaucher phenotypes have been identified, but a prominent group is shown in Figure 6.5. The large number of known mutations in Gaucher disease has led to another interesting contrast to sickle cell disease: Many affected individuals are not true homozygotes but "compound heterozygotes"; they have a different mutant allele of the β-glucosidase gene on each chromosome.

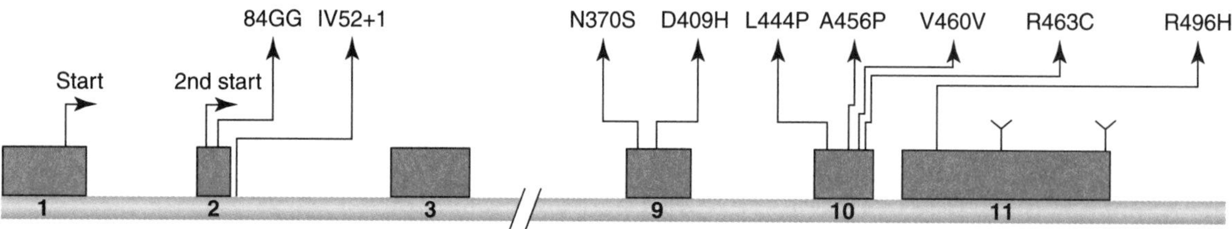

Figure 6.5 *Sites of several mutations in the β-glucosidase gene. Homozygotes and compound heterozygotes for these and other changes show the clinical phenotype of Gaucher disease. Note that the mutations are of different types and occur at multiple sites. (Modified from Horowitz M, Am J Med Genet 53:921, 1993.)*

Because of the differences in β-glucosidase mutations, several different clinical presentations have been distinguished (see Table 6.1). The common feature of all of these clinical pictures is that they are related to defective recycling of membrane glycolipids. The substrate is glucocerebroside (glucosylceramide), derived from leukocyte membranes in the blood and from neuronal tissue in the nervous system.

Type I Gaucher disease (OMIM # 230800) generally occurs in adults. In many it is a remarkably well-tolerated condition. Adult Gaucher disease is notable for gradual enlargement of the spleen due to the sequestration of membrane degradation products within macrophages. A characteristic finding in adult Gaucher disease is the Gaucher cell (see Figure 6.6). This cell is large and often seen in the bone marrow. It contains multiple intralysosomal layers of membranes that represent partially degraded molecules and whose final degradation and recycling has been either slowed or inhibited completely by the enzyme mutation(s). These cells accumulate in the spleen, and the resultant splenic enlargement gradually leads to platelet entrapment and thrombocytopenia. Despite the thrombocytopenia, many patients do not have severe bleeding difficulties and can go for many years without problems. Gaucher cells continue to accumulate, however, with complications

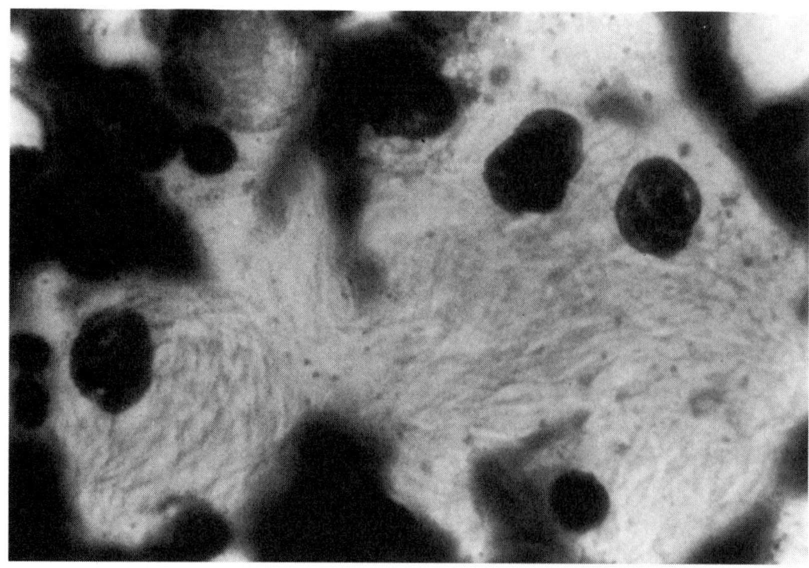

Figure 6.6 *Gaucher cells in bone marrow aspirate. These cells have pale purple cytoplasm on conventional staining and contain multilayered membranous inclusions. The cells can be very large—some are 100 μm in diameter. They are derived from macrophages. See also Plate 3.*

involving the bone marrow. The accumulation of Gaucher cells in bones leads to spontaneous fractures because of progressive cortical weakness as well as degeneration of adjacent joints. Thus, the initial presentation of Gaucher disease in the formerly asymptomatic adult may be as a spontaneous fracture, often in the femur. Only after that event has occurred are splenomegaly and mild thrombocytopenia detected. The clinical course of Gaucher disease in adults varies, because there is such a broad spectrum of mutations underlying the phenotype. Adult ("Type I") Gaucher disease is relatively common (see Table 6.1). It is more frequent in individuals with Eastern European Jewish heritage, but it is by no means confined to that population. Even within the Ashkenazic Jewish community, multiple mutations have been identified.

"Type II" Gaucher disease (OMIM # 230900) has been called the "infantile" form. This condition, fortunately quite rare, causes not only enlargement of the liver and spleen but also relatively rapid neurologic degeneration. This condition is generally lethal within the first several years of life, with progressive loss of neurologic function. There is little time for bone involvement to develop, and thus that is not an important part of the clinical picture in Type II Gaucher disease.

"Type III" Gaucher disease (OMIM #231000) also is characterized by liver and spleen enlargement but has a more gradual onset and also is associated with loss of neurologic function. These individuals can reach the second decade of life but usually have severe neurodegeneration by that time. Types II and III are remarkably rare compared with the broad distribution of Type I Gaucher disease.

The treatment of individuals with Gaucher disease has undergone considerable study and represents an important joining of information about cell biology, enzymology, and protein chemistry. Because the β-glucosidase enzyme can be taken up by cells from their environment, early efforts at treatment employed intravenous administration of relatively purified enzyme. Although some results were promising, they did not reach the point of clinical applicability. More recently, the

Table 6.1 Gaucher disease phenotypes*

Type	Designation	Clinical features	Populations affected
I	Adult	Splenomegaly, thrombocytopenia, bone marrow involvement, fractures	20+-fold increased in Ashkenazi Jews but seen in all populations
II	Infantile	Hepatosplenomegaly, rapid neurological deterioration, death	All populations
III	Juvenile	Hepatosplenomegaly, slow but relentless neurological degeneration	All populations

* All reflect homozygous or compound heterozygous mutations in the β-glucosidase gene.

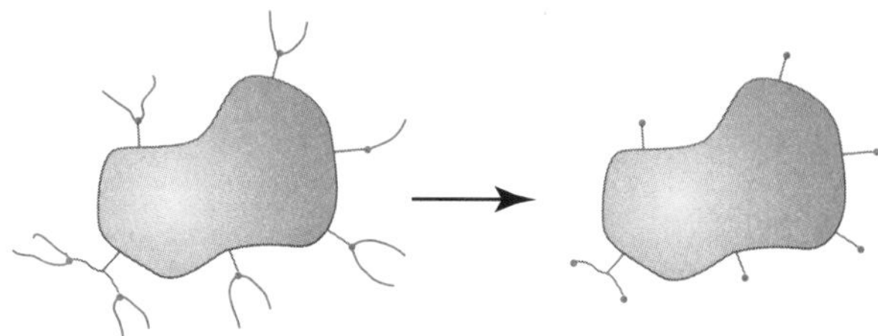

Figure 6.7 *Preparation of Ceredase, a derivative of β-glucosidase that can be administered to individuals with Gaucher disease and taken up by macrophage receptors, leading to resumed enzyme activity and reduced symptoms and signs of Gaucher disease. As isolated from the placenta, the enzyme contains complex carbohydrate side chains. Much of the carbohydrate is removed, but mannose-6-phosphate residues are left; these are essential for targeting the product to macrophages. Such a preparation is useful only in peripheral tissues and organs; it does not enter the central nervous system or ameliorate problems there.*

development of reliable enzyme replacement treatment has changed the management picture. For the newer treatment, large amounts of enzyme have been purified from placentas. As isolated, this enzyme has a complex set of carbohydrate additions (see Figure 6.7). These carbohydrates change the chemical character of the enzyme and complicate readministration. However, partial hydrolysis of the carbohydrate adducts yields a modified enzyme that can be administered intravenously and taken up reliably by macrophages in the spleen, liver, bone, and elsewhere. Recently, recombinant enzyme has become available. Once taken up, the enzyme, known as "Ceredase," functions appropriately to degrade accumulated glucocerebrosides, leading to both an improvement of bone disease and a reduction in spleen and liver size. The treatment is not simple, because recurrent infusions are needed. Currently, this treatment is very expensive, and Ceredase has "orphan drug" status (see Chapter 14).

A particular concern with regard to the success of this approach in treating accumulations in bones, liver, and spleen is the fact that the treatment fails to penetrate the central nervous system. Thus, the irony is that although the treatment can be remarkably effective for patients with Type I disease, the adult form of the condition (for whom brain involvement is not a problem), it cannot prevent continued neurologic degeneration in individuals with Type II or III disease. Thus, the latter individuals can have somatic improvement despite progressive loss of neurologic function. Obviously, this cannot be considered an acceptable treatment for individuals with Type II or III Gaucher disease.

As described above for sickle cell disease, it is feasible to consider bone marrow transplantation for patients with adult Gaucher disease, but this is rarely performed, both because of the relatively recent availability of the enzyme replacement and because of the difficulties inherent in transplantation. Replacing resident bone marrow with unaffected stem cells can control many manifestations of Type I disease, but its long-term effect on neurologic function is not known, so it cannot be recommended for individuals with Types II and III disease.

As noted, there is no known difficulty with being a carrier for any of the mutations in Gaucher disease. Apparently, the presence of one normal allele provides sufficient enzymatic activity so that reasonable metabolism can be achieved. Prenatal or presymptomatic diagnosis by DNA studies is possible for pregnancies and individuals at risk for Gaucher disease, but it is complicated by the existence of different mutations that can cause the same phenotype. Thus, prior to considering such diagnostic procedures, it is essential to determine the gene change(s) in an affected individual in the kindred. If the mutation(s) is(are) not known, it is possible that several mutations could be excluded by the testing, while the mutation(s) for which the pregnancy or the individual was at risk would not be recognized. This situation is a striking contrast to that in sickle cell disease, in which the mutation is always the same, but it is now becoming recognized in many other recessive conditions, as will be considered in detail below.

D. CYSTIC FIBROSIS

Cystic fibrosis (CF) (OMIM #219700), a disorder of membrane transport, is the most frequently encountered recessive condition in people of Caucasian and northern European descent. The understanding of the clinical spectrum of CF has been altered by the continuing discovery of older individuals with the disease. Conventional descriptions of CF involve complications from difficulty with managing secretions in infancy. This may present as intestinal blockade in the newborn due to meconium ileus, arising from firm intestinal secretions that resist peristalsis. This problem represents a combination of causes but especially reflects insufficient pancreatic secretions. The problem of pancreatic secretions is now manageable, however, with the use of oral enzyme replacements. Thus, many of the digestive and intestinal problems can be circumvented or minimized.

A more impressive and chronic difficulty is related to pulmonary secretions. Here the problem is largely that of clearing lung contaminants from the mucosal surfaces. The increased tenacity of the pulmonary secretions in CF reduces the rate of bronchial clearance, leading to the potential for establishment of chronic sites of bacterial contamination and, ultimately, infection. These poorly cleared areas ultimately allow damage to the underlying mucosal surfaces and the basic architecture of the lung, compromising gas exchange. The picture progresses from chronic bronchitis through changes of florid bronchiectasis to ultimate pulmonary insufficiency. This has been the pattern commonly recognized for untreated individuals.

More recent studies have revealed a much broader spectrum of clinical presentations for CF. Some individuals in the fourth and fifth decades of life with chronic lung disease have now been shown to carry mutations in the CF gene. These people may have had no difficulty with pancreatic exocrine secretions, and, in contrast to the more severely affected people manifesting problems earlier in life, they may have had relatively mild clinical courses.

The gene whose mutations are responsible for CF has been isolated. It encodes a very large protein known as the "cystic fibrosis transmembrane regulator" (CFTR). Many mutations have been identified in the CFTR gene (see Figure 6.8). These undoubtedly underlie the broad spectrum of clinical presentations. In addition, some CFTR mutations may not have very severe clinical effects; others may represent relatively neutral changes, detectable only as molecular polymorphisms. The function of CFTR is still being clarified, but it clearly is related to ion transport across cell membranes. In particular, transport of the chloride ion, with its attendant sodium cation, is regulated by CFTR. Because ion transport obligates water movement as well, the defective movement of ions in areas such as the pancreas, intestine, and lung reduces the moisture in the secretions and consequently increases their viscosity. It is this thick layer of secretions that resists physiologic clearance mechanisms.

Although the spectrum of mutations in cystic fibrosis is broad, several are relatively common. The most common of these mutations represents an interesting three-base-pair deletion. This deletion neatly removes a single codon for phenylalanine. The mutation has thus achieved the designation "ΔF508" (OMIM #219700.0001). The absence of this

Figure 6.8 *The remarkable spectrum of mutations (and polymorphisms) in the CFTR gene. The density of the lower line emphasizes the broad spectrum of changes and the difficulties that can develop in attempting testing on the basis of suspected mutation(s). (Adapted from Scriver CR, Beaudet AL, Sly WS, Valle D (eds): The Metabolic and Molecular Basis of Inherited Disease, New York, McGraw-Hill, chap. 127, p. 3826.)*

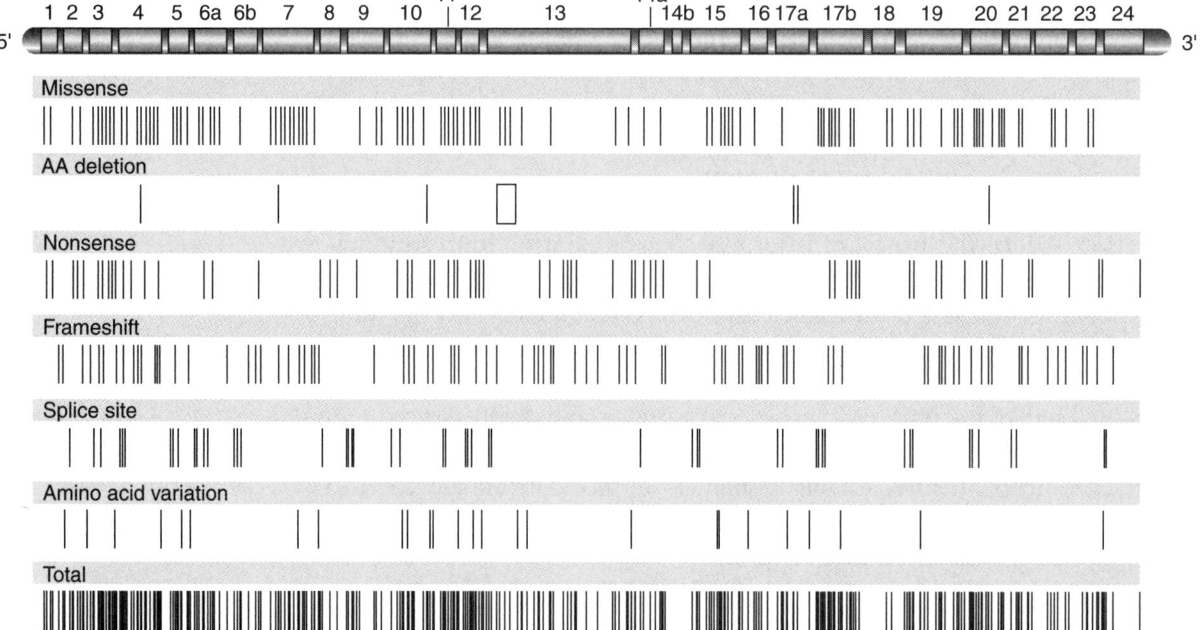

single amino acid at position 508 in the protein is sufficient to cause the entire clinical picture.

The diagnosis of CF begins with suspicion based on the clinical presentation. Obviously, that suspicion must now be broadened to include young and middle-aged adults with recurrent respiratory difficulties. The laboratory test most frequently used when CF is suspected clinically is known as the "sweat chloride test." In this test, the relative concentration of chloride (and hence the conductivity) in skin secretions is measured. Affected individuals show increased sweat chloride levels. Although frequently possible, studying the CFTR gene for specific mutations is more time-consuming and expensive. Furthermore, the large number of potentially responsible mutations can make this a difficult approach (although essential for prenatal and presymptomatic diagnosis in a given kindred).

Management of the care of individuals with CF has undergone considerable improvement, which has been responsible for increases in lifespan and in quality of life. The pediatric approaches to managing meconium ileus and intestinal motility are now generally successful, and replacing pancreatic enzymes with oral supplements can permit reasonable intestinal function, digestion, motility, and absorption. In contrast, the respiratory problems have been more of a challenge. Appropriate antibiotic use has reduced the level of complications early in life. Nevertheless, the chronic bacterial colonization, particularly with *Pseudomonas* species, and difficulties with managing secretions generally lead to the loss of gas exchange surface and to chronic lung disease later in life. While oxygen supplementation may be helpful, it does not fundamentally affect the loss of gas exchange surface.

Because pulmonary problems are so prominent in the long-term care of individuals with CF, the development of treatments based on biotechnology has received considerable emphasis. Because part of the difficulties with clearing secretions is their tenacity, and because much of this gel-like material represents cell debris and high-molecular-weight molecules, the introduction of agents that degrade these polymers has been considered as an option for improving pulmonary toilet. One approach that has had some success has been to introduce topical nucleases through pulmonary aerosols. These enzymes assist in degrading the DNA that is one of the highest-molecular-weight components of the secretions. Reducing the length of these DNA polymers assists in clearing the residual material by reducing its viscosity.

The fact that the respiratory epithelium is so frequently affected in CF has made it an obvious target for efforts at gene replacement. Although this will be discussed in more detail in Chapter 14, the notion of replacing the CFTR gene within the respiratory epithelium by aerosol delivery systems has been appealing. Delivery systems for pulmonary treatment often have been based on viruses, such as adenoviruses, that are known to have a host range limited to respiratory tissues. Attempts to develop successful gene replacement strategies using recombinant adenoviruses are currently under intense investigation. Unfortunately, the fact that the respiratory epithelium is constantly being shed at the surface and renewed from basal layers means that these treatments must be repeated at reasonable intervals so that the newly developed surfaces will continue to have a functional gene. These experimental efforts are promising and will be the basis of future, more successful gene-based treatment.

As noted above, CF is the most common recessive genetic condition in Caucasians. Nevertheless, the frequency of carriers for mutant alleles is not as high in this population as that for sickle cell anemia in African-descended populations. It needs to be reemphasized that sickle cell disease represents a unique and consistent mutation, in contrast to the many mutations that can cause a CF phenotype (recall Figure 6.8). Thus, it is quite likely that many individuals presenting with, and receiving the diagnosis of, CF do not represent homozygotes but rather compound heterozygotes, similar to the situation with Gaucher disease. We already have considered clinical variations in the presentation and severity of sickle cell disease. We noted that the level of fetal hemoglobin, controlled by a gene unlinked to the globin locus, can substantially affect the clinical outcome. While such epistasis may appear complicated when just the single sickle cell mutation is involved (see above), the problem of how compound heterozygotes might respond to epistatic or environmental factors is very difficult to formulate. Thus, it is currently difficult to determine the underlying mutation on the basis of the clinical presentation in CF. As discussed above, this makes gene-based diagnosis more difficult because one cannot know a priori what mutations to test for. Cystic fibrosis resembles Gaucher disease with respect to the presence of a relatively common mutation (ΔF508) complicated by that of multiple other gene changes.

There is evidence that CFTR is essential for the entrance of *Salmonella typhi* into epithelial cells of the gastrointestinal tract. Thus, CFTR mutations may have provided some protection against this important pathogen and hence could have been valuable for survival in earlier populations.

E. PHENYLKETONURIA

Phenylketonuria (PKU) (OMIM # 261600) is an important cause of mental retardation. PKU is caused by defective function of the phenylalanine hydroxylase (PAH) gene. This metabolic defect causes an increase in the level of the amino acid phenylalanine in the serum, the main chemical feature of the condition. PKU is relatively prominent in Caucasians, with a prevalence of about 1 in 10,000. It also is recognized in other populations throughout the world. The most prominent clinical characteristic of affected individuals is mental retardation. Few other characteristic clinical features have been de-

Phenylalanine Hydroxylase (PAH)

Phenylalanine → Tyrosine

Tyrosine → Thyroxine, Epinephrine, Melanin

Figure 6.9 *The metabolic pathway involved in phenylketonuria. Absence or malfunction of phenylalanine hydroxylase (PAH) creates both a reduction of the levels of tyrosine (and its important metabolites) and an increase in phenylalanine concentrations behind the barrier.*

fined, although some individuals do have seizures and abnormal postural changes.

The biochemical change in PKU is an example of a metabolic blockade. The normal metabolism of phenylalanine involves its hydroxylation by PAH, as shown in Figure 6.9. The absence or ineffective function of PAH causes a characteristic increase in the concentration of the substrate. Thus, high blood levels of phenylalanine develop. Human observations and animal studies indicate that prolonged high blood levels of this single amino acid cause neurologic damage, presumably explaining the mental retardation.

The diagnosis of PKU takes advantage of screening techniques to detect high blood levels of phenylalanine. In the United States since 1963, the Guthrie test has been performed by collecting a single blood specimen from the heel of a newborn (see Figure 6.10). This screening test detects elevated blood levels of phenylalanine. Because of the screening program in the United States, individuals with high blood levels are regularly detected. Although not all of them have classic PKU, this screening approach identifies infants who need further testing. The Guthrie test is one of the simplest and most broadly used genetic screening procedures. As will be discussed later, its design permits detection of infants affected with PKU, as well as those whose phenylalanine levels are elevated because of other causes. Some of these elevated levels may be transient; others reflect even more rare genetic disorders.

The PAH gene has been studied in considerable detail. The mutations are varied and spread throughout the gene, resembling the pat-

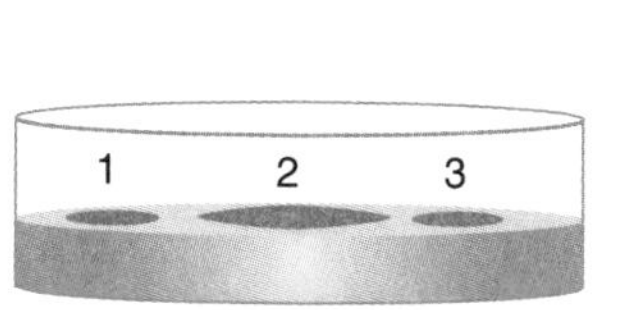

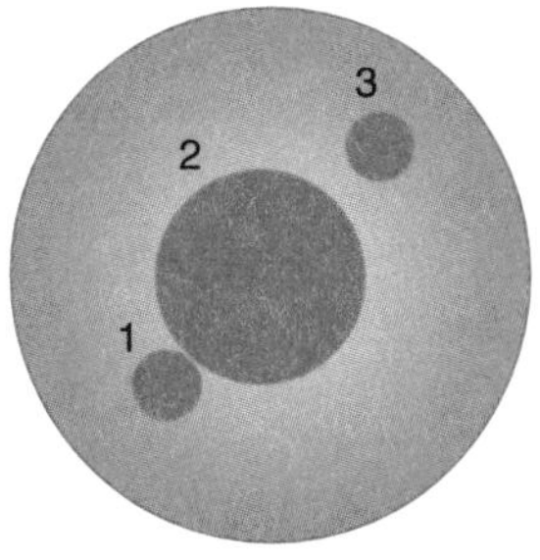

Figure 6.10 *The Guthrie test. A blood spot on filter paper establishes the potential for competition between phenylalanine in the blood and β-2-thienylalanine in the agar. High levels of phenylalanine permit bacterial growth, as shown. Note that this is not diagnostic for PKU; it was designed to screen for high levels of phenylalanine.*

terns noted in Gaucher disease and CF. No single mutation can uniformly explain the presence of PKU. Furthermore, as might be expected, non-Caucasian populations have different PAH mutations.

Because of the severe mental retardation characteristic of individuals with PKU, treatment is begun early. The basis of treatment is limitation of dietary phenylalanine. Reduced amounts of this essential amino acid decrease blood levels of phenylalanine. Long-term clinical studies have established the efficacy of this dietary treatment, and the mental function of treated adults is substantially improved. Nevertheless, the treatment is not without its difficulties. In the first place, it is difficult to maintain; infants and children need to be kept away from other foods. In the second place, the diet is not particularly tasty, making compliance a difficult challenge. In the third place, the treatment has led to a new question regarding the appropriate management for phenylketonuric adults who were treated as infants and children. It is not currently possible to predict the long-term implications (neurologic or otherwise) of feeding a regular diet to an adult whose PKU was treated in childhood. As anticipated, resuming a nonrestricted diet causes a predictable elevation in blood phenylalanine levels. What is not known is how these levels will affect the individual over an adult lifetime. One perspective is that the elevated levels might cause damage only in the developing nervous system. On the basis of this hypothesis, high adult levels should not create difficulties. On the other hand, if chronic high blood phenylalanine levels result in subtle neurotoxicity, there may be slow neurodegeneration in these individuals. Long-term observations and clinical measurements will be necessary to resolve this important question.

There is one situation, however, in which adults treated for their PKU in childhood need special attention: pregnancy. A woman with treated PKU who becomes pregnant and is no longer maintaining dietary control exposes her developing fetus to high levels of phenylalanine at a critical time in its neurologic development. Studies in animals have shown significant fetal damage due to high maternal phenylalanine levels. Thus, the care of pregnant individuals who themselves have PKU involves continued (or resumed) dietary control. The fetuses of such women necessarily are at least heterozygous for a mutation in the PAH gene and thus may be less able to tolerate high phenylalanine levels than those with no such mutation.

PKU has received considerable attention. Originally believed to represent an opportunity to "cure" mental retardation by biochemical manipulation, PKU is now recognized as a condition affecting different populations with different mutations, whose long-term neurologic effects after treatment are still unknown. The problems of normal diets in affected adults and especially the diet for mothers during pregnancy are still the subjects of long-term investigations that should clarify the picture. Nevertheless, the multiplicity of mutations and degrees of clinical severity make predictions difficult for individuals.

F. TREATMENT

Recessive disorders, as noted earlier, have provided many opportunities for understanding metabolic pathways. Significantly, studies of some of these disorders originally helped to define these pathways. Treatments have been considered, as described in the examples above. Several different approaches have been brought to bear on the metabolic abnormalities of different recessive conditions. No single approach is applicable to all conditions, however, often for specific biochemical and clinical reasons. The relatively simple approach of limiting the dietary content of a potentially toxic metabolite has proved very useful for PKU, galactosemia, and even more rare metabolic problems. This is not successful for Gaucher disease, however, because the molecular substrates involved are synthesized by the body rather than derived from exogenous sources. For this reason, the notion of enzyme replacement has arisen and already has proved effective in Gaucher disease and several other conditions. As described above, however, this involves commitment to long-term, repeated treatment. Furthermore, whether this treatment will be tolerated (for instance, whether antibodies to the enzymes will develop) and remain effective for decades is not known. The possibility of performing longitudinal clinical observations in these treatment populations, although of fundamental importance, will likely be complicated by the interim development of newer and even more effective treatments. Thus, it may be very difficult to obtain and compare extended follow-up data about cohorts managed with different treatment protocols.

The notion of activating a molecular switch for synthesizing an alternative hemoglobin has considerable appeal for the treatment of individuals with sickle cell disease. The effectiveness of this approach already has been shown and has encouraged both patients and physicians. Nevertheless, the long-term consequences of such treatment, which may itself be potentially hazardous, are not known.

The possibility of replacing a defective gene has driven much thinking and research, particularly in relation to recessive disorders, but it also is applicable to other sorts of conditions transmitted by mendelian inheritance. Several clinical trials of treatment using gene replacement in CF are under way. The effectiveness of such protocols over both the short and the long term is not known. Nevertheless, the fact that genes are being introduced into the bronchial epithelium means that the treatments will have to be repeated periodically as these tissues are renewed. Permanent gene replacement is the goal of other protocols currently being studied. These will be reviewed in more detail in Chapter 14. Of course, for a number of these conditions the option of bone marrow transplantation remains. This clearly can be effective, but, as recognized, it is not without its own difficulties. This obviously can be of particular value for individuals with conditions related to blood-forming elements and in whom the bone marrow contains stem cells responsible for producing affected blood cells. It has not been effective where the nervous system is involved, however.

The evolving thoughts and plans for treating individuals with recessive and other mendelian conditions generate considerable interest and justifiable hope. These must be tempered with an awareness of the need for very long-term follow-up to be certain that unexpected consequences do not develop. Furthermore, successful treatments, such as dietary management for infantile and childhood PKU, have led to the emergence of an entirely new category of adults. There is no natural history, and there are no long-term data regarding the prognosis and potential medical peculiarities that may be associated with adults who have been successfully treated for PKU earlier in life. As these individuals are followed, new observations undoubtedly will be made that will influence their care. Ultimately, some of these may lead to dietary modifications, while others may reveal challenges peculiar to PKU in adults. We already have discussed the importance of pregnancy in adults with PKU, but, clearly, other common conditions to which such individuals are susceptible (e.g., hypertension, diabetes) may have unique features in terms of presentations and pharmacologic and clinical responses.

G. IMPLICATIONS OF THE CARRIER STATE

One of the most conspicuous features of recessive disorders is their rarity. It is relatively unlikely that any physician will encounter many homozygotes (or compound heterozygotes) during the course of a lifetime in general medical practice. For this reason, among others, considerations about these conditions are frequently lost from memory. A few simple calculations, in a relationship originally developed by Hardy and Weinberg, are quite enlightening, however (see Figure 6.11). These workers began with the simple assumption that the population is at genetic equilibrium. They assumed that there would be only two alleles at a locus of interest, occurring at frequencies p and q, where $p = 1 - q$. Obviously, the frequency of homozygosity (i.e., two copies) for either allele will be p^2 or q^2. If one

Two alleles of frequencies p,q

■ Frequency $= \frac{1}{10{,}000}$ $(= p^2)$, $\therefore p = \frac{1}{100}$

$\therefore q = 1 - p = \frac{99}{100}$

□ Frequency $= \frac{9801}{10{,}000}$ $(= q^2)$

◧ Frequency $= \frac{198}{10{,}000}$ $(= 2pq) \sim 2\%$

$p + q = 1 \qquad p^2 + 2pq + q^2 = 1$

Figure 6.11 *The Hardy-Weinberg relationship assumes that the population is genetically stable and considers a simple case of a two-allele system in which the sum of the frequencies of the two alleles $(p + q)$ must equal 1. This example begins with a recessive condition with a frequency of homozygotes of 1 in 10,000 (10^{-4}).*

takes a frequency for homozygotes of 1/10,000, such as is seen for PKU, this becomes p^2; p then becomes 1/100. The alternative is q^2, which represents the frequency of homozygosity for the normal allele at the same locus. This equals 9801/10,000. Because $q = 1 - p$, it becomes 99/100. Although this might appear to end the discussion because p seems quite small, a more interesting number reflects the frequency of heterozygotes, or carriers, to be expected on the basis of these same relationships. The frequency of heterozygosity is $2pq$. It is simple to show that this equals $2 \times 1/100 \times 99/100$. This is approximately 1/50. Thus, for a condition whose homozygous incidence is one in 10,000, one out of every 50 individuals (2%) would be expected to be a heterozygous carrier for the mutant allele. Similar calculations can be made for disorders of higher or lower frequency.

The important conclusion from the Hardy-Weinberg equilibrium assumptions is that even for rare recessively inherited diseases, the frequency of carriers of the mutant allele is relatively high. In any clinic population of reasonable size, there will be a number of individuals heterozygous for a PKU mutation. For a condition such as sickle cell disease, with an incidence of homozygotes in Africa of one in ~600, the carrier frequency is obviously considerably higher. (You might wish to try the math.)

These simple calculations and the fact that several thousand recessive disorders have been identified immediately indicate that any given individual is a carrier for several mutant alleles. The reason that homozygotes (or compound heterozygotes) for recessive phenotypes are rare in the United States reflects the largely outbred characteristics of our population. As noted earlier in this chapter, consanguinity and geographic, religious, or ethnic isolation can change the incidence of homozygosity simply by bringing together more heterozygotes.

As noted, the clinical status of heterozygotes is usually normal. Carriers of the sickle cell mutation have been studied extensively, and no problems have been identified. Nevertheless, it is not known what the effect of heterozygosity for mutant alleles of multiple genes might be. The notion of epistatic gene effects (introduced above in our consideration of sickle cell disease) cautions us about simplistic interpretations of these data. It is not at all difficult to propose situations in which heterozygosity for one mutant allele in a metabolic pathway could alter the substrate level for another enzymatic reaction also operating at a heterozygous level and lead to overt or subtle clinical difficulties. A current area of considerable interest is elucidation of the interaction of alterations in different metabolic pathways and their potential clinical effects. The potential for heterozygosity in thousands of individual genes establishes a formidable base of genetic and clinical variation (see also Chapter 12).

Current efforts at medical treatment for individuals manifesting different recessive disorders represent a new event in the population history of many mutations. It is obvious that, prior to these interventions, many recessive conditions were lethal; many still are. Even untreated PKU, although not immediately lethal, is associated with severe mental retardation, and affected individuals would be unlikely to reproduce (thus, it is lethal in a genetic sense). Similarly, until relatively aggressive management in recent years, many individuals homozygous for sickle cell disease did not reproduce. For sickle cell disease, as noted earlier, the fact that the heterozygous state confers relative resistance to falciparum malaria could serve as an important positive selection factor, and this certainly is consistent with the distribution of sickle cell disease in areas where malaria is endemic. Without such selection pressure, however, the frequency of individuals homozygous for rare metabolic defects would be expected to fall unless one of two events occurred. First, new mutations could continue to rise as old ones were eliminated by homozygotes who could not pass them on, leading to genetic lethality. This can be tested for many genes by comparing the chromosomal haplotypes within which the mutant alleles are found. The notion of haplotypes was introduced in Chapter 1. If new mutations were occurring and replacing old ones, the new mutations should be in equilibrium in terms of frequency with the surrounding haplotypes, but they are not. Thus, the possibility remains and must be considered strongly that at some time in the past, possibly well before the current era, heterozygosity for at least some of these mutant alleles provided some selective advantage. We may never identify the survival advantages imparted by such mutant alleles, but it is important to consider that they might provide differential responses to the environment or to infections, trauma, or degenerative conditions.

Because of the multiplicity of mutant alleles that frequently underlie individual recessive phenotypes, as discussed above, it often is difficult to either establish or exclude heterozygosity for an individual. Thus, the clinician, geneticist, or genetic counselor often is forced to work in reverse. To understand the affected individual, it is necessary to know the molecular details of the responsible mutations. For what we now recognize as a relatively atypical situation in the case of sickle cell disease, this is a trivial undertaking, because all carriers have the identical mutant allele at the β-globin locus. This makes prenatal diagnosis for sickle cell disease a very direct task.

In contrast, the possibility of compound heterozygosity for many other recessive conditions creates a formidable set of difficulties for prenatal and presymptomatic diagnosis. Unless both mutant alleles are identified in the compound heterozygote, it can be very difficult to make any predictions, either in the prenatal clinic or for a presymptomatic but potentially affected sibling. The current approach often is based on considering the frequency with which known mutations have been encountered within the general population; thus, one can consult a ranked list of likely mutant alleles. Unfortunately, not finding any of the known mutations does not exclude the possibility of a very rare, possibly as yet undescribed, mutation. For this and other reasons, screening tests based on metabolic abnormalities, such as the Guthrie test for PKU, have an important advantage over DNA-based testing for the same disorder. Measuring the end product of the metabolic pathway that is at risk for variation provides a more broad-based screening approach than DNA techniques that depend on being able to identify a specific mutation.

Because of the possibility of compound heterozygosity, there exist variations in the clinical severity and presenting features within the same diagnostic category. Thus, as noted above, Gaucher disease presents three identifiable clinical phenotypes, each of which involves gene changes within the same genetic locus. Similarly, CF now is being recognized in adults of middle age who have had relatively minor difficulties earlier in life and might, for instance, present with no gastrointestinal disturbances, or at least none that have been sufficiently severe to cause any problems. Awareness of such broad variation adds considerable challenge to the general practice of medicine in any age group, and this will only become more prominent as the broad range of expression of compound heterozygosity is explored. More aspects of these problems will be considered in Chapters 12 and 13.

CASE STUDY

Hilda was a frail but remarkably active 82-year-old retired teacher. Although she always tried to take extra dietary calcium, she had noted mild kyphosis. While walking on a rainy day, she fell against the curb and fractured her right hip. She was brought to the emergency room and admitted for definitive management. Her vital signs were stable. She had a prominent bruise over the right hip but no bruising elsewhere and no pallor or petechiae. Aside from general findings consistent with her age, her examination showed a spleen tip palpable at the left costal margin; liver was normal. Her chemistry and urinalysis findings were normal. Hgb = 12.8; WBC = 17,500; platelets = 72,000. She was taken to surgery for replacement of the hip. When the femoral neck was exposed, a thick but not viscous yellow material was found exuding from the fracture site; it was not purulent. A sample of the material was placed on a microscope slide and showed the picture seen in Figure 6.12. The surprising finding was sheets of pale cells with voluminous cytoplasm and multiple intracellular layered structures. Based on this finding, the surgeons were alerted to the presence of adult Gaucher disease, and further examination of the femur showed virtually complete replacement of the marrow cavity with similar cells. A long prosthesis was used to strengthen the femoral shaft. The patient recovered well.

COMMENT

This remarkable presentation surprised both the patient and her physicians. It is an example of one end of the spectrum of adult Gaucher disease, and enzyme analysis confirmed the defective function of β-glucosidase. Although thrombocytopenia had likely been present for many years, she had accommodated to it well and no one suspected the abnormality. Her spleen was enlarged, although not strikingly so, and it had not been detected earlier in her life. Her femoral bone marrow was replaced by Gaucher cells and subsequent x-rays of the other leg showed cortical thinning as well, placing her at risk for additional fractures. Her course did not indicate a need for Ceredase treatment and she has done well through 3 more years of follow-up.

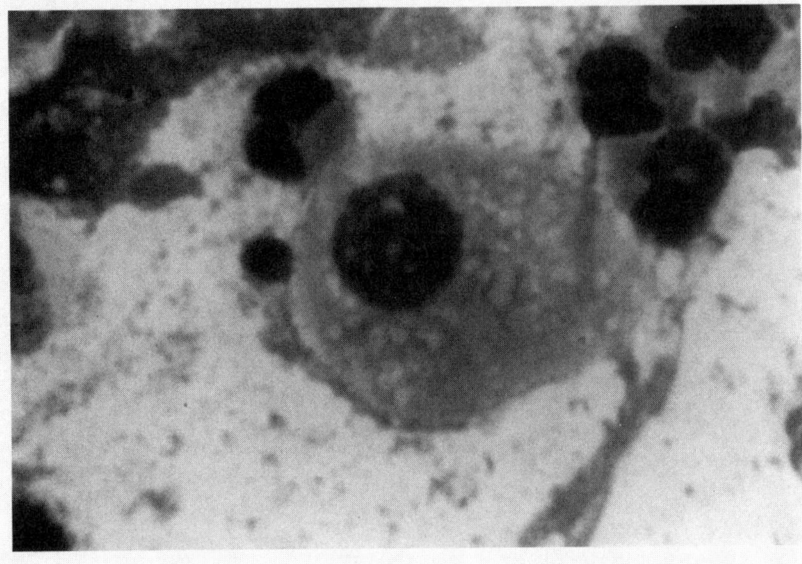

Figure 6.12 *Gaucher cells in exudate from femoral fracture. Compare with Figure 6.6.*

STUDY QUESTIONS

1 As described earlier, sickle cell disease always reflects homozygosity for the identical mutation. Consider the mechanisms promoting variations in clinical severity. Can any of these be established through testing?

2 Your patient Raymond is 24 and has sickle cell disease. His symptoms have been mild, and he considers himself to be doing well. He plans to marry Tanya, who has been healthy. They have come to you for counseling about a future pregnancy outcome. Tanya and Raymond are not related; both are African-American. You tell them:

   a The likelihood of their having a child with sickle cell disease is low because the mother is unaffected.
   b Any child affected with sickle cell disease would be expected to have a mild course, similar to Raymond's.
   c Tanya should have a test for sickle cell carrier status.
   d Prenatal diagnosis is necessary.
   e Any child must at least be a carrier.

3 A couple had a child with Type II Gaucher disease who died at 18 months of age. The uncle of the father of the child is found to have splenomegaly at age 58; his past history is remarkable for good health. The uncle is your patient. You say:

a This is most likely a lymphoma.
b The spleen should become smaller with Ceredase treatment.
c He needs a platelet count and bone marrow aspiration.
d He may be a compound heterozygote for Type I Gaucher disease.
e He may be a homozygote for Type I Gaucher disease.
f He needs a splenectomy.

4 Untreated homozygotes for PKU often have fair skin and light brown hair. Can you offer an explanation based on the metabolic changes?

FURTHER READING

Barany F. Genetic disease detection and DNA amplification using cloned thermostable ligase. *Proc Natl Acad Sci USA* 88:189, 1991.

Beutler E. Gaucher disease as a paradigm of current issues regarding single gene mutations of humans. *Proc Natl Acad Sci USA* 90:5384, 1993.

Charache S et al. Hydroxyurea and sickle cell anemia: Clinical utility of a myelosuppressive agent. *Medicine* 75:300, 1996.

Collins FS. Cystic fibrosis: Molecular biology and therapeutic implications. *Science* 256:774, 1992.

Eisensmith RC and Woo SLC. Molecular basis of phenylketonuria and related hyperphenylalaninemias: Mutations and polymorphisms in the human phenylalanine hydroxylase gene. *Hum Mutat* 1:13, 1992.

Hubbard RC et al. A preliminary study of aerosolized recombinant human deoxyribonuclease I in the treatment of cystic fibrosis. *N Engl J Med* 326:812, 1992.

Mistry PK et al. Genetic diagnosis of Gaucher's disease. *Lancet* 339:889, 1992.

Perrine SP et al. A short-term trial of butyrate to stimulate fetal-globin-gene expression in the beta globin disorders. *N Engl J Med* 328:81, 1993.

Pier GB. *Salmonella typhi* uses CFTR to enter intestinal epithelial cells. *Nature* 393:79, 1998.

Rouse B et al. Maternal phenylketonuria collaborative study (MPKUCS) offspring: Facial anomalies, malformations and early neurological sequelae. *Am J Med Genet* 69:89, 1997.

Stern RC. The diagnosis of cystic fibrosis. *N Engl J Med* 336:487, 1997.

USEFUL WEB SITES

HELIC (INCLUDES GENLINE)

http://www.hslib.washington.edu/helix
Medical Genetics Knowledge Base. Electronic textbook under development relates testing to diagnosis, management, and counseling. Directory of laboratories providing testing for genetic disorders.

OMIM (ONLINE MENDELIAN INHERITANCE IN MAN)	http://www.ncbi.nlm.nih.gov/omim Basic catalog listing mendelian genes and traits. Clinical descriptions, references, and links to other databases.
HEALTHLINKS	http://healthlinks.washington.edu/ Broad access to many related areas of basic and clinical genetics. Includes "Genome Machine II" database of genes and diseases mapped to a particular chromosome location. Also chromosome information.
MARCH OF DIMES	http://modimes.org/ Resource for families and professionals about birth defects. Professional education section.
NATIONAL ASSOCIATION FOR RARE DISORDERS (NORD)	http://www.NORD-rdb.com/~orphan Primary nongovernmental clearinghouse for information about rare disorders. Over 1100 disease reports. Broad resource guide. Includes Rare Disease Database, a wide range of information for families and professionals. Listings by symptoms, causes, affected populations, treatments. Listing of support groups. Includes orphan drug designation database.
OFFICE OF RARE DISEASES (NIH)	http://cancernet.nci.nih.gov/ord/index Broad-based NIH index including support groups, clinical research database, investigators, and extensive glossary.

CHAPTER 7

X-Linked Inheritance

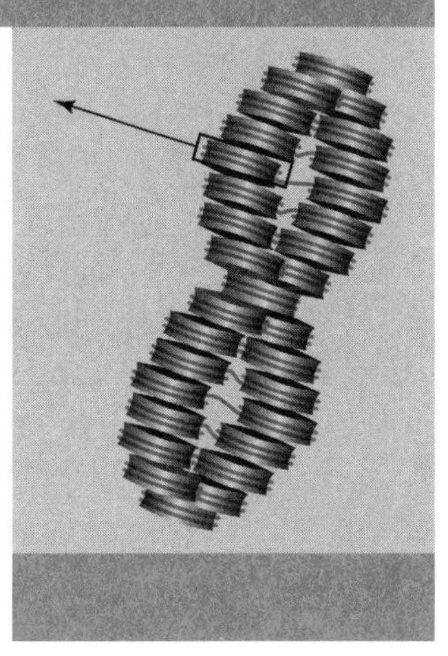

A. OVERVIEW

Disorders due to abnormalities on the X-chromosome follow a characteristic clinical pattern due to the biology of this chromosome, as reviewed in Chapter 3. The clinical manifestations are notable for their obligate expression in males (who have only a single X-chromosome and hence must express the genes on that chromosome). In females, the Lyon hypothesis (see Chapter 3) predicts random inactivation of one of the X-chromosomes early in embryonic development. The lineage of each of the embryonic cells retains this X-inactivation pattern. Because the inactivation is initially random, females become mosaic in terms of which X-chromosome is expressed in which cell. This mosaicism means that the effect of a mutant allele on a single X-chromosome may be minimally detectable, not apparent at all, or obvious, depending on whether the X-chromosome carrying the mutant allele has been inactivated in the organ or organ system in which that gene is normally expressed. Obviously, the degree of involvement in females may be subtle and detectable only with detailed testing. This is important in terms of the general pattern of transmission of X-linked disorders.

Figure 7.1 shows several pedigrees characteristic of X-linked inheritance. Note that the symbol for a carrier female is the presence of a large dot within the circle. As before, affected males are marked by

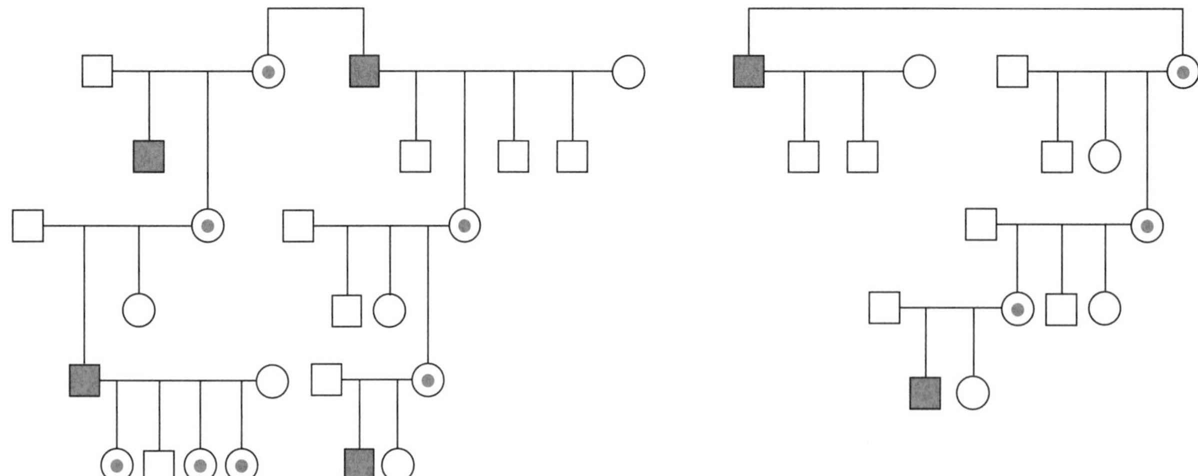

Figure 7.1 *Two pedigrees showing X-linked inheritance. Note how several generations may appear unaffected if carrier females are not detected. Affected males from earlier generations may not be available for examination or may not have had their diagnoses established by current clinical or laboratory criteria. Thus, careful questioning with a broad understanding of potential phenotypic variation(s) may be needed to clarify the family pattern and identify those members at risk.*

solid squares. Several important observations are obvious:

- Females are not affected as severely as males and, thus, may be considered normal on initial clinical encounters.
- An affected male cannot transmit the trait to his sons, because the trait is on the X-chromosome, and the father must necessarily transmit his Y-chromosome to a son.
- All of the daughters of an affected male must be carriers, because the only X-chromosome that the father can give to a daughter contains the mutation.

These features, which are direct consequences of the biology of the X-chromosome, impart a pattern to these pedigrees that some have called "diagonal." This is because the affected males have only carrier daughters and unaffected sons. These sons are not at risk for further transmission of the mutant allele. The carrier daughters, however, are at 50% risk for having an affected son or a carrier daughter. Obviously, because carrier females may go undetected, many generations can pass with clinically silent transmission through females until an affected male appears. Testing then may demonstrate a pervasive picture of female carriers and significantly alter the known recurrence risks for other members of the kindred.

Unlike the situation seen in recessive conditions (see Chapter 6), the phenotype frequency in males equals the frequency of the mutant allele. This follows from the facts that in a two-allele system the mutant allele must be either present or absent, and, if present, it will be expressed.

The X-chromosome has been particularly well studied in many organisms. Because there is little opportunity for meiotic exchanges between the X- and Y-chromosomes (although a small amount of recombination can occur in isolated regions), the genes on the X-chromosome have been remarkably preserved in evolution. (This phenomenon has been called "Ohno's law.") This means that the complement of genes on the X-chromosome and much of their organization is similar in humans and other mammals, presenting many opportunities for comparative genetic, metabolic, and therapeutic studies.

B. HEMOPHILIA A

Hemophilia A (OMIM # 306700) is a disorder of blood coagulation due to absence or abnormality of factor VIII in the clotting cascade. It is one of the best-known X-linked conditions and has gained considerable notoriety.

The gene for factor VIII is quite large and is located on the distal part of the large arm of the X-chromosome, as shown in Figure 7.2. Many mutations have been characterized in this gene, helping to explain the variations in clinical severity and in the presentations encountered in individuals with hemophilia A. The typical picture in hemophilia A is an otherwise healthy boy or young man with a history of either spontaneous hemorrhage or continued bleeding after minimal trauma. This bleeding may be into soft tissue, such as muscle, or it may be into joints, leading to the complication of hemarthrosis. Because the clotting factor is either defective or absent, these episodes of bleeding may be quite severe. In some episodes the bleeding stops slowly and spontaneously, presumably because the pressure of accumulated blood acts to tamponade it. Other situations may lead to life-threatening hemorrhage. These recurrent bleeding episodes are particularly disturbing to the patient and his family and can lead to anxiety, reduced mobility, and general avoidance of activities.

Figure 7.2 *A simplified map of the X-chromosome, showing the loci described in this chapter. Note the close proximity of the glucose-6-phosphate dehydrogenase deficiency (G6PD), color blindness (CB), and hemophilia A (HemA) loci on the end of the long arm. The precise physical relationships among these have been established using molecular genetics. PEX and the Duchenne muscular dystrophy (DMD) gene are found on the short arm. The FraX site is the position of the FMR-1 gene.*

Because hemophilia A represents a deficiency of biologically active factor VIII, it is treated by administering the active clotting factor. The quality and properties of the preparations used have changed from early treatment with fresh plasma, to cryoprecipitated plasma, to the point where recombinant factor VIII has become available. These products arrest bleeding by replacing the defective clotting factor. Because clinical manifestations and severity differ, the treatment protocols differ as well. Some individuals need prophylactic administration of relatively small doses, while others do not need treatment except after trauma that causes bleeding. These management protocols are usually established relatively early in life. For many reasons, including diagnostic confusion and because the individuals themselves are less able to avoid minor trauma, the symptoms of hemophilia A are more often severe in younger boys than in adolescents and adults.

The mutant allele for hemophilia A is passed through asymptomatic carrier females. It can pass through many generations. A particularly notable pedigree developed from the fact that Queen Victoria of England was a carrier and had one affected son (Leopold, Duke of Albany) as well as a several carrier daughters. One of her granddaughters, Alexandra, also a carrier, married Nicholas II, the last Romanoff Czar in Russia; the illness of their son, born with hemophilia A, created considerable intrigue and political difficulties toward the end of the Romanoff era.

C. FRAGILE X SYNDROME

It has been known for many years that a disproportionate number of males have mental retardation. The complete basis for this has not been established, but recent studies have helped to clarify an entire group of disorders.

Among mentally retarded males there is a population whose karyotypes often show broken chromosomes when their cells are cultured in media lacking folic acid. This is usually due to fragmentation (failure to condense during mitosis) of the X-chromosome at a specific location, Xq27 (see Figure 7.3); this "fragile site" is the FMR-1 gene (see Figure 7.2). Despite the ability to detect this important chromosomal landmark, little progress was made until the nature of the underlying genetic

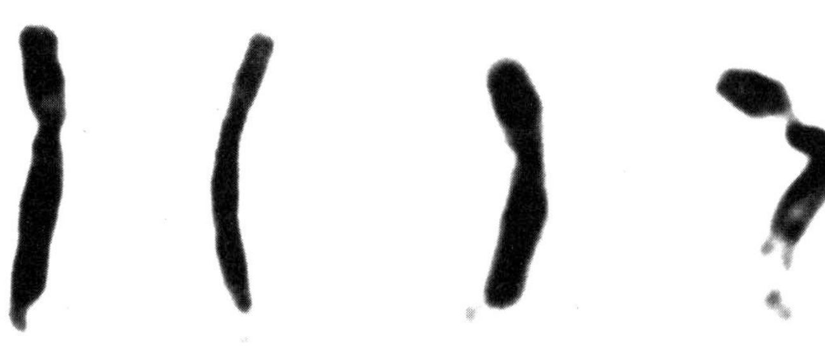

Figure 7.3 *Several X-chromosomes showing fragmentation at the fragile site. (Photograph courtesy of Dr. George Thomas.)*

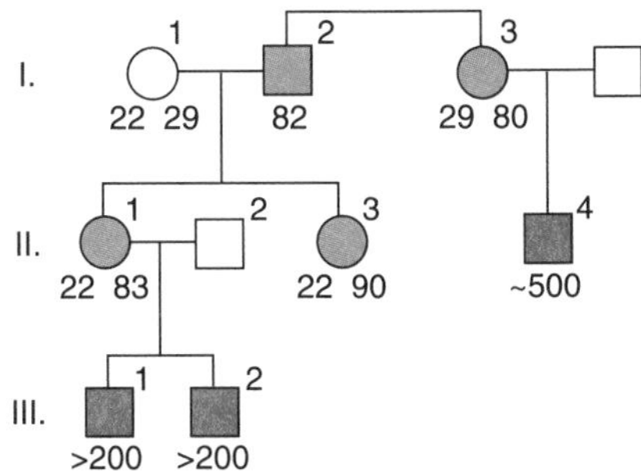

Figure 7.4 *A Fragile X syndrome kindred. The numbers of triplet repeats are shown beneath the symbols. Note the normal transmitting male (I-2) and his sister (I-3), who have similar numbers of repeats in the premutation range but no symptoms. Sons of females with repeats at the premutation level (II-4, III-1,2) can receive X-chromosomes with greatly expanded repeats, in which case they will show the clinical features of Fragile X. Note the absence of significant expansion of the repeats in females II-1 and II-3, who received the premutation from their father. (Adapted from Caskey et al, Science 256:784, 1992.)*

change was clarified. Fragile X syndrome (OMIM #309550) was the first disorder characterized by a triplet repeat in the DNA. We already have considered the implications of such repeats in our review of myotonic dystrophy in Chapter 5. Fragile X syndrome involves expansion of a CGG trinucleotide 5' to the coding region of the FMR-1 gene. No gene product is produced in this amplified situation. Fragile X syndrome is remarkably common, affecting about one in 1500 males. However, unlike the situation in hemophilia A, one third of females with the expansion are mentally impaired (although usually not severely so).

An additional anomaly in Fragile X syndrome is seen in its transmission (see Figure 7.4). A threshold exists, with individuals who have 50-200 repeats being considered to have premutations. A male with such a premutation is usually clinically normal and called a "normal transmitting male." His daughters (who must inherit the premutation) are usually asymptomatic as well and have little change in their repeat numbers. Both male and female children of these females are at risk for receiving greatly increased numbers of repeats (250-4000 the "full" mutation), as well as symptoms. Thus, expansion of this triplet repeat is largely confined to female meioses (since clinically affected males rarely reproduce). It is not quite as clear, however, whether clinical severity is related to the degree of expansion of the repeat sequence.

The degree of mental retardation varies, but the average IQ of affected males is about 40. Males with the full mutation can show behavior changes resembling those of autism, with delayed language skills and poor coordination. These males generally have coarse facial features, with elongation and a thickened bridge and filtrum of the nose, as well as large testicles. Unfortunately, however, the visual impression of males with Fragile X syndrome is not always distinctive; thus, the presence of mental retardation in a male justifies a search for Fragile X.

Evaluation of kindreds once a male with a full mutation has been detected often identifies individuals who have triplet expansions to premutation levels but are unaffected themselves. There currently is no treatment for affected individuals, so detection through prenatal and presymptomatic studies can be useful for families.

D. DUCHENNE MUSCULAR DYSTROPHY

Duchenne muscular dystrophy (DMD) (OMIM # 310200) is the most common form of muscular dystrophy in males. It has a relatively early age of onset, usually before age 6. Affected boys exhibit problems with walking and a waddling gait, sometimes mistaken for clumsiness. This apparent clumsiness or weakness may appear anomalous, because of muscular enlargement, particularly early in the disease; this has been termed "pseudohypertrophy." Weakness of pelvic, paraspinal, and shoulder girdle musculature is often present early as well. Knee and hip extensor weakness makes rising from a seated position difficult. To accomplish this maneuver, affected individuals usually flex their hips and push themselves up by placing their hands on their knees and moving them up the thighs (Gowers' sign).

These problems are progressive, with the development of muscular atrophy leading to flaccid limbs and prominence of bones as the overlying muscles become thinner. Involvement of the heart muscle is common, and congestive heart failure is frequent as a terminal event in those few individuals who live beyond the third decade. Some patients show a decline in IQ, but this is not usually severe. In general, those developing symptoms earliest have the worst prognosis.

There is a remarkably high apparent spontaneous mutation rate for DMD; as many as 1 in 3000 to 4000 liveborn males are affected. This may reflect the fact that the gene itself is very large (~2.3 mb), presenting a large target for mutations. The gene is located on the short arm of the X-chromosome (Xp21.2; see Figure 7.2). This gene encodes an enormous protein called dystrophin (>400 kDa), found in cardiac and skeletal muscle. It is now possible to measure dystrophin levels in muscle biopsies of individuals suspected of having the condition.

A broad range of mutations has been found in individuals with DMD; about 60 to 70% of affected individuals show a deletion in the gene. Some of these mutations present as what has been called "Becker type muscular dystrophy," but this is now recognized to be part of the milder aspects of the clinical and mutational spectrum of DMD.

The progressive weakness and lack of mobility of affected males gives them reduced reproductive fitness. However, the carrier status of their mother and other female relatives is an important piece of information for the entire family. For most of these families, a specific DNA-based study can establish the presence of the underlying gene change in female carriers as well as in pregnancies and presymptomatic boys. Interestingly, a few carrier females have muscle weakness, presumably due to some expression (or lack thereof) of the mutant gene in muscles through lyonization, as discussed in Chapter 3.

As proposed many years ago on the basis of population studies, DMD behaves like an X-linked lethal disorder, because affected males do not reproduce. Haldane formulated an instructive relationship in regard to such disorders. He noted that in a population of $2n$ persons at equilibrium, there should be $3n$ X-chromosomes. About one-third of all DMD mutations will be found on the n X-chromosomes of affected males, and about two-thirds would be on the $2n$ X-chromosomes of carrier females. The mutations on the n male X-chromosomes would be lost in each generation because of the genetic lethality. Thus, to maintain equilibrium in the distribution of mutations across $3n$ X-chromosomes, about one-third of the mutations must arise per generation to replace those lost. This formulation permits an estimate of the mutation rate, because about one-third of affected males should represent new mutations. Therefore, obviously, the other two-thirds of patients have mothers who are carriers. This high number of carriers means that family screening studies can be valuable in counseling unsuspecting and clinically unaffected females who are at risk for having affected sons. Although there is no effective treatment currently, isolation of the gene and characterization of the dystrophin protein have led to attempts at gene replacement in order to restore muscle function. This area of research has not yet proved useful clinically but has considerable promise for affected individuals.

E. COLOR BLINDNESS

Defective color vision is one of the most frequent inherited conditions in males, present in approximately 8% of the population. Whether this can be called a true "disease" is somewhat subjective, but affected individuals may have considerable difficulty with perception, and aspects of the problem demonstrate several important features. "Color blindness" does not describe a single condition; there are many ways to alter color perception. Of specific interest to considerations in this chapter are those forms of color vision deficiency that are related to abnormalities of the visual pigment genes on the X-chromosome (OMIM #303800/303900).

Each cone cell of the human retina contains one of three visual pigments—for absorption maxima in the blue, green, or red wavelength range. The gene for the blue color vision pigment is located on chromosome 6, but the genes for pigments absorbing in the red and green maxima are located in tandem on the X-chromosome, very close to

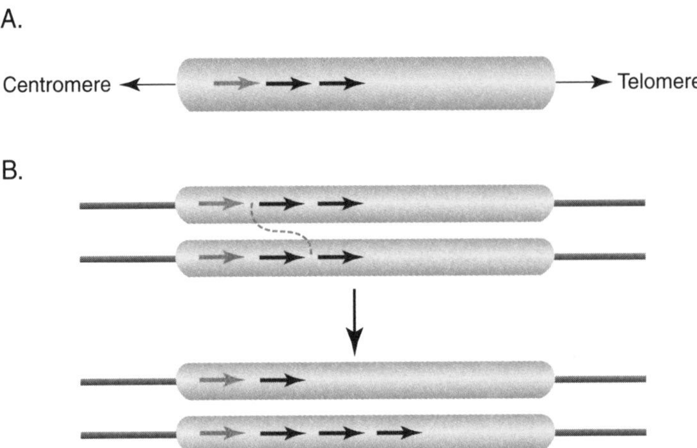

Figure 7.5 *(A) The basic organization of the human color vision gene cluster on Xq28. Note the head-to-tail arrangement, with one red (absorption maximum ~560 nm) and a pair of identical green (absorption maximum ~530 nm) pigment genes. (B) Many variations in this basic organization are possible as a result of aberrant pairing in meiosis, because the genes are 96% identical. This can lead to changes in gene number. (After Nathans et. al, Science 232:203, 1986.)*

the hemophilia A gene (see Figures 7.2 and 7.5). The structure of the visual pigment genes undoubtedly has contributed to the likelihood that defective genes will develop. As shown in Figure 7.5*A*, the visual pigment genes exist in a cluster of very similar genes arranged in a head-to-tail array. These may show variations in the numbers of normal genes present. Because normal color vision is achieved by expressing a single red and a single green pigment gene from the X-chromosome, simply changing the numbers of these genes does not necessarily change visual sensitivities. Such changes—duplications—most likely are the result of aberrant crossing-over at meiosis. Because the genes are so very similar, abnormal pairing can easily generate additional copies, as shown in Figure 7.5*B*. Note that entire genes are duplicated. This is distinct from the amplification of trinucleotides in Fragile X and myotonic dystrophy.

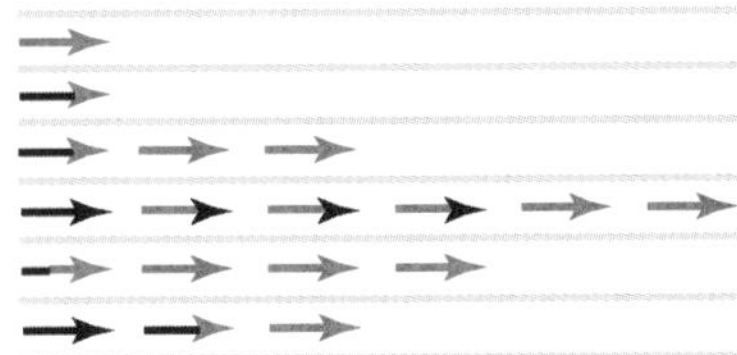

Figure 7.6 *Hybrid visual pigment genes can be derived from crossing-over within coding regions. This can lead to genes containing sequences from both red and green pigments that may or may not be functional. (Sack and Morrell, Inv Ophthal Vis Sci 34:2634, 1993.)*

In addition to changing the number of normal genes, aberrant crossing-over within this gene cluster also can lead to the loss of genes and/or to their mutation in such a way that the protein is not expressed at all or the protein that is expressed is abnormal (see Figure 7.6). Having these highly homologous genes in a contiguous array on the X-chromosome undoubtedly has established the potential for many genetic variations. In addition, most affected males have few problems and are not at a selective disadvantage genetically. In fact, many men are unaware that they have difficulties with color perception, and community conventions minimize the immediate consequences of color vision confusion. For example, traffic lights always have the upper light red and the lowest light green, and many street signs can be distinguished on the basis of their shape rather than their color.

In contrast to males, heterozygous females rarely have difficulty with red/green color distinctions, and it is difficult to detect females carrying an abnormal visual pigment gene simply by routine color vision testing. Of course, DNA testing can reveal the gene changes, but it is rarely indicated.

Color blindness arising from genetic changes in the tandem pigment gene cluster on the human X-chromosome is very instructive. It provides examples of the consequences of aberrant meiotic pairing and crossing-over, as well as of a change(s) within a contiguous array of virtually identical genes. The fact that this tandem gene arrangement contributes substantially to the frequency of the changes is substantiated by noting the rarity of changes in the pigment gene with a blue absorption maximum found isolated on chromosome 6. In addition, chromosome 6 gene mutations are generally present only in heterozygotes, so that homozygosity for changes in blue visual pigment genes is extremely rare (but can be encountered in both males and females). The tandem arrangement of the α- and β-globin gene clusters, shown in Figure 1.5, provides a similar opportunity for aberrant crossing-over. Many anomalous hemoglobins have been produced by this method, although they are collectively less frequent, probably because the globin genes are less similar and because the consequences include chronic hemolytic anemia.

F. GLUCOSE-6-PHOSPHATE DEHYDROGENASE (G6PD) DEFICIENCY

Glucose-6-phosphate dehydrogenase (G6PD) deficiency (OMIM # 305900) is an enzyme defect associated with nonspherocytic hemolytic anemia. G6PD deficiency occurs worldwide and has been recognized and studied for many years. There is remarkable variation in its clinical spectrum, undoubtedly reflecting the fact that over 300 different genetic variants of the G6PD protein are known. Males deficient for the enzyme have reduced resistance to oxidative stress; this leads to altered erythrocyte stability. The most frequent form of oxidative stress is due to drug exposure. The characteristic picture is that an individual is given the drug and within several days develops an episode of acute hemolysis, with its characteristic pain, anemia, and urinary abnormalities (see below). Interestingly, such hemolysis usually is confined to the older red cells in the circulation, because of the relative resistance of younger red blood cells to oxidative stress.

G6PD deficiency represents an important interaction between the host and the environment. As will be discussed in more detail in Chapter 10, G6PD deficiency is largely a drug sensitivity. There are several important categories of drugs to which deficient individuals are particularly susceptible; some of these drugs are indicated in Table 7.1. Exposure to any of these agents can lead to acute hemolysis. Rarely, some individuals present with chronic hemolysis without any identifiable environmental cause. Some individuals have sensitivities to food; an Italian population was identified whose hemolysis developed after eating fava beans. Their condition was known as "favism," but it was really just a special case combining G6PD deficiency with exposure to a dietary oxidant.

G6PD deficiency is undoubtedly the most common hemolytic anemia throughout the world, but the broad range of variants makes its clinical manifestations hard to predict. In addition to favism, notable in Italian communities, large populations of individuals in areas endemic for malaria, particularly in Africa and Southeast Asia, have hemolysis induced by primaquine. Interestingly, G6PD deficiency received more attention following racial integration of the U.S. armed forces; some African-American GIs in the Korean War developed hemolysis after receiving malaria prophylaxis. Table 7.1 emphasizes that a broad range of drugs must be considered when treating sensitive individuals. Despite this, there also is a large group of agents that are safe to use.

The obvious treatment for individuals deficient in G6PD is to avoid contact with inducing agents. This certainly requires circumspection in prescribing and also some dietary considerations. Because acute hemolysis can be severe, it is important to document G6PD deficiency in any individual once it has been identified so that future problems can be avoided. Testing for G6PD deficiency is usually based on enzyme activity, which is detected by a simple laboratory test; DNA-based diagnosis is not commonly used. (Again, there is a broad range of mutations in the responsible gene.)

Table 7.1 ▶ **Drugs and chemicals causing problems in individuals with G6PD deficiency**

Acetanilide	Niridazole
Doxorubicin	Nitrofurantoin
Furazolidone	Phenazopyridine
Methylene blue	Primaquine
Nalidixic acid	Sulfamethoxazole

Drugs tolerated by individuals with G6PD deficiency (and without nonspherocytic hemolytic anemia)

Acetaminophen	Isoniazid	Quinidine
Ascorbic acid	Phenacetin	Quinine
Aspirin	Phenylbutazone	Streptomycin
Chloramphenicol	Phenytoin	Sulfamethoxypyridazine
Chloroquine	Probenecid	Sulfisoxazole
Colchicine	Procainamide	Trimethoprim
Diphenhydramine	Pyrimethamine	Tripelennamine

(After Beutler, *N Engl J Med* 324:169, 1991.)

G. THE FEMALE CARRIER

As mentioned earlier, males with an X-linked genetic change necessarily have difficulties, because their only copy of the responsible gene is abnormal. Because of random X-inactivation and mosaicism in females, there is an excellent chance that a carrier female will have no difficulties at all. Occasionally, the cells responsible for a particular product encoded on the X-chromosome may be derived from cells in which the normal gene has been inactivated. In such an individual it is possible to identify features of the disorder. Women with clinical hemophilia A have been recognized, and, as noted, some female carriers of the full mutation for the Fragile X syndrome also show cognitive defects. The relative rarity of these "manifesting heterozygotes" attests to the validity of the Lyon hypothesis and the whole idea of random X-chromosome inactivation. It also makes family studies particularly valuable, because the absence of symptoms can create a kindred with many asymptomatic female carriers who are at 50% risk for having affected male children. As clinical and laboratory studies are refined, more subtle details are often appreciated in carrier females who formerly were considered to be asymptomatic. The main point of identifying such women remains counseling and assistance with detecting and caring for symptomatic males.

H. X-LINKED DOMINANT INHERITANCE: HYPOPHOSPHATEMIC RICKETS

There is a group of disorders whose genes are found on the X-chromosome but for whom the transmission pattern differs from those discussed above. These have been called "X-linked dominant" disorders to distinguish their pedigree patterns. In this group of conditions, both males and females are affected and can transmit the trait to offspring. Pedigrees for X-linked dominant traits show that females transmit the trait to 50% of both sons and daughters. (Each has a 50% chance of receiving the mutant X-chromosome.) Males, however, transmit the trait to none of their sons and all of their daughters.

One of the best-studied examples of this transmission pattern is seen in X-linked hypophosphatemic rickets (OMIM #307800), the most common inherited form of rickets in the United States. Affected individuals are usually short, with bowed legs and osteomalacia. They often develop ankylosis of the spine and large joints and can have hearing loss.

Because of defective renal phosphate transport, patients have phosphate wasting and low serum phosphate levels. They also have abnormal metabolism of 1,25-dihydroxyvitamin D. Treatment with both phosphate and vitamin D often leads to improvement in bone abnormalities.

The responsible gene (called PEX) is located at Xp22.1 and shows homologies to endopeptidases, but its exact function has not been determined. As might be expected because of the conservation of genes on the X-chromosome (recall Ohno's law) a homologous gene has been found on the X-chromosome of the mouse.

CASE STUDY

Charles is 17. He is active in school but generally avoids contact sports. He has known hemophilia A (OMIM #306700) and had several serious hemarthroses as a young child. The second of these led to reduced mobility of his left shoulder, but he managed to work around it and, being right-handed, considers this a minor problem. He is a good student and is comfortable with more quiet activities. He has recently assumed management of his own factor VIII administration and had learned to administer appropriate amounts of recombinant protein when necessary. He has a detectable level of circulating factor VIII activity, and he has not found it necessary to administer doses at regular levels. Over the last 2 years he has had only three episodes of bleeding, and he has considered them minor. His uncle (his mother's brother) who also had hemophilia A died last year from AIDS.

COMMENT

This student shows the progress in the care of individuals with hemophilia A.

- He has learned to adjust the amount of recombinant factor VIII in a reasonably appropriate relationship to bleeding or trauma and has had few problems.
- He has adjusted well to the care of his condition and has stated that "it's no different than taking care of diabetes."
- He remains at considerable risk in the case of serious trauma, but he carries identification to notify emergency care personnel of the need for factor VIII.
- His uncle is an unfortunate example of someone who became infected with HIV from a contaminated factor VIII preparation nearly 15 years ago. This complication has been recognized for a group of individuals receiving blood- or tissue-based products prior to the recognition of and testing for HIV. It is a complication eliminated by the use of recombinant material.

STUDY QUESTIONS

1 You are working on a temporary assignment in a remote Pacific island community where several males with Fragile X have been identified. Because of your medical background, you have been asked to help identify normal transmitting males. Unfortunately, there are no resources for DNA or chromosome testing, but you did bring a few items in an examining bag. These items include a hematocrit tube, a tuning fork, Ishihara plates, some test kits for G6PD, and an IQ test protocol designed for this community. What testing can you perform? What are the limits to your testing? What will you do with the results?

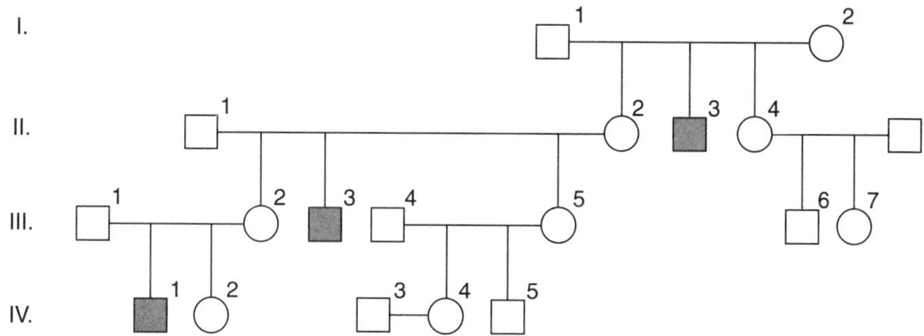

Figure 7.7 *Three affected males have been identified in this kindred.*

2 Concerned parents have brought in their 2-year-old son, who has developmental problems that no one has been able to clarify. They are young and healthy and are interested in having more children. What can you suggest?

a Their recurrence risk is 50%.
b Their recurrence risk is 25%.
c Their recurrence risk is ~3%.
d They should have a karyotype performed on the child.
e They should have IVF and preimplantation sex determination.
f There is a low risk of recurrence in females.

3 You are planning a linkage study on the family in Figure 7.7. Three males have been identified with a serious, late-onset X-linked disorder. Although the gene has not been isolated and no metabolic tests are available, you have identified a highly polymorphic DNA marker that is tightly linked to the phenotype. Unfortunately, individuals II-2 and III-5 are not available for testing.

a Can you determine the carrier status of IV-2 and IV-4?
b Can you offer prenatal diagnosis to IV-3 and/or IV-4?
c Can you offer presymptomatic diagnosis to IV-5 and III-6?

FURTHER READING

Beutler E. Glucose-6-phosphate deficiency. *N Engl J Med* 324:169, 1991.

Caskey CT et al. Triplet repeat mutations in human disease. *Science* 256:784, 1992.

Fu Y-H. Variation of the CGG repeat at the fragile X site results in genetic instability: Resolution of the Sherman paradox. *Cell* 67:1047, 1991.

Harper PS. *Practical Genetic Counseling*, 4th ed. Bristol, Wright, 1993.

Holm IA et al. Mutational analysis of the PEX gene in patients with X-linked hypophosphatemic rickets. *Am J Hum Genet* 60:790, 1997.

Nathans J et al. Molecular genetics of inherited variation in human color vision. *Science* 232:203, 1986.

Roberts RG et al. Searching for the 1 in 2,400,000: A review of dystrophin gene point mutations. *Hum Mut* 4:1, 1994.

Rousseau F et al. A multicenter study on genotype-phenotype correlations in the fragile X syndrome, using direct diagnosis with probe StB12.3: The first 2253 cases. *Am J Hum Genet* 55:225, 1994.

Sutherland GR and Mulley JC. Fragile X syndrome and other causes of X-linked mental handicap, in Rimoin DL, Connor JM, Pyeritz RE (eds). *Emery and Rimoin's Principles and Practice of Medical Genetics*, 3d ed. New York, Churchill Livingstone, 1997, p 1745.

Tennyson CN et al. The human dystrophin gene requires 16 hours to be transcribed and is cotranscriptionally spliced. *Nat Genet* 9:184, 1995.

USEFUL WEB SITES

HELIC (INCLUDES GENLINE)

http://www.hslib.washington.edu/helix
Medical Genetics Knowledge Base. Electronic textbook under development relates testing to diagnosis, management, and counseling. Directory of laboratories providing testing for genetic disorders.

OMIM (ONLINE MENDELIAN INHERITANCE IN MAN)

http://www.ncbi.nlm.nih.gov/omim
Basic catalogue listing mendelian genes and traits. Clinical descriptions, references, and links to other databases.

HEALTHLINKS

http://healthlinks.washington.edu/
Broad access to many related areas of basic and clinical genetics. Includes "Genome Machine II" database of genes and diseases mapped to a particular chromosome location. Also chromosome information.

MARCH OF DIMES

http://modimes.org/
Resource for families and professionals about birth defects. Professional education section.

NATIONAL ASSOCIATION FOR RARE DISORDERS (NORD)

http://www.NORD-rdb.com/~orphan
Primary nongovernmental clearinghouse for information about rare disorders. Over 1100 disease reports. Broad resource guide. Includes Rare Disease Database, a wide range of information for families and professionals. Listings by symptoms, causes, affected populations, treatments. Listing of support groups. Includes orphan drug designation database.

CHAPTER 8

Mitochondrial Disorders

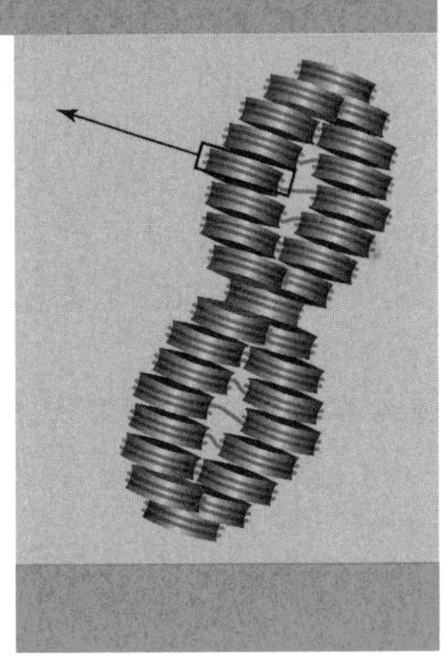

A. OVERVIEW

Mitochondrial DNA has been referred to as the "25th chromosome." It is a circular, double-stranded DNA molecule whose complete sequence of 16,569 nucleotides has been determined. All genes in mitochondrial DNA have been defined (see Figure 8.1). They are responsible for encoding 22 transfer RNAs necessary for mitochondrial protein synthesis, two ribosomal RNAs (12s and 16s), and 13 proteins essential to oxidative phosphorylation. Thus, it is not surprising that changes in mitochondrial genes affect cell processes related to energy and mitochondrial oxidative functions and that tissues with particularly high energy requirements, especially brain and muscle, often are the first to show the effects of mitochondrial mutations. There is at least one copy (and often many) of mitochondrial DNA in each of the hundreds of mitochondria in the cell; this means that there are many copies of the mitochondrial "genome" in every cell.

These features underlie five important aspects of mitochondrial disorders, which can be summarized as follows:

- The portion of the sperm that penetrates and fertilizes the egg generally contains no mitochondria. Thus, the population of mitochondria in the developing embryo is derived from the mother (in virtually all studied cases) and must reflect the structure of the

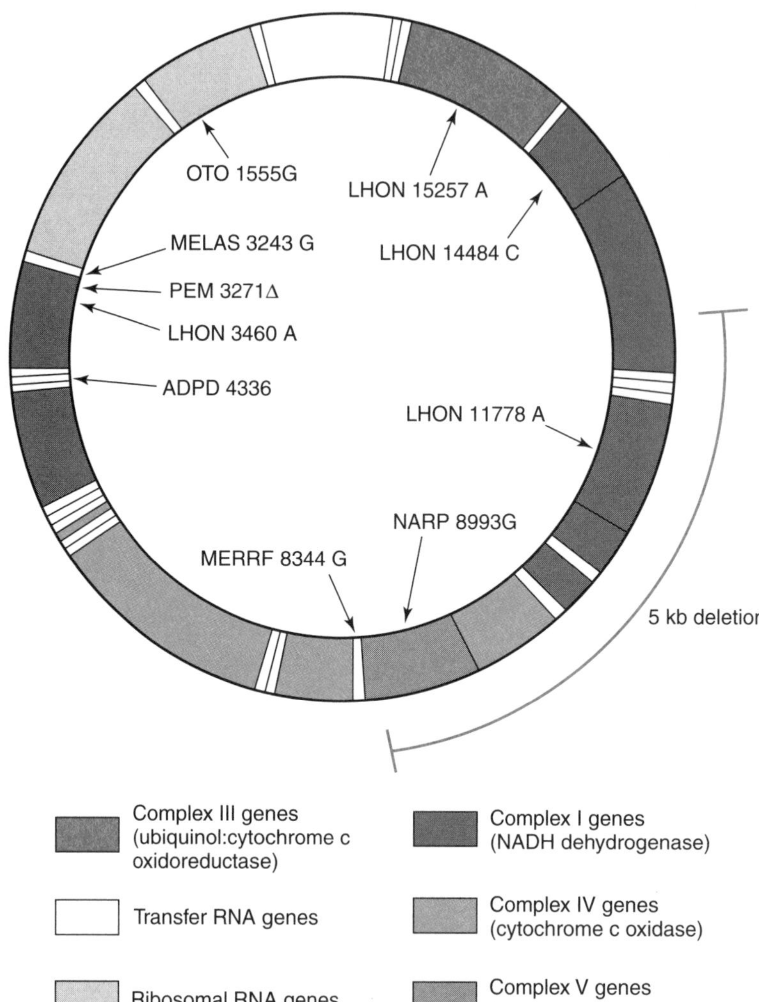

Figure 8.1 *Map of the mitochondrial genome (the entire sequence is known). The positions of identified mutations associated with several specific phenotypes are shown. See Table 8.1 for identification of abbreviations. The position of a common 5-kb deletion associated with ocular myopathy and aging is shown. The positions of genes related to energy production are indicated. (After Wallace, Am J Hum Genet* 57:201, 1995.)

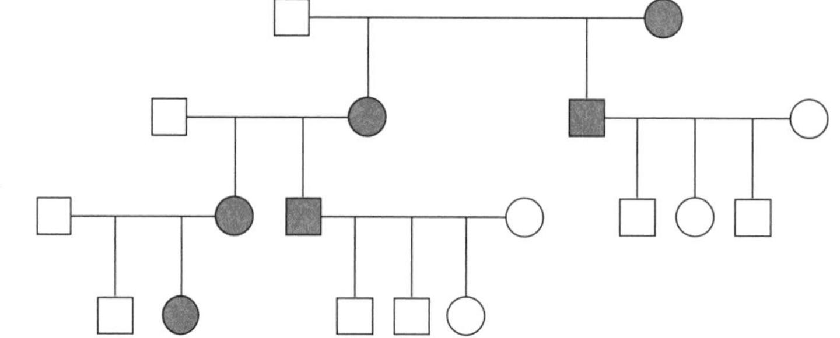

Figure 8.2 *Pedigree showing transmission and expression of a mitochondrial trait. Note that transmission occurs only through females.*

mitochondrial DNA in the fertilized oocyte. Thus, mitochondrial disorders must be inherited from the mother, producing a characteristic pedigree, as shown in Figure 8.2. Either sex can be affected, but the male cannot transmit the disorder because of the lack of mitochondrial transmission via sperm. Such inheritance is said to be "cytoplasmic" and does not follow the pattern of distribution of chromosomes in the nucleus of the cell.

- Mutations in mitochondrial DNA may lead to the condition of "heteroplasmy," in which there is more than one population of mitochondrial DNA molecules in a cell. (When all mitochondria in each cell have the same DNA, the cell is said to be "homoplasmic.") The proportions of heteroplasmic and homoplasmic mitochondrial DNA may differ in different tissues and may change with age, complicating the clinical picture. (Because there are sequences in nuclear DNA homologous to those in mitochondria, using PCR on total cell DNA also may detect nuclear DNA sequences. Thus, it is important to evaluate claims of heteroplasmy critically to be certain that changes are being detected in mitochondrial DNA. This confusion is less likely if mitochondrial DNA is purified prior to study.)
- The physiologic effect of defective mitochondrial function depends on the energy requirements of the cell. Thus, not surprisingly, nerves and muscle often show clinical changes first.
- Mutations in mitochondrial DNA occur more frequently (as much as tenfold faster) than those in nuclear genes involved in oxidative phosphorylation.
- Oxidative phosphorylation integrity declines with aging, paralleling the accumulation of mitochondrial DNA mutations in somatic cells.

B. DISORDERS

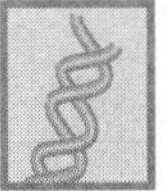

Three facts complicate the phenotypes of mitochondrial DNA mutations:

1. Heteroplasmy permits the relative proportions of normal and abnormal mitochondrial DNA to differ among cells.
2. Most products of the mitochondrial genome do not act alone but work in connection with proteins encoded by nuclear genes. Because nuclear genes are responsible for encoding most mitochondrial proteins, their mutations can produce mitochondrial dysfunction with a conventional autosomal mendelian inheritance pattern. In contrast, mitochondrial dysfunction due to mitochondrial DNA mutations is distinctive, particularly when the family pattern is considered.
3. Mitochondria in somatic cells can acquire and accumulate mutations with age. Although these are not transmissible, they may affect the rate and severity of clinical changes.

Table 8.1 ▶ **Disorders with defective mitochondrial function and mitochondrial DNA mutations**

Disorders	Mutations*	OMIM#
Leber hereditary optic neuropathy (LHON) (also early-onset dystonia)	Multiple	535000 516000
Leigh disease and neurogenic muscle weakness, ataxia, and retinitis pigmentosa (NARP)	T→G 8993; T→C 8993	516060
Aminoglycoside ototoxicity (OTO)	A→G 1555	580000
■ *tRNA mutations*		
Progressive encephalomyopathy (PEM)	$\Delta\beta$271	590050/60
Myopathy, encephalomyopathy, lactic acidosis, stroke (MELAS) (also adult-onset diabetes mellitus and sensorineural hearing loss)	tRNA leu A→G 3243	540000
Myoclonic epilepsy and ragged red fibers (MERRF)	tRNA lys A→G 8344	545000
Late onset Alzheimer disease, Parkinson disease	tRNA gln A→G 4336	502500
■ *Structural DNA changes*		
Kearns-Sayre syndrome (KSS)	Deletions	530000
Chronic progressive external ophthalmoplegia (CPEO)	Duplications	555000
Ocular myopathies	Rearrangements	
Inherited cardiomyopathies		

* This is a representative list; other mutations also have been reported. See references and OMIM.

Table 8.1 summarizes several diseases caused by mitochondrial mutations. A substantial majority of the mitochondrial DNA in a given cell must be abnormal for clinically detectable problems to develop. Generally, this threshold is about 85%. In addition, as noted above, the proportion of abnormal mitochondrial genomes may change with the age and number of divisions of the host cell. This becomes a particular concern in tissues such as brain, where increases in the number of mitochondrial mutations can lead to progressive deterioration in energy metabolism and brain function. Thus, in aging, cumulative neurologic effects of mitochondrial mutations may be prominent (see below). Table 8.1 indicates that different clinical presentations may be related to the same mutation. This reflects the proportion of abnormal mitochondrial DNA molecules as well as the age of the individual. In some patients, the full myopathic picture may never develop despite a suspicious family history; this can be confusing if the characteristics of mitochondrial disease are not considered.

Disorders associated with mitochondrial gene mutations are related to muscle or nerve function and often are enigmatic in their presentation. Several of these conditions are now recognized. They are generally distinguished on the basis of findings on muscle biopsy and in vitro studies of mitochondrial function. They often are known by their abbreviations, including MERRF—myoclonic epilepsy and ragged red fibers (OMIM #545000)—and MELAS—mitochondrial myopathy, encephalopathy, lactic acidosis, and stroke-like episodes (OMIM #540000). These conditions usually have a delayed clinical onset and always are heteroplasmic. Different family members (each with the same mutation) may have very different presentations in terms of age of onset of symptoms, organs involved, and clinical progression. From a clinical perspective, the two critical determinants of symptoms appear to be the amount of mutant mitochondrial DNA the individual inherited and the age of the individual. Aging is likely to be an important contributor to the clinical picture (and in itself may be associated with mitochondrial DNA changes; see below). There also are forms of maternally inherited deafness and diabetes.

Leigh disease is often lethal in infancy or childhood due to degeneration of basal ganglia. It is associated with the same mutations as NARP, which is characterized by retinitis pigmentosa, macular degeneration, olivopontocerebellar atrophy, and retardation. The particular manifestations are difficult to predict, reflecting both heteroplasmy and the likelihood that the mutations often arise de novo in different families.

A peculiar condition known as Leber hereditary optic neuropathy (LHON; OMIM #535000) appears related to mitochondrial integrity as well. This is an optic nerve degeneration usually seen in young adults and can be associated with peripheral neuropathy and cardiac arrhythmias. LHON has a maternal pattern of inheritance, but more males than females are affected according to clinical studies. At least 19 different mitochondrial DNA mutations have been characterized in LHON, five of which appear essential in leading to illness. Interestingly, these five mutations are associated with different degrees of clinical severity; some cause other neurologic symptoms as well. Not all maternal relatives of individuals with LHON have visual loss, and not all individuals with the same mitochondrial mutation have the same clinical course. It is likely that the clinical outcome is affected by other variations in the mitochondrial DNA, the proportion of heteroplasmy, the status of nuclear-encoded genes related to mitochondrial function, and the age and sex of the patient.

Because of the small size and complete characterization of the mitochondrial genome, it is possible to establish the nature of any genetic changes with certainty. Specialized laboratories can thus proceed directly to mitochondrial DNA analysis in equivocal cases so that the diagnosis can be based on precise molecular information.

C. MITOCHONDRIA AND AGING

An important aspect of mitochondrial biology is related to changes of mitochondrial DNA in somatic cells during aging. For example, a 5-kb deletion in the mitochondrial genome, virtually never seen in normal hearts before age 40, accumulates in the hearts of older individuals. Studies of this same deletion in brain have shown that it is rarely seen before age 75 but that it accumulates in the cortex and basal ganglia, reaching ~3% and ~11%, respectively, by age 80. (Such changes are not seen in the cerebellum.) The cumulative effects of this and other somatic mutations could change the efficiency of mitochondrial metabolism in older individuals, possibly unmasking the effects of mutations elsewhere in the mitochondrial or nuclear genomes. It offers additional perspective on the delayed onset of some of the disorders related to energy metabolism. Of course, such effects would

CASE STUDY

Martha was always healthy and enjoyed sports through high school and college. Her job as an architect/manager involved considerable travel with often daily visits to construction sites for building supervision, and she had no difficulties. At age 27 she began to notice "jerking" in her legs. Although at first infrequent, these sensations became notable at least once a week. The problem progressed to the point at which she was unwilling to climb stairs and ladders at building sites because she was fearful of instability. After about 10 months, she was working at her office one day and fell into a seizure. Although she recovered quickly and had no injuries, she was sent to neurologic evaluation. She had a slight amount of recent memory loss, but nothing that was diagnostic. She had a sensitivity to movement of her legs; when her right knee was flexed, she showed tremors throughout the leg. Routine laboratory tests were normal, and nerve conduction testing was normal as well. Over the next 3 months she had more difficulty walking and two episodes of what her housemate described as "little seizures." Upon reexamination she had mild but significant bilateral leg weakness. There was mild elevation of the creatinine phosphokinase in the blood. A muscle biopsy revealed multiple areas of fiber degeneration with poorly-defined borders.

DNA studies revealed heteroplasmy for the A → G 8344 mutation characteristic of MERRF syndrome (OMIM #545000; see Table 8.1).

COMMENT

This woman had several features characteristic of mitochondrial disorders:

- The onset was late and progressed once symptoms were recognized.
- There were no known affected individuals in her family (according to her report; none were available for examination).
- Her presentation involved neuromuscular symptoms.
- Muscle fibers showed structural changes on biopsy.

These features are consistent with a heteroplasmic situation in which only a fraction of mitochondrial genomes shows mutation(s). Symptoms usually develop only after ~85% of mitochondrial DNA is involved. The high energy demand of muscles often leads to their showing dysfunction earliest.

not necessarily be expected to have implications for the offspring of affected individuals.

Because of the underlying difficulties with oxidative phosphorylation and energy metabolism, the management of mitochondrial myopathies and neural degeneration has not been successful to date.

STUDY QUESTIONS

1 You see a 28-year-old woman for a routine examination; she appears normal. Her brother has been diagnosed with LHON at age 20. You tell her:

- a You would like to see her mother.
- b She should remain free of problems because nothing has been seen in her by this age.
- c She is unlikely to be homoplasmic for an LHON mutation.
- d She should have mitochondrial DNA studies done.
- e The possibility of her developing symptoms is related to the percentage of heteroplasmy in her brother.
- f Her risk of having affected children is low.

2 A 72-year-old man has been referred to you with muscle weakness of recent onset. He tells you that his sister died with muscular dystrophy 35 years ago. His parents died in an accident at young ages. You tell him:

- a He may have Becker muscular dystrophy.
- b A muscle biopsy may be useful.
- c His children are at low risk because this is likely to be a recessive disorder.
- d His children should be examined.
- e Mitochondrial DNA studies may suggest a course of treatment or clarify the prognosis.

3 You are evaluating a patient with muscle weakness, and the pathologist calls to tell you that the mitochondria appear normal in a recent muscle biopsy. You think:

- a The sample studied must have been small enough to include only normal mitochondria due to heteroplasmy.
- b It is unlikely that structural changes would have occurred yet based on the clinical findings.
- c What you really need are studies of mitochondrial *function*.
- d You need to find another pathologist.
- e It would be a good idea to do electron microscopy on the specimen.

FURTHER READING

Corral-Debrinski M et al. Mitochondrial DNA deletions in human brain: Regional variability and increase with advanced age. *Nat Genet* 2:324, 1992.

Harding AE et al. Prenatal diagnosis of mitochondrial DNA$^{8993T\rightarrow G}$ disease. *Am J Hum Genet* 50:629, 1992.

Holt IJ et al. Deletions of muscle mitochondrial DNA in patients with mitochondrial myopathies. *Nature* 331:717, 1988.

Holt IJ et al. A new mitochondrial disease associated with mitochondrial DNA heteroplasmy. *Am J Hum Genet* 46:428, 1990.

Howell N. Mitochondrial gene mutations and human diseases: A prolegomenon. *Am J Hum Genet* 55:219, 1994.

Kelly DP and Strauss AW. Inherited cardiomyopathies. *N Engl J Med* 330:913, 1994.

Shoffner JM and Wallace DC. Mitochondrial genetics: Principles and practice. *Am J Hum Genet* 51:1179, 1992.

Shoffner JM et al. Mitochondrial DNA variants observed in Alzheimer disease and Parkinson disease patients. *Genomics* 17:171, 1993.

Wallace DC. Mitochondrial DNA variation in human evolution, degenerative disease, and aging. *Am J Hum Genet* 57:201, 1995.

USEFUL WEB SITES

HELIX (INCLUDES GENLINE)

http://www.hslib.washington.edu/helix
Medical Genetics Knowledge Base. Electronic textbook under development relates testing to diagnosis, management, and counseling. Directory of laboratories providing testing for genetic disorders.

OMIM (ONLINE MENDELIAN INHERITANCE IN MAN)

http://www.ncbi.nlm.nih.gov/omim
Basic catalogue listing mendelian genes and traits. Clinical descriptions, references, and links to other databases.

HEALTHLINKS

http://healthlinks.washington.edu/
Broad access to many related areas of basic and clinical genetics. Includes "Genome Machine II" database of genes and diseases mapped to a particular chromosome location. Also chromosome information.

MARCH OF DIMES

http://modimes.org/
Resources for families and professionals about birth defects. Professional education section.

NATIONAL ASSOCIATION FOR RARE DISORDERS (NORD)

http://www.NORD-rdb.com/~orphan
Primary nongovernmental clearinghouse for information about rare disorders. Over 1100 disease reports. Broad resource guide. Includes Rare Disease Database, a wide range of information for families and professionals. Listings by symptoms, causes, affected populations, treatments. Listing of support groups. Includes orphan drug designation database.

OFFICE OF RARE DISEASES (NIH)

http://cancernet.nci.nih.gov/ord/index
Broad-based NIH index including support groups, clinical research database, investigators, and extensive glossary.

CHAPTER 9

Congenital Changes

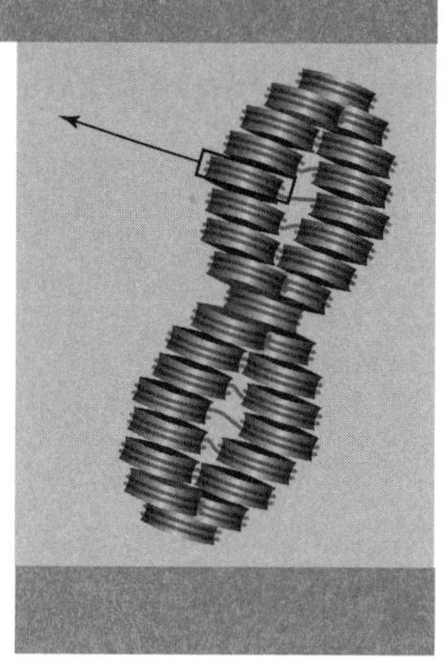

Approximately one in 50 newborn infants has a recognizable abnormality. The spectrum of such changes is broad, ranging from serious, life-threatening malformations to trivial features. Nevertheless, whatever its severity, the appearance of anything irregular causes concern and raises questions about diagnosis, etiology, management, prognosis, and potential genetic implications. It is especially important to establish the appropriate diagnosis for these conditions, which by definition are "congenital." Also, it is important to recall that congenital means "present at birth"; it says nothing about etiology. Currently, most congenital changes are unexplained. In the discussion below and in interactions with patients and their families, it is essential to distinguish inherited and/or heritable congenital problems from those without genetic implications. This can be confusing when affected individuals are first encountered. In addition, as the affected individual grows older, certain problems that may have generated concern earlier in life may either have been forgotten or no longer be considered to be of much consequence; this can affect diagnosis, counseling, and future management.

Many congenital changes are visible and obvious to parents, families, and medical personnel. Others may be detected only later in life; it is not unusual for both patients and physicians to be surprised by the detection of a congenital problem in an older person. We will consider only several representative categories of congenital changes.

A. HEMANGIOMAS

The development of interconnections between the arterial and venous circulations can occur in any part of the body. Clusters of such connections are frequently called "hemangiomas." Other names include A-V malformation (AVM), port-wine stain, strawberry mark, nevus flammeus, and stork bite (see Figure 9.1). In extreme cases, A-V malformations can lead to high-output cardiac problems because of substantial blood shunting. They also may be warm to the touch due to the large amount of blood they contain. The appearance

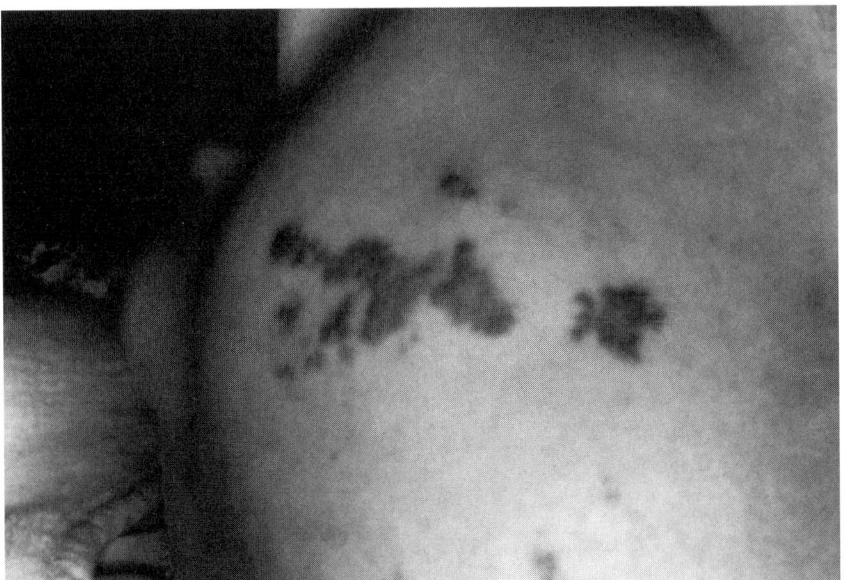

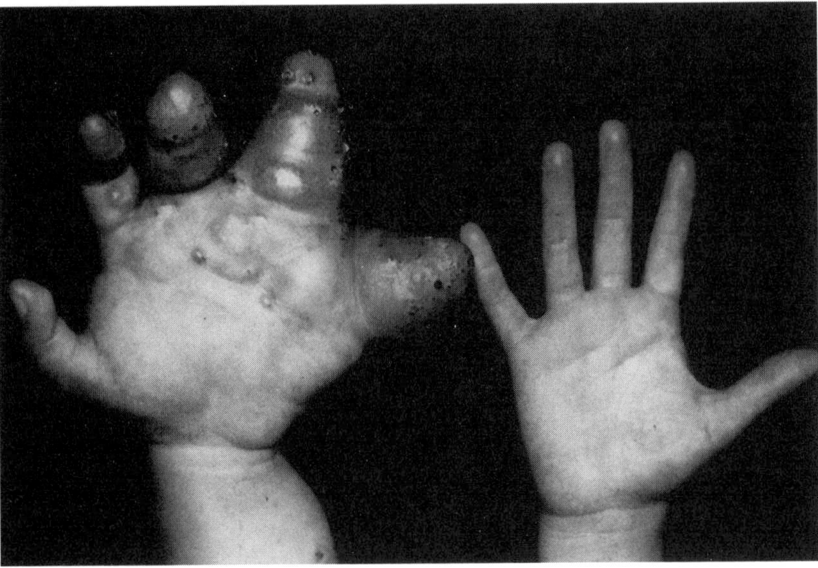

Figure 9.1 *Hemangiomas. These lesions can range from superficial "port-wine stains" (top) to complex, deep changes with high degrees of vascularity (bottom). See also Plate 4. (Photographs courtesy of Dr. Grant Anhalt and Dr. Gaylord L. Clark, Jr.)*

Table 9.1

Syndromes with hemangiomas*

Name	OMIM#	Name	OMIM#
Maffucci	166000	Klippel-Trenaunay-Weber	149000
Riley-Smith	153480	Kasabach-Merritt	141000
Neurofibromatosis	162200	Port-wine stain/Strawberry mark	163000/100
Tuberous sclerosis	191100	Sturge-Weber	185300
von Hippel-Lindau	193300	Turner (XO)	

* Other, less frequently encountered, associations also are recognized

(color, size, and texture) of hemangiomas can change with time, emotions, posture, and development.

Some hemangiomas may be part of a recognizable larger syndrome. These more complex lesions frequently have other associated clinical findings that suggest specific diagnostic classifications (see Table 9.1). Hemangiomas may cause complications by trapping platelets; large hemangiomas in infants and children can cause thrombocytopenia due to such entrapment (e.g., the Kasabach-Merritt syndrome [OMIM #141000]). Still other hemangiomas cause difficulties largely through cosmetic effects. There is no question that the striking appearance of some of these lesions can create self-consciousness and social difficulties. Although individuals may become less sensitive to these features later in life, the lesions can continue to present a source of discomfort and anxiety. Some hemangiomas can affect the development of contiguous organs or extremities in infancy and childhood. The basis for this is not always known, but it is likely that altered blood flow can either stimulate or repress regional development. Thus, what might begin as a local vascular lesion can lead to a more prominent developmental problem, with discrepancies in the size, length, or shape of a leg, arm, or hand.

Because of the prominent appearance of hemangiomas, various approaches have been considered for their management. The approach or approaches chosen must reflect the location and severity of each problem as well as the intrinsic complexity of each lesion. Direct surgical removal of hemangiomas often is difficult because of problems obtaining vascular stasis in such complex vascular arrays. Because of this recognized difficulty, various techniques can be used to reduce the vascularity of hemangiomas. Interventional radiologists can selectively embolize some hemangiomas. Such embolization can reduce vascularity substantially, not only decreasing the amount of blood in the lesion (and sometimes its clinical prominence) but also making subsequent surgical excision safer.

Fortunately, most hemangiomas do not require excision. The desire for cosmetic improvement remains, however, and one particularly useful approach is to use laser therapy with the laser beam adjusted to the absorption wavelength of hemoglobin. Appropriately directed laser beams can heat and coagulate the contents of superficial hemangiomas, ultimately reducing their bulk. Some hemangiomas can be largely eliminated by this approach; others can be reduced in size and visual prominence.

The fact that most hemangiomas and AVMs do not imply genetic risk is often of considerable comfort to affected individuals and their families. Nevertheless, affected individuals do require special attention to embrace developmental, cosmetic, and circulatory aspects of the lesions themselves. Infection of or trauma involving a hemangioma may create a serious problem with complex management requirements. Often, considerations of embolization and surgical management arise only after such events.

B. CLUBFOOT

The appearance of an infant with a clubfoot deformity (talipes equinovarus, OMIM #119800) is usually diagnostic (see Figure 9.2). This deformity represents a spectrum of changes involving the ankle, distal lower leg, and foot. These changes usually reflect intrauterine effects such as pressure or fetal positioning. They can be part of the larger spectrum of arthrogryposis, which may involve other joints as well. Because of the need for integrity of the foot and of the ankle joint for walking as well as because of cosmetic considerations, clubfoot deformity needs to be corrected.

As is the case for hemangiomas and other congenital problems, clubfoot may be part of a larger developmental or systemic problem (see Table 9.2). However, the majority of instances represent isolated events.

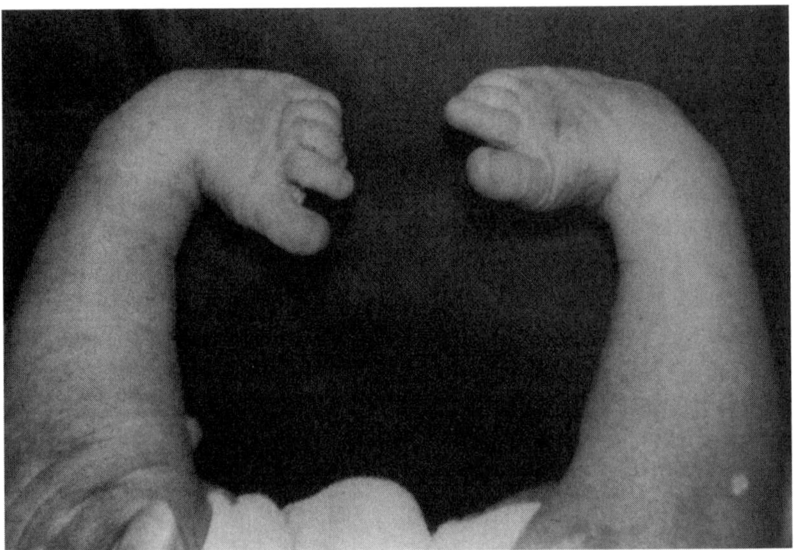

Figure 9.2 *Clubfoot deformity. Despite some variation in presentation, the basic presentation is consistent. (Photograph courtesy of Dr. Gaylord L. Clark, Jr.)*

Table 9.2

Syndromes with clubfeet*

Name	OMIM#	Name
Diastrophic dwarfism	222600	Aminopterin toxicity
Larsen	245600	13 Trisomy
Freeman-Sheldon	193700	18 Trisomy

* Other, less frequently encountered, associations also are recognized

The management of individuals with clubfoot should begin early and is the province of pediatricians and orthopedic surgeons. The abnormal positions of the foot and ankle generally respond to moderate but continued pressure. This can be accomplished by several mechanisms. For certain individuals, appropriately designed shoe and ankle supports can accomplish this directly. For others, bracing with slow, continual adjustment of the pressure and alignment is required. Surgery is needed infrequently for managing uncomplicated clubfoot; a combination of bracing and splinting usually achieves the required repositioning if maintained conscientiously and for a sufficient time.

The clubfoot deformity in its many forms represents the sort of congenital malformation that is largely reversible. It is essential that management should begin early to minimize interruptions in weight bearing and walking. Once the appropriate alignment has been achieved, it may be necessary to maintain bracing during early growth and walking. Usually, however, a successful treatment program is followed by virtually normal foot, ankle, and leg development and a good gait.

C. HAND CHANGES

Because they are so important in all aspects of life and because they are so prominent visually, the hands always receive special attention during the examination of newborns. The array of changes that can involve the hands is extensive, ranging from relatively uncomplicated and inconsequential features to total absence. Because hands represent the end point of a complicated set of developmental interactions, it sometimes is possible to establish at least some of the developmental pathways that have been altered in individual cases. For instance, abnormal development of the radius can affect all of the structures on the lateral side of the hand and forearm. A "radial ray" absence or deformity has characteristic changes. Other changes are more complicated, however, and it is frequently impossible to dissect the embryologic basis for a very obvious clinical change. Several important categories of hand malformations are shown in Figure 9.3.

Because the hands can be affected in many ways, numerous hand malformations are important features of well-known syndromes (Table

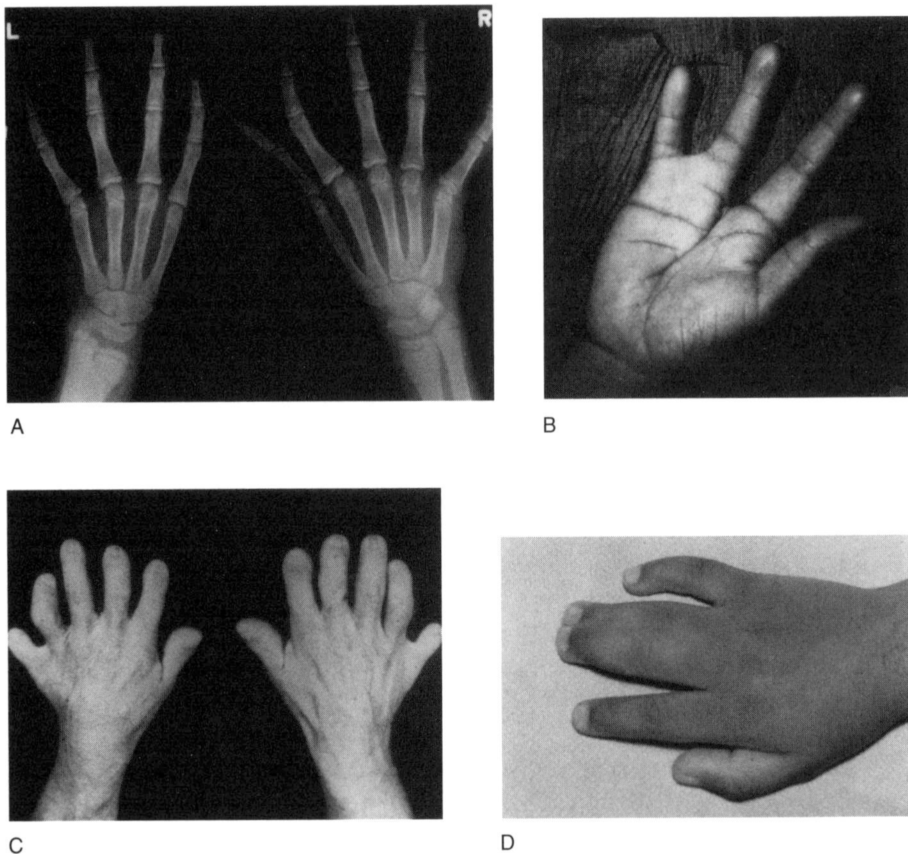

Figure 9.3 *Congenital hand malformations: (A) Radiograph showing radial ray deformities in Holt-Oram syndrome: absent thumb on left hand, small thumb "digitalized" on right hand. (B) "Split hand" deformity. (C) Postaxial polydactyly in Ellis-van Creveld syndrome. (D) Fused digits. (B and D courtesy of Dr. Gaylord L. Clark, Jr.)*

9.3). Some of these syndromes also involve developmental problems in other organ systems, and thus the final array of clinical changes may be complex.

The first step in managing an individual with hand changes is to ascertain the extent of the change(s). This is done by careful clinical examination and frequently also by radiography. In the newborn this may be difficult, because some bony centers are not yet visible radiographically. Later, the bone growth patterns become more clear. Although concern about hand structure and function is justifiably important and usually raised from the first day the changes are recognized, it is important to emphasize that many individuals with visibly abnormal hands can function remarkably well. Furthermore, some surgical repairs and reconstructions are appropriate only after development and growth have proceeded.

Syndromes with hand malformations*

Table 9.3

Name	OMIM#	Name	OMIM#
Holt-Oram	142900	Thrombocytopenia-absent radius	274000
Fanconi	227650	Cornelia De Lange	122470
Diastrophic dwarfism	222600	Carpenter	201000
Ellis-van Creveld	225500	Apert	101200
Rubinstein-Taybi	180849	Smith-Lemli-Opitz	270400
Oculodentodigital	164200	Oral-facial-digital	258850
13 Trisomy		18 Trisomy	

* Other, less frequently encountered, associations also are recognized

An example of the sort of problem that may be resolved early is polydactyly. As shown in Figure 9.4, rudimentary digits may be excised early in life. Such digits frequently have no bones and never would develop any; their removal prevents difficulties later and also can limit self-consciousness and potential damage to an unstable appendage. Such an approach often leaves the affected individual with minimal scars later in life, and he or she may be unaware that surgery was performed earlier. Because polydactyly can have syndromic and genetic implications, however (see Table 9.3), when genetic counseling is considered, it is important to be aware that such rudimentary digits were present earlier in life.

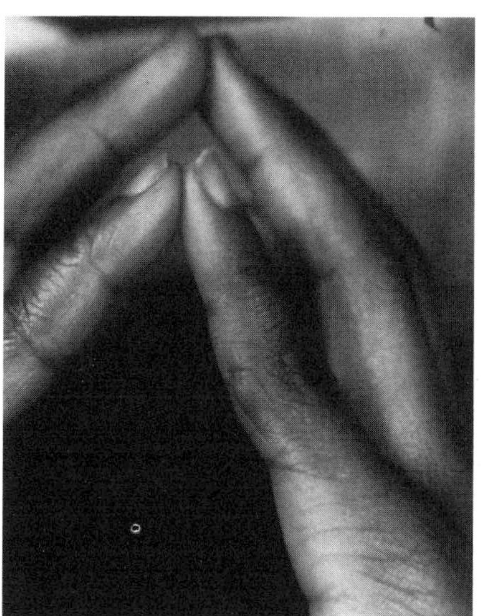

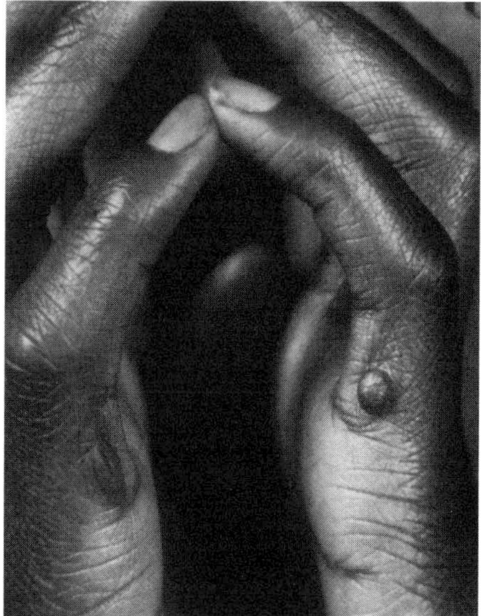

Figure 9.4 *Early removal of postaxial digits may leave minimal scars (left) or only residual tissue (right).*

A more complex group of malformations involves those with dysplastic or absent thumbs. An effective pincer grip between the thumb and other fingers is central to hand function. The absence of this function causes significant difficulties in activities of daily living and the operation of various mechanical devices. For this reason, in the absence of a functional thumb an effort often is made to create the possibility of a pincer grip. This is surgically complex, but specialized surgeons can achieve remarkably effective results. Although the cosmetic features will not be normal, the function can be quite satisfying. Because such reconstructions often involve multiple stages with intervening physical therapy, patients needing them often are referred to specialized centers.

As indicated in Table 9.3, hand changes may be signals of other abnormalities, including those of the heart and skeleton, and the genetic implications present a challenge to the examining physician. At times the full extent of the systemic involvement may not be clear; the associated changes may become apparent only later in life. Nevertheless, the development of the hands and digits occurs during periods of embryonic development when other major organ systems are being formed. This fact justifies heightened suspicion about possible developmental changes in other systems and should lead to their thorough investigation.

D. HEART DISEASE

Congenital heart disease (CHD) is common, involving nearly 1% of liveborn infants in North America. As many as 50% of these individuals will have some degree of ventricular septal defect (VSD), but, remarkably, only a relatively small fraction of these show persistence of the problem in later infancy. Thus, there must be a mechanism for late closure of these defects in many individuals. Given the remarkable complexity of the heart, its valves, and the great vessels, it should not be surprising that a broad range of defects can occur. This is made even more striking by the relatively narrow time window for the formation of specific parts of the heart and great vessels. Although most CHD is detected appropriately in childhood, a significant, although small, fraction is not detected until adulthood.

Because of the complex developmental interactions involved in forming the heart and great vessels, it is not surprising that heart defects can be prominent in individuals with chromosomal abnormalities. As discussed in Chapter 3, both Down syndrome and Turner syndrome have important cardiovascular components. Rarer aneuploidies and other types of chromosomal anomalies also are associated with CHD (see Table 9.4). Usually, however, their clinical presentations involve multiple unusual features in different organ systems, and the clinician recognizes that heart disease is only one of a group of changes.

As shown in Table 9.5, both DiGeorge syndrome (OMIM #188400) and velocardiofacial (OMIM #192430) syndrome are associated with

Table 9.4

Chromosomal syndromes with congenital heart and vascular disease*

21 trisomy	13 trisomy
18 trisomy	XO (Turner)
18q-	4p-
XXXXX	XXXXY
22q11(del)	13q-
Triploidies	

* Only the most frequent chromosomal associations are listed

CHD; several individuals with these syndromes have been found to have visible chromosome deletions in 22q11. This finding led to further studies of individuals with CHD who lack other syndromic features that revealed a significant fraction of 22q deletions in individuals from families with more than two affected members. This kind of careful chromosome study has become a useful tool in evaluating individuals who have CHD; abnormalities have been found even in isolated individuals with CHD and no family history.

Again, because of the complex developmental interactions involved, many mendelian conditions are associated with CHD in addition to other manifestations. Thus, one of the most important steps in evaluating an individual with CHD involves determining the family history and investigating the possibility of a complex herritable syndrome or evidence for mendelian transmission of any sort. Table 9.5 shows only part of the very long list of mendelian disorders associated with cardiovascular changes. Obviously, these have different transmission patterns. Because the presentation of heart or vascular disease may be confusing in the face of multiple other clinical problems, it is especially important to consider the entire clinical picture before reaching a diagnosis, and certainly before making any suggestions about recurrence risks or prognosis.

Table 9.5

Mendelian syndromes associated with congenital heart disease*

Name	OMIM#	Name	OMIM#
Holt-Oram	142900	Zellweger	214100
Williams	194050	Myotonic Dystrophy	160900
Noonan	163950	Ellis-van Creveld	225500
Carpenter	201000	Cornelia De Lange	122470
Fanconi	227650	Velocardiofacial	192430
Rubinstein-Taybi	180849	Thrombocytopenia-absent radius	274000
Weill-Marchesani	227600	Alagille	118450
Smith-Lemli-Opitz	270400	Beckwith-Wiedemann	130650
Ivemark	208530	DiGeorge	188400

* Only the most frequent Mendelian associations are listed.

Another important consideration in CHD is the fact that developmental changes in this category have a known association with teratogens (see Table 9.6). These may be drugs, environmental elements, infections, or other diseases that in themselves are associated with an increased risk of CHD. One of the best studied and most prominent of these was the tragic experience with the use of the drug thalidomide. In addition to its prominent effects on skeletal malformations with shortened and abnormal limbs (phocomelia), thalidomide use during pregnancy has been associated with a broad range of CHD manifestations. Another important category of illness where cardiovascular effects are well recognized is rubella, or German measles. As many as one-third of infants born after a maternal rubella infection during pregnancy have CHD in addition to other problems. Chapter 6 introduced the challenging situation in which children of mothers homozygous for PKU whose blood levels of phenylalanine are uncontrolled during pregnancy have a high risk of cardiac septation defects in their children. Finally, there is a significant association of CHD with maternal alcohol use (fetal alcohol syndrome).

The approach to the patient with CHD can be particularly challenging because of the broad range of entities that may be associated with it. As suggested by Tables 9.4–9.6, genetic, environmental, toxic, and infectious factors, as well as other maternal conditions, must be considered. Despite the risks due to any of these underlying factors, however, it is frequently the case that no obvious predisposing factor can be identified. In such cases, other family members may be interested in and concerned about the risk of recurrence.

When all recognized predisposing factors can be eliminated, it is necessary to offer counseling on the basis of empiric risks. CHD has been particularly well studied in this regard, and the likelihood of recurrence can be estimated based on the experience of large collections of patients and their families. It is important to realize that, although these risks have been derived empirically based on large numbers of patients, they show remarkable similarity between different popula-

Table 9.6 ▶ **Teratogens associated with congenital heart and vascular disease***

Drugs		
Alcohol	Dilantin	Phenytoin
Lithium	Trimethadione	Valproic acid
Retinoic acid	Amphetamines	Thalidomide
Maternal Disorders		
Rubella	Phenylketonuria	Diabetes
Thyroid disease	Connective tissue disease	

* Only the most frequent associations are listed. The use of any drugs during pregnancy must be considered carefully.

Table 9.7

Empiric risk figures for congenital heart disease*

Condition	Suggested risk (%) One affected sibling	Suggested risk (%) One affected parent
Ventricular septal defect	3	4
Patent ductus arteriosus	3	4
Atrial septal defect	2.5	2.5
Tetralogy of Fallot	2.5	4
Pulmonic stenosis	2	3.5
Coarctation of aorta	2	2
Aortic stenosis	2	4

* From Nora and Nora, *Circulation* 57: 205, 1978.

tions. Table 9.7 presents an example of empiric risk figures derived from large studies by Nora and Nora. Obviously, the risks are relatively low. No individual has a risk greater than 4%, the risk for the offspring of an affected parent. Interpreting such a table requires circumspection, however. Large clinical studies have shown that VSDs have the highest likelihood of recurrence. Furthermore, as noted earlier in this section, individuals may be born with VSD, but the defect may close during infancy. If such an individual is examined only as an adult, the fact that the defect was present earlier (although asymptomatic) may never be identified, and this could distort the risk prediction.

The use of empiric risk figures for counseling about CHD is an example of the usefulness of large clinical data sets for estimating risks. Nevertheless, such lists need to be used only after careful consideration of the family history. For instance, there are families in which a clearly mendelian pattern of transmission appears likely even though no identifiable syndrome is present. Counseling is such that a kindred must take into consideration the mendelian pattern of CHD in the rest of the family and the possibility of such changes as the chromosome deletions in 22q11 described earlier. Failure to do so ignores potentially critical data that could change the recurrence risk from under 5% to as high as 50%.

The remarkable success of cardiac surgery in the repair of CHD has permitted more and more affected individuals to live healthy lives. As these individuals plan families of their own, it will be important to document their own history of CHD and to consider it strongly in prenatal and possibly preconception counseling. As will be discussed in Chapter 12, it is likely that further genetic marker studies will permit refinement of empiric risk figures for selected kindreds. Obviously, this will change the entire perspective on counseling for affected individuals and their families and may identify family members in whom problems have not been detected who are at increased risk for having affected children.

CASE STUDY

Maria appeared normal at birth and showed good early growth and development. On a routine pediatric visit at age 4 months she was suspected to have a heart murmur. She was evaluated by a cardiologist who discovered a small ventricular septal defect and recommended careful observation. The child continued to do well and another complete cardiologic evaluation at age 3 showed no murmur, but there was a mild interventricular conduction defect on her electrocardiogram. Subsequent growth and development were normal.

Although her parents were said to be normal, examination of her father showed an unusual, faint heart murmur. Further evaluation showed a tiny atrial septal defect. He also had a minor interventricular conduction abnormality on his electrocardiogram. Chromosome studies on father and daughter were normal.

COMMENT

As noted in Table 9.7 the risk for the daughter's having a congenital cardiac anomaly was increased by the finding in her father (but, of course, this was not known at the time of her conception). The empiric risk prediction in this situation would have been ~4%. This also raises the chance of recurrence in any of Maria's children. This situation also emphasizes the fact that one must examine the parents carefully if subtle changes are expected; the statement "everything is normal" may be misleading, especially in the case of subtle changes. Furthermore, the father's defect might have closed completely, and then no one could have put the entire picture together. For Maria herself, who now has no murmur, she needs to be aware of her personal and family history in order to assess the recurrence risk if she has children.

STUDY QUESTIONS

1 Your brother has called you to say that he was rejected from track events in college because someone detected a heart murmur. He has always been considered healthy. You suggest:

- a He has a recurrence risk of 3% for any children.
- b He needs a stress test.
- c Someone should examine your parents.
- d He needs an echocardiogram.
- e You need an echocardiogram.

2 Holt-Oram syndrome (OMIM #142900) is inherited as an autosomal dominant trait. It is associated with a variety of changes in the hands as well as cardiac malformations. The former show an impressive left-sided predominance. What explanation(s) can you suggest?

3 Parents have brought in their child with a pigmented skin macule on her chest. You examine the child and discover eight more in various parts of

the body. The parents have had no similar lesions. They have heard about laser photoablation and have asked your opinion. You tell them:

a Most of these spots will become less prominent with time.
b Laser treatment is usually most appropriate for spots on the face and arms.
c You would like to have the girl seen by an ophthalmologist.
d Microembolization may be more effective for treatment.
e New spots may continue to appear, so treating those present now is not a good idea. They should wait until after puberty.

FURTHER READING

Cohen MM Jr. *The Child with Multiple Birth Defects*, 2d ed. London, Oxford University Press, 1996.

Hall JG et al. The distal arthrogryposes: Delineation of new entities—review and nosologic discussion. *Am J Med Genet* 11:185, 1982.

Nora JJ. Recurrence risks in children having one parent with a congenital heart disease. *Circulation* 53:701, 1976.

Nora JJ and Nora AH. The evolution of specific genetic and environmental counseling in congenital heart diseases. *Circulation* 57:205, 1978.

Samuel M and Spitz L. Klippel-Trenaunay syndrome: Clinical features, complications and management in children. *Br J Surg* 82:757, 1995.

Smith AT et al. Holt-Oram syndrome. *J Pediatr* 95:538, 1979.

Stevenson RE et al (eds). *Human Malformations*. London, Oxford University Press, 1993.

Warkany J. *Congenital Malformations*. St. Louis, Year Book, 1971.

Webb DW. The cutaneous features of tuberous sclerosis: A population study. *Br J Dermatol* 44:307, 1992.

Wilson DI. Deletions with chromosome 22q11 in familial congenital heart disease. *Lancet* 340:573, 1992.

Winter R and Baraitser M (eds). *London Dysmorphology Database 2.1 on CD-ROM, London Neurogenetics Database 2.1 on CD-ROM, Dysmorphology Photo Library 2.0 on CD-ROM*. London, Oxford University Press, 1998.

Wynne-Davies R. Family studies and the cause of congenital talipes equinovarus, talipes calcaneo-valgus and metatarsus varus. *J Bone Joint Surg* 46B:445, 1964.

USEFUL WEB SITES

HELIX (INCLUDES GENLINE)

http://www.hslib.washington.edu/helix
Medical Genetics Knowledge Base. Electronic textbook under development relates testing to diagnosis, management and counseling. Directory of laboratories providing testing for genetic disorders.

MARCH OF DIMES	http://modimes.org/ Resources for families and professionals about birth defects. Professional education section.
NATIONAL ASSOCIATION FOR RARE DISORDERS (NORD)	http://www.NORD~rdb.com/~orphan Primary nongovernmental clearinghouse for information about rare disorders. Over 1100 disease reports. Broad resource guide. Includes Rare Disease Database, a wide range of information for families and professionals. Listings by symptoms, causes, affected populations, treatments. Listing of support groups. Includes orphan drug designation database.

CHAPTER 10

Drug Sensitivities (Pharmacogenetics)

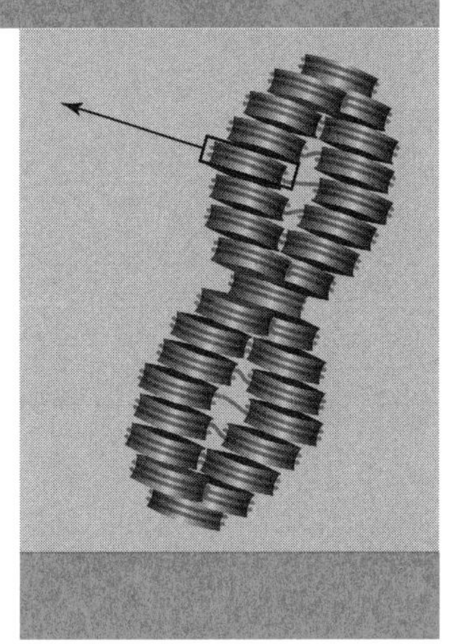

The notion of a "drug allergy" is often mentioned. This term should imply a true allergic reaction due to either a direct sensitivity to the drug or the drug's acting as a hapten. In general, such responses are not inherited. In contrast, an inherited "drug sensitivity" is based on Mendelian principles. Inherited drug sensitivities have several unusual features, however. The first is that they may never be detected simply because the individual never comes in contact with the critical agent. This feature leads to the second, which is that, despite the fact that such sensitivities are inherited, there may be no family history of difficulties. Thus, an individual with an inherited problem may be newly diagnosed in a kindred with no apparent antecedents. An immediate view of this situation might lead to the conclusion that the problem is not inherited or, at the worst, that the affected individual represents a new mutation. If testing is available, however, a very different view can be developed based on evidence of genetic transmission.

One inherited drug sensitivity already was introduced in Chapter 7. G6PD deficiency (OMIM #305900) often is manifest by an acute hemolytic crisis following exposure to a sensitizing drug. Such a situation may be relatively common in a given population due to the broad distribution of G6PD mutations. For example, in areas endemic for malaria, identifying of individuals sensitive to primaquine may not be particularly surprising, because the drug is likely to be used frequently. G6PD deficiency also has the general property of being manifested in only males (reflecting X-chromosome mosaicism in females). As described earlier, G6PD deficiency generally causes hemolysis in older red cells; thus, profound anemia is uncommon on this basis alone, because younger erythrocytes survive. Other forms of drug sensitivities, however, can cause serious problems indeed.

A. DRUG METABOLISM

One of the earliest demonstrations of a human physiologic polymorphism was made in the study of hepatic metabolism of isoniazid (OMIM #243400). Isoniazid is hydroxylated and inactivated in the liver. Clinical studies confirmed an earlier impression that populations can differ in their rates of isoniazid inactivation. Some individuals do it rapidly, others slowly. It became clear that such a pattern was derived from a simple genetic polymorphism and that the different metabolic rates reflected the distribution of two alleles (see Figure 10.1). Rapid inactivators are either heterozygotes or homozygotes for the normal allele, while slow inactivators are homozygotes for the mutant allele of the enzyme arylamine *N*-acetyl transferase (EC 2.3.1.5). The distribution of phenotypes differs in different populations. Slow inactivators are uncommon among Asians, while up to two-thirds of Caucasians may be in this category.

The paradigm of isoniazid metabolism is now recognized for other drugs as well. As mentioned earlier in this chapter, such polymorphisms in metabolic efficiency may cause no clinical problems and remain completely undetected until the individual is challenged. Fortunately, the different rates of isoniazid inactivation do not generally require adjustments in treatment. The use of standard doses generally leads

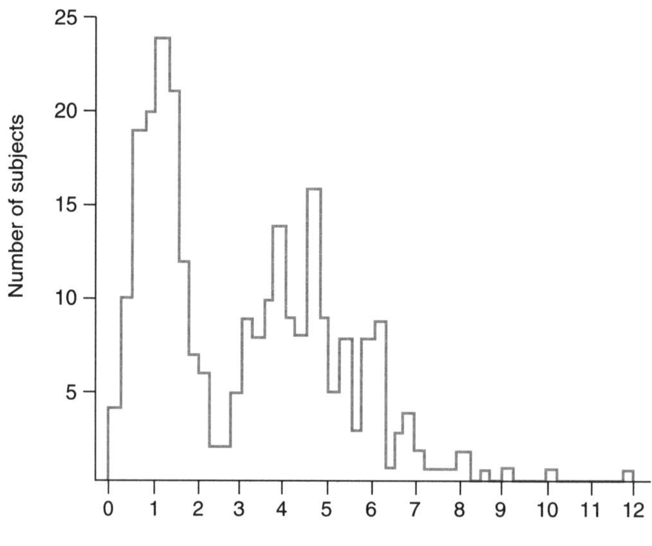

Figure 10.1 *Plasma concentrations of isoniazid 6 hours after drug ingestion (~9.5 mg/kg) by 267 members of 53 complete family units. (Adapted from Evans and McKusick, Br Med J* 2:485, 1960.)

to blood levels sufficient for satisfactory management of mycobacterial infections, and there is a broad range of therapeutic effectiveness for isoniazid. With drugs whose therapeutic and toxic ranges overlap or are far more stringent, however, such an underlying polymorphism can have very serious effects and require careful monitoring during drug administration.

B. ACUTE INTERMITTENT PORPHYRIA

Acute intermittent porphyria (AIP) (OMIM #176000) is a broadly distributed condition in which heterozygosity for a mutant allele often is without clinical consequence. Challenges that exceed the limited metabolic capacity of such individuals can have serious results, however (see Figure 10.2). The AIP mutation occurs in the cyclization step in porphyrin biosynthesis. Uroporphyrinogen-1-synthase (EC 4.3.1.8) mediates the cyclization of a single pyrrole into the first tetrapyrrole, which then is further modified to form heme and other porphyrin-containing groups. Obviously, because this is the only pathway for synthesizing heme in humans, a complete blockade would be lethal. This is a likely reason that individuals homozygous for nonfunctional mutations in this gene have not been reported.

The problem for individuals with AIP develops when the flux through the heme pathway exceeds the capacity of the limited amount of enzyme. The consequences are an increase in the metabolites ahead

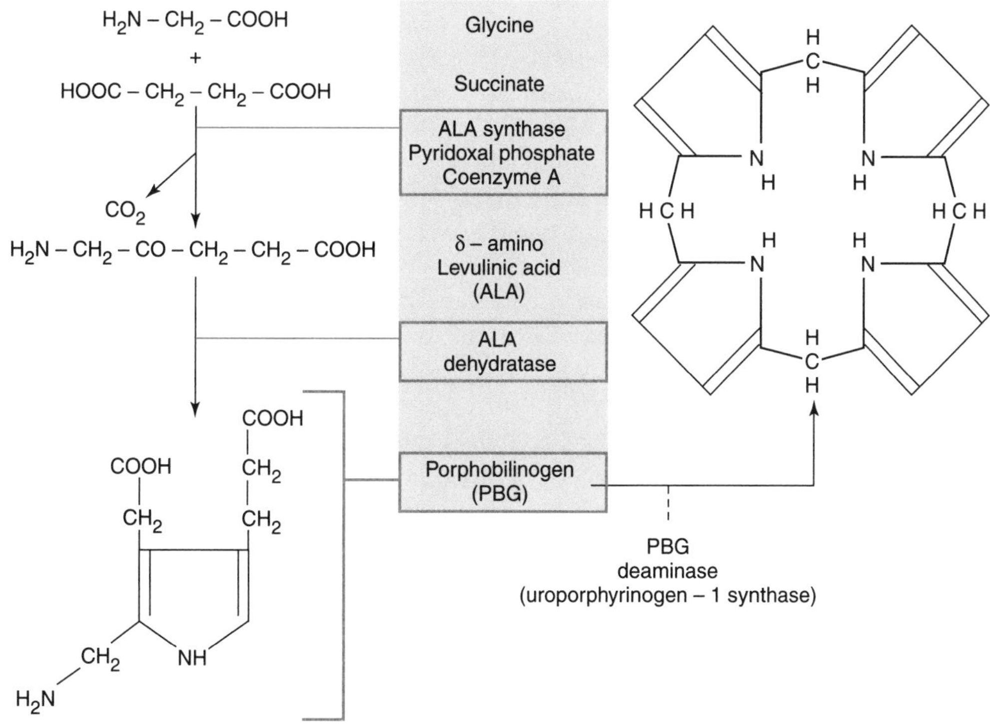

Figure 10.2 *Porphyrin biosynthesis pathway. ALA synthase catalyzes the first committed step and is subject to considerable allosteric control. Mutations in the gene for uroporphyrinogen-1-synthase lead to accumulation of metabolites when flux through the pathway is increased. (Adapted from Sack, JAMA* 264:1290, 1990.)

of the enzyme defect in the synthesis pathway (resembling phenylalanine accumulation in the recessive disorder PKU; recall Chapter 6). Clinical problems develop because at least one of these metabolites is potentially neurotoxic. Acute attacks of AIP are manifested by pain and sometimes paralysis and weakness, reflecting the neurotoxicity of these elevated levels of normal metabolites. Not only are these metabolites capable of causing neuropathy, but they also can be detected easily, particularly in the urine, using a simple colorimetric test. Thus, estimating the concentration of porphobilinogen in the urine can be helpful both diagnostically and therapeutically in individuals with AIP.

The main reason for the clinical development of AIP symptoms is related to the striking inducibility of the first committed enzyme in this metabolic pathway. At basal levels, this enzyme, δ-aminolevulinic acid synthase (ALA synthase), does not provide more substrate than individuals heterozygous for an AIP mutation can handle. When ALA synthase is induced, however, the level of substrate flow through this pathway may rise over a hundredfold, exceeding the limitations imposed by the AIP mutation.

Induction of ALA synthase can occur by many means. Unfortunately, many of these involve drugs. The prototypic drugs for inducing ALA synthase are barbiturates. Administering any of these for sedation, during anesthesia, or for seizure control can very quickly lead to serious clinical problems with pain and neurologic damage. The damage from an acute attack of AIP can cause transient and sometimes permanent neurologic damage and paralysis. Death can occur because of respiratory muscle paralysis.

Fortunately, the biochemical physiology of the heme pathway also provides a treatment option. By administering derivatives of heme, the final product of the pathway, it is possible to repress the activity of ALA synthase. Heme derivatives can reduce ALA synthase activity (by repressing its induction) and thus reduce the quantity of metabolites confronted by the reduced capacity of uroporphyrinogen-1-synthase in AIP. This helps control an attack by minimizing the accumulation of potentially neurotoxic metabolites.

Once an affected individual has been identified, it is imperative that special care be directed to preventing future exposure to inducing drugs. This can be done through the use of a MedicAlert bracelet or other means of identification, but it is essential in long-term management. Many drugs are capable of inducing AIP attacks; Table 10.1 summarizes the general categories.

AIP thus represents a serious and potentially life-threatening situation in which a biochemical change coupled with relevant drug exposure causes severe neurotoxicity. The enzyme defect in AIP can be quantitated by measurements performed on erythrocytes. This can establish whether someone is affected. Performing such studies in large kindreds

Table 10.1

Drugs inplicated in causing AIP attacks

Barbiturates	Sulfonamides	Griseofulvin
Diphenylhydantoin	Chlordiazepoxide	Ergots
Estrogens		

* This short list only emphasizes drugs commonly used. Before using any agents in individuals with AIP, it is important to consider the advice of a pharmacist.

has often led to surprising conclusions. Figure 10.3 presents an example of a kindred identified by a female proband. Note that in three different generations no one had ever been symptomatic to anyone's knowledge. A comparison of the kindred without laboratory data with that of the same kindred after erythrocyte enzyme measurements shows that other individuals clearly have the same enzymatic background and are potentially at risk for an AIP attack. Obviously, if everyone knew his or her enzyme profile, physicians and pharmacists could be alerted to avoid exposure to problem drugs. However, it is generally the case that data such as those from the fully studied kindred in Figure 10.3 are not available. Thus, the AIP mutation may be transmitted simply as an asymptomatic metabolic polymorphism that becomes clinically conse-

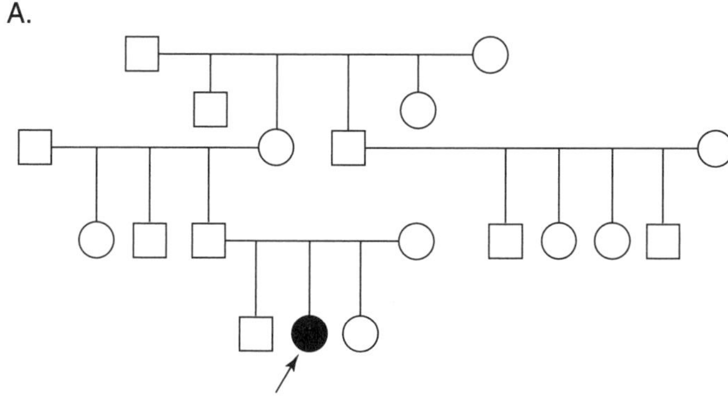

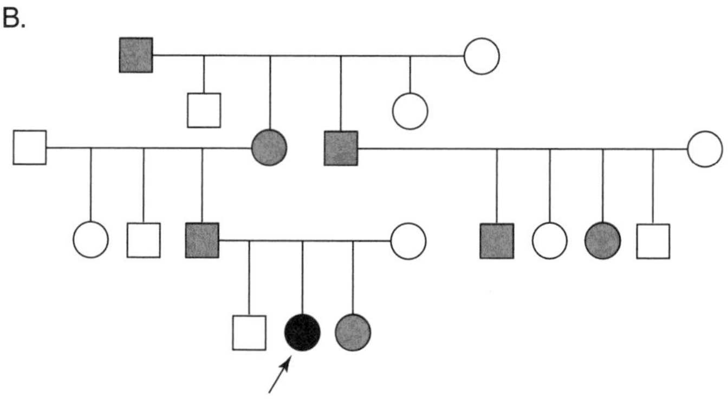

Figure 10.3 *(A) Extensive kindred of a woman newly diagnosed with AIP and claiming no other affected family members on the basis of symptoms. (B) The same kindred following determination of uroporphyrinogen-1-synthase levels. Note the remarkable number of asymptomatic individuals who remain at risk for an attack.*

quential only after exposure to an inducing agent. This observation emphasizes the importance of studying "asymptomatic" family members after an affected individual is identified.

AIP also presents differently depending on the sex of the individual. Although it is certainly not a sex-limited condition and is clearly transmitted as an autosomal dominant trait, far more young (premenopausal) women are symptomatic than men or postmenopausal women. The explanation for this also has been enlightening. Estrogens can induce ALA synthase. Thus, cyclic variations in estrogen levels can cause recurrent attacks. Such situations may occur in a kindred in which enzymatically affected males have never had clinical difficulties. Thus, we are forced to consider clinical complexities due to the interaction of a genetic mutation with normal physiological variation.

C. ANESTHETIC REACTIONS

Many people never require surgery. Among those who do, different anesthetic agents may be used. These two facts provide an important background perspective that makes appreciation of complex genetic differences in anesthetic susceptibility more clear. It is easy to encounter an extensive kindred with a susceptibility that has never been detected simply because of lack of exposure. Two important examples of anesthetic susceptibility illustrate important lessons and the potential for serious illness.

A serum protein that hydrolyzes choline esters has been called *pseudocholinesterase* [also called *butyrylcholinesterase* (EC 3.1.1.8)]. Although the normal function of this serum protein is not completely understood, it does serve to inactivate such drugs as succinylcholine, a muscle-relaxing agent widely used during anesthesia. Failure to inactivate this drug rapidly after surgery obviously prolongs the duration of high serum levels of the drug and leads to continued muscle paralysis. The manifestations of pseudocholinesterase deficiency (OMIM #177400) are easy to predict. A patient who has undergone a surgical procedure with succinylcholine muscle relaxation is found to have prolonged postoperative respiratory paralysis and requires extended artificial ventilation. Gradually, the drug is removed by other detoxification mechanisms, and normal breathing returns. Fortunately, such instances are rare, reflecting the relative infrequency of the mutant allele, as well as the relative rarity of use of the drug succinylcholine.

An individual's susceptibility to such drugs can be measured using an agent called dibucaine to inhibit the activity of serum pseudocholinesterase. The so-called dibucaine number thus becomes a measure of the amount of inhibitor needed for a particular serum sample (see Table 10.2). When an individual has been identified with this deficiency, other family members can be screened through this relatively simple blood enzyme inhibition assay, and their individual dibucaine numbers can be determined. Based on this information, risk assessments can

Variations in pseudocholinesterase levels*

Table 10.2

Dibucaine number	Frequency in Europeans	Enzyme activity	Suxamethonium sensitivity
80	95%	Normal	None
62	3.5%	Slightly decreased	±
49	1:2,500	Slightly decreased	+++
22	1:11,000	Decreased	+++
0	1:170,000	Absent	++++

* Adapted from WHO Technical Reports Series 524, Geneva, 1973.

be made. The simple form of pseudocholinesterase deficiency is an autosomal recessive condition in which only homozygotes are susceptible to clinical difficulties. Unfortunately, the practical measurement of dibucaine numbers can be confused by the presence of other mutations; about one-fifth of individuals who are susceptible to succinylcholine have no measurable serum cholinesterase activity. This presumably reflects a third allele at which the enzyme activity is zero by the standard assay.

An even more rare problem encountered in anesthesia is known as "malignant hyperthermia" (OMIM #145600). In individuals susceptible to this reaction, serious, generalized muscle contractions and remarkably high fevers [up to 42°C (107.6°F)] can develop rapidly after anesthetic exposure. This susceptibility, which can be lethal if uncontrolled, is transmitted as an autosomal dominant trait. Malignant hyperthermia reflects any of several mutations in the ryanodine receptor in skeletal muscle.

D. AMINOGLYCOSIDE OTOTOXICITY

Hearing loss is recognized as a potential side-effect of prolonged use of aminoglycoside antibiotics such as streptomycin. However, some kindreds show increased sensitivity to what in others would be moderate doses of these drugs. The transmission pattern of this sensitivity follows a pattern of maternal inheritance, and an A→G transition at position 1555 of the 12s ribosomal RNA gene in mitochondria has been identified in several families (OMIM #580000; see also Figure 8.1 and Table 8.1). This mutation may affect ATP synthesis in the hair cells of the cochlea; alternatively, this mutation may permit the equivalent in humans of the known ability of this class of antibiotics to reduce the fidelity of translation in bacterial ribosomes.

These susceptibilities to exogenous agents—drugs, hormones, and anesthetics—may never become clinically manifest in the individual at risk. Nevertheless, such individuals can easily transmit these susceptibilities, particularly in disorders with dominant inheritance, such as

AIP and malignant hyperthermia. Individuals with known drug sensitivities are at risk for serious complications themselves. In addition, detecting their susceptibility justifies family studies to detect and inform others who are at risk for similar difficulties. Although encountered infrequently, such susceptibilities can be lethal unless recognized early, so detection efforts are important for counseling and preventive medicine. They also reflect an important background of genetic polymorphisms that may affect the expression of other genes, traits, or biological responses, as discussed in Chapter 12.

CASE STUDY

Ronaldo was a 38-year-old Italian seaman whose container ship had docked the previous evening. He had developed a fever of 38.8°C (102° F) with abdominal pain and dysuria 2 days before arrival and thus was taken to a local clinic where the presence of a urinary tract infection was confirmed. This was his first episode of such an infection and he was given a sulfonamide antibiotic and returned to his ship. When he awoke the next morning he felt worse with continuing fever and cramping pains in his arms and legs. These pains grew worse through the day and he developed dark urine. He was taken back to the clinic where he was found to be more ill and complaining of generalized muscle pain. His laboratory studies showed hemoglobinuria and his blood smear showed polychromatophilia and scattered nucleated erythrocytes. He was treated with rest and intravenous fluids; after 3 more days his episode had resolved completely. A measurement of his G6PD enzyme level showed ~1% of normal. He was given a card identifying him as G6PD deficient (OMIM #305900) and advised about specific drug and food interactions. He was returned to his crew (which, by then, had reached another port).

COMMENT

This picture of acute hemolysis following the use of a drug (in this case, a sulfonamide) is characteristic of individuals with G6PD deficiency. Language difficulties and his lack of previous history complicated the management but, as noted earlier, this condition is remarkably common in certain population groups (e.g., Italians and others from the Mediterranean area) and should at least have been considered by those who saw this man on his first visit.

- Often there is no antecedent history, but there may be numerous female carriers.
- Affected males may develop problems late in life or not at all, depending on drug, food, or environmental exposure.
- Although most episodes are self-limited, the presentations can be frightening and, occasionally, severe.
- Measuring the enzyme level establishes the diagnosis unequivocally.

STUDY QUESTIONS

1 It has been hypothesized that one's isoniazid inactivation status might affect one's predisposition to developing systemic lupus erythematosus. How would you investigate this potentially important relationship?

2 You are the director of anesthesiology for a large surgical hospital. Would you establish a program to detect pseudocholinesterase deficiency and/or malignant hyperthermia? What tests would you perform? What would you do with the results?

3 You are an internist whose patient is a 62-year-old with aggressive, metastatic prostate cancer, and you are discussing treatment options with your colleagues in surgery and oncology. The patient had been well all of his life, but one of his sisters died in her early twenties with what was said to be a "porphyria attack." What would you do about this information? Would it influence the treatment you might suggest (or rule out) for your patient?

FURTHER READING

Ball SP and Johnson KJ. The genetics of malignant hyperthermia. *J Med Genet* 30:89, 1993.

Evans DP and McKusick VA. Genetic control of isoniazid metabolism in man. *Br Med J* 2:485, 1960.

LaDu BN et al. Phenotypic and molecular biologic analysis of human butyrylcholinesterase variants. *Clin Biochem* 23:423, 1990.

MacLennan DH and Phillips MS. Malignant hyperthermia. *Science* 256:789, 1992.

Prezant NR et al. Mitochondrial ribosomal RNA mutation associated with both antibiotic-induced and non-syndromic deafness. *Nat Genet* 4:289, 1993.

Puy H et al. Molecular epidemiology and diagnosis of porphobilinogen deaminase gene defects in acute intermittent prophyria. *Am J Hum Genet* 60:1373, 1997.

Roychoudhury AK and Nei M. *Human Polymorphic Genes: World Distribution*. New York, Oxford University Press, 1988.

Stein JA and Tschudy DP. Acute intermittent prophyria: A clinical and biochemical study of 46 patients. *Medicine* 49:1, 1970.

USEFUL WEB SITES

GENECARDS

http://bioinformatics.weizmann.ac.il/cards/index.html
Presentation of integrated data from large genetics databases in a concise screen of data. Data include gene name, synonyms, gene locus, protein product(s), associated diseases. Searching can be extended from basic cards to other sites.

HELIX (INCLUDES GENLINE)

http://www.hslib.washington.edu/helix
Medical Genetics Knowledge Base. Electronic textbook under development relates testing to diagnosis, management, and counseling. Directory of laboratory providing testing for genetic disorders.

OMIM (ONLINE MENDELIAN INHERITANCE IN MAN)	http://www.ncbi.nlm.nih.gov/omim Basic catalog listing mendelian genes and traits. Clinical descriptions, references, and links to other databases.
NATIONAL ASSOCIATION FOR RARE DISORDERS (NORD)	http://www.NORD-rdb.com/~orphan Primary nongovernmental clearinghouse for information about rare disorders. Over 1100 disease reports. Broad resource guide. Includes Rare Disease Database, a wide range of information for families and professionals. Listings by symptoms, causes, affected populations, treatments. Listing of support groups. Includes orphan drug designation database.
OFFICE OF RARE DISEASES (NIH)	http://cancernet.nci.nih.gov/ord/index Broad-based NIH index including support groups, clinical research database, investigators, and extensive glossary.

CHAPTER 11

Genetics and Immune Function

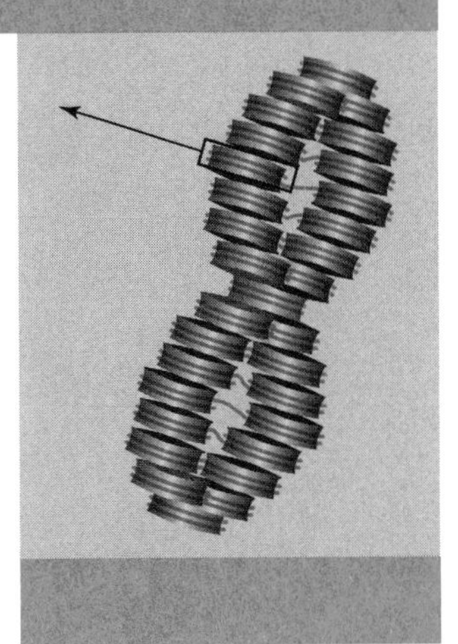

The biologic distinction between "self" and "nonself" is mediated through complex interactions that involve differences at cell surfaces as well as in soluble molecules. Dissecting these pathways has involved molecular and cell biology studies that can only be summarized here. As noted below, individual mendelian disorders illustrate the consequences of defects in individual parts of the immune defense system. Details of this important and fascinating field can be found through the "Further Reading" listings. The discipline of "immunogenetics" begins with the basic mechanisms of mammalian genetics, including polymorphisms, gene families, and recombination. The additional features of somatic mutation and cell differentation add considerable complexity, as well as adaptability, to immunologic responses but are superimposed on the basic processes.

A. THE MAJOR HISTOCOMPATIBILITY COMPLEX

Although formidably complicated at first glance, the genetics of the major histocompatibility complex (MHC) are surprisingly straightforward. Most of the complexity arises from the enormous amount of polymorphism at the responsible loci. This complexity underlies many of the challenges of tissue and organ transplantation, because the protein products of genes of the MHC appear on most cell surfaces.

The genes of the human MHC constitute rather typical gene families; all are clustered on chromosome 6 (see Figure 11.1). As shown, this

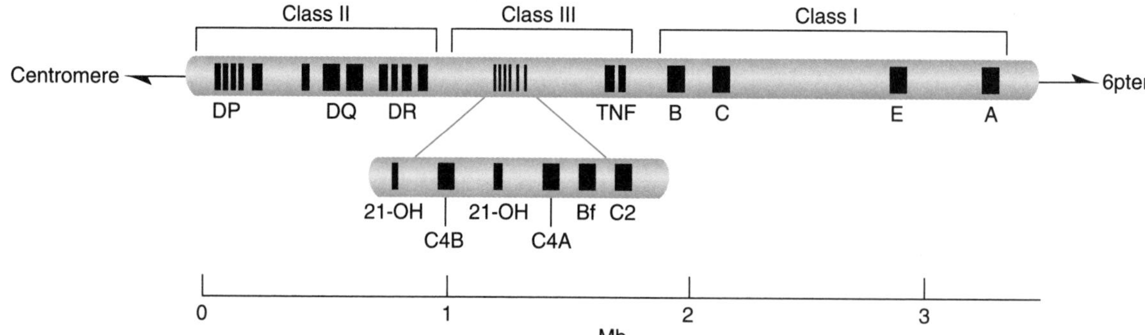

Figure 11.1 *Outline genetic map of the major histocompatibility complex (MHC) on chromosome 6. As described in the text, the Class III genes contain additional genes, including those for 21-hydroxylase (21-OH), complement factors (e.g., C2, C4A, C4B), properdin factor B (Bf), and tumor necrosis factors alpha and beta (TNF-α, TNF-β).*

large cluster, spanning about 4.2 mb, can be separated into three groups, or "classes." Classes I and II contain genes encoding human leukocyte antigens (known as "HLA antigens"). These antigens are structurally distinct, however.

Class I antigens are formed by combining β_2-microglobulin, a small (12 kDa) invariant protein unlinked to the MHC (its gene is found on chromosome 15), with a 44-kDa heavy chain protein encoded by a Class I gene (see Figure 11.2, left side). There are ~15 Class I genes.

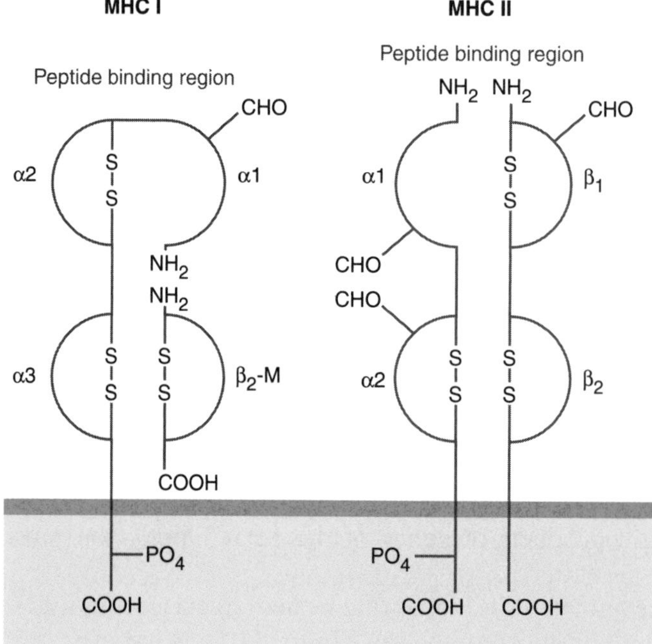

Figure 11.2 *Diagrams of MHC molecules. (Left) Class I MHC; the groove between the α_1 and α_2 domains is the region of specific peptide binding. The β_2-microglobulin is intimately associated with this complex. (Right) Class II MHC; the peptide-binding groove is at the top. No β_2-microglobulin is present.*

Class I antigens have two protein domains outside the cell that, as a complex, form a peptide binding site. Within the cell there are two additional domains that are homologous to the immunoglobulin superfamily (see below). Class I antigens are found on all nucleated cells.

Class II antigens do not involve β_2-microglobulin. Instead, they are heterodimers comprising two subunits, a 30–32 kDa heavy chain, called α, and a 27–29 kDa light chain, called β, each of which is encoded by Class II genes. There are approximately 23 Class II genes. Class II antigens are characteristic of the plasma membranes of B cells, activated T-cells, and macrophages. Most other cells lack Class II antigens, but many can be induced to express them by agents such as tumor necrosis factor α (TNF-α) and interferon γ (IFN-γ). Each of the chains has two extracellular domains, a short membrane-spanning domain and an intracellular domain (see Figure 11.2, right).

Class II genes are organized into three main clusters, encoding isotypes DR, DQ, and DP (see Figure 11.1); each cluster has one α-chain gene and one or more β-chain genes. Usually, Class II antigens are formed by α and β chains of the same isotype, but some interisotypic pairs also are found. DR Class II antigens have only β-chain polymorphisms (hence, there can be only two DR types—one from each parental chromosome). In contrast, DQ and DP antigens can be formed from *cis* and *trans* heterodimers. This mixing of the products encoded by maternally and paternally derived chromosomes can lead to four possible types of DP and DQ Class II antigens.

Based on structural and DNA sequence details, the Class I and Class II genes are related to the T-cell receptor and immunoglobulin genes and often are grouped as the "immunoglobulin gene superfamily."

Physically located between the Class I and Class II gene clusters, Class III genes are not themselves HLA genes. Rather, they encode several complement factors (C2, C4A, C4B, and properdin factor B), as well as other proteins related to immune function and host defense, including TNF-α and TNF-β. Other genes not obviously related to immune processes, such as the enzyme 21-hydroxylase, also are found in this region. Polymorphism is not a conspicuous feature of genes in the Class III region, although several important and frequent mutations for hemochromatosis (OMIM #235200) are recognized in the HFE gene.

The Class I and Class II genes (and hence their protein products) are very polymorphic, and their alleles are codominant. This means that any individual can express two alleles for each Class I and Class II DR locus (one from the copy of chromosome 6 inherited from each parent). As noted, the possible mixing of α- and β-chain proteins encoded by both chromosomes 6 gives four possible DP and DQ antigens. This situation could theoretically lead to over 10^7 combinations, but many of these have not been detected. Because there is very little recombination in the MHC region, haplotypes are generally transmitted intact. This means that a parent and a child share one HLA Class I haplotype. By the same logic, there is a one in four chance that any two sibs will have inherited the same haplotype from both the parents (and hence will be said to be "HLA identical"). Similarities in HLA haplotypes have been central to tissue and organ transplantation.

As might be expected, the distribution of HLA haplotypes is not random and linkage disequilibrium across the MHC is notable. Concentrations of HLA haplotypes in specific populations are common. For example, Class I HLA-A group A 24 is not found in blacks or Asians, while it is recognized in Caucasians.

B. HLA AND DISEASE

Because of the high degree of polymorphism in the HLA system, it is not surprising that associations have been detected between certain HLA alleles and relatively common disorders. The pathophysiologic mechanisms underlying these associations are generally unclear, however. Of course, one would expect an association between a mutation in a Class III gene, such as that for 21-hydroxylase, and mutations in flanking Class I and Class II alleles simply on the basis of the linkage disequilibrium across the region (little crossing-over is expected). Consistent with this pattern, similar relationships have been noted for hemochromatosis (OMIM #235200), whose gene, HFE, is found in the MHC, as noted above.

Other conditions with HLA associations are important but not so easily explained. Some sort of immunological reactivity has been implicated, however. Table 11.1 presents a group of diseases and their associ-

Table 11.1 ▶ **HLA-disease associations***

Disorder	HLA Allele	Relative Risk
Ankylosing spondylitis	B27	87.4
Seropositive juvenile rheumatoid arthritis	Dw4/Dw14	116
	Dw14	47
	Dw4	26
Narcolepsy	DR2	49
Reiter syndrome	B27	37
Goodpasture syndrome	DR2	15.9
Dermatitis herpetiformis	DR3	15.4
Adrenal hyperplasia (congenital)	B47	15.4
Pemphigus (in Jews)	DR4	14.4
Subacute thyroiditis	B35	13.7
Psoriasis vulgaris	Cw6	13.3
Membranous nephropathy	DR3	12
Celiac disease	DR3	10.8
Acute anterior uveitis	B27	10.4

* Only those with at least a tenfold relative risk are shown. Adapted from Vogel and Motulsky, *Human Genetics*, 3rd ed., Springer-Verlag, Berlin, 1997; and Koopman, *Arthritis and Allied Conditions*, 13th ed., Williams & Wilkins, Baltimore, 1997.

ations with Class I and Class II. An important consideration in viewing these data is that they are expressed as "relative risks." This represents the following relationship:

$$\text{Relative risk (RR)} = ad/bc$$

where

a = proportion of patients who have the allele
b = proportion of patients who lack the allele
c = frequency of the allele in control individuals
d = frequency of absence of the allele in control individuals.

The most prominent of the conditions shown in Table 11.1 is ankylosing spondylitis. In Caucasians, 88 to 95% of affected individuals have the HLA-B27 allele. As shown, an 85 to 90-fold increase in the risk of developing ankylosing spondylitis has been estimated for individuals with the B27 allele. Unfortunately, the interpretation is not as straightforward as might be anticipated. Approximately 8% of Caucasians have the B27 allele, and ankylosing spondylitis is not that frequent a condition. Only about 4% of Caucasians with the B27 allele will develop the disorder. Thus, the B27 allele usually is associated with, but by itself is not a sufficient explanation for, the disease; the other contributing factor or factors have not been identified.

Although these associations are not always understood from a physiologic perspective, recognizing them can be useful in diagnosis and management. For example, finding the B27 allele can add a clearer diagnostic perspective for an individual with early symptoms of spondyloarthropathy and supports the likelihood of later development of ankylosing spondylitis.

Although they differ considerably, disorders linked to HLA have a number of common features:

- They have familial associations (see also Chapter 12).
- There is no clear-cut inheritance pattern.
- Penetrance is weak.
- Many have been considered autoimmune diseases and have associated immunologic abnormalities.
- They are generally chronic (or at least subacute) and seldom affect reproductive fitness.
- Details of their pathophysiology are not understood.
- The HLA association is *not* absolute, and thus the "relative risk" notion is appropriate (see Table 11.1).
- Finding an HLA association for a subgroup of individuals with a disorder can at least *suggest* a different basis for the condition.

This group of characteristics has led to several hypothetical explanations of HLA-disease associations. These are not mutually exclusive:

- The actual cause of the disease is a virus for which the HLA molecule acts as a receptor. Many viruses are known to interact with specific cell surface proteins. For example, HIV interacts with the CD4 protein.
- A specific polypeptide is the etiologic agent, and only certain HLA molecules can accommodate this protein in their extracellular binding domain (recall Figure 11.2).
- The specificity of recognition originates in the T-cell antigen receptor, which recognizes only a specific HLA/polypeptide complex. In this case, the primary recognition derives from the T cell, not the HLA protein.
- There is defective intracellular transport of polypeptides from the cytoplasm to the endoplasmic reticulum, leading to reduced formation of HLA Class I-peptide complexes within the cell and on its surface. The presence of fewer such complexes on the cell surface could blunt an effective immune response to the virus. Alternatively, there could be empty Class I molecules on the cell surface, capable of binding high numbers of extracellular viral polypeptides and leading to a T-cell response. Finally, a low surface density of Class I molecules leads to reduced self-tolerance and an autoimmune response.
- The "molecular mimicry" hypothesis suggests that the associated HLA antigen shares immunologic features with the etiologic HLA-viral peptide combination. This could lead either to failure to recognize the HLA-etiologic virus complex as foreign (permitting it to propagate) or to recognition of the associated HLA antigen as foreign, triggering an autoimmune response.
- A sixth possibility relates only to diseases associated with Class II HLA molecules. In this situation, some initial insult (for example, a viral infection) to a cell not normally expressing Class II molecules leads to inflammation of the cell and synthesis of IFN-γ, which then stimulates expression of Class II molecules; an aggressive immune response follows.

The considerable interest in HLA-B27 disease associations has led to informative models in transgenic rats expressing the human HLA-B27 gene. Such rats develop signs of spondyloarthropathy, indicating that the HLA-B27 gene and not a linked gene is responsible for the inflammatory disease in this system.

An instructive example of the value of HLA associations in distinguishing between clinically similar disease presentations has come from studies of diabetes. Ninety-five percent of individuals with Type I (juvenile-onset) diabetes have DR3 and/or DR4 antigens. This association is not seen for individuals with Type II (adult-onset) diabetes. While there often are other clinical features that help with these distinctions, finding specific HLA associations can suggest pathologic relationships as discussed above. More discussion of this problem is presented in Chapter 12.

C. IMMUNOGLOBULINS

The formation of immunoglobulins involves recombination, mutation, and localized nucleotide changes, all of which lead to the production of specific and unique proteins. As is explained in detail in immunology textbooks, formation of the mature product of the heavy-chain gene cluster on chromosome 14 involves several events:

- the choice among multiple V, D, and J gene regions
- imprecision in joining these regions
- apparently random insertion of nucleotides at the joints between the regions
- somatic mutation in the J region

Once the heavy-chain gene product is synthesized, it can pair with multiple light-chain product subtypes in the structure shown in Figure 11.3.

There are extensive families of heavy- and light-chain genes. The somatic reorganization and, ultimately, the expression of the reorganized product occurs on either the maternal or the paternal allele (never on both) in each primordial B cell. For heavy chains this is

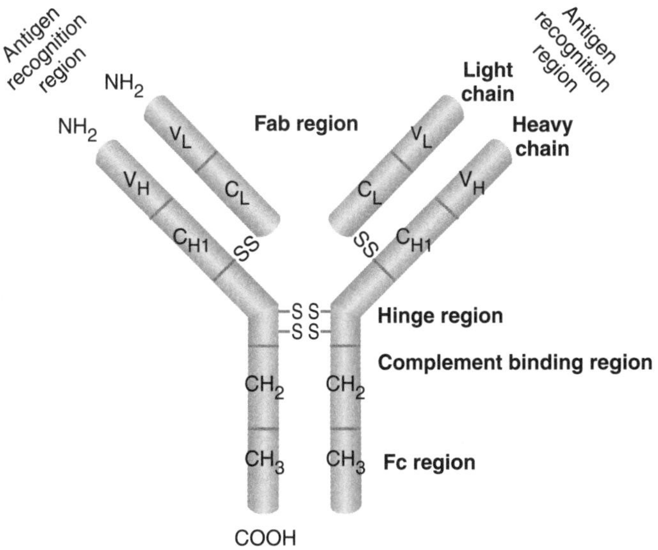

Figure 11.3 *Simplified view of immunoglobulin G (IgG) molecule. Note that it is formed from two heavy and two light chains, joined by disulfide bonds. The antigen recognition regions are at the N-terminal parts of the* V_L and V_H regions.

referred to as isotypic exclusion; for the light chains, it is called allelic exclusion.

D. T-CELL RECEPTORS

Forming the membrane-bound T-cell antigen receptors also involves assembling different parts (V, D, and J) and thus has important similarities to immunoglobulin synthesis. Unlike the situation with immunoglobulins, however, formation of T-cell receptors (TCRs) does not involve somatic mutations. Most T-cell receptors represent heterodimers of α and β chains (designated TCR α:β), but another group (TCR γ:δ) also exists; its function is less clear. The genes for the members of the TCR family are dispersed but still locally organized as families. TCR α genes are on chromosome 14; TCR β genes are on the long arm of chromosome 7; TCR γ genes are on the short arm of chromosome 7. The same general organization is seen for each cluster.

E. IMMUNOGLOBULIN GENE SUPERFAMILY

The basic structural similarity of a number of gene groups permits them to be referred to collectively as the "immunoglobulin (Ig) gene superfamily." Table 11.2 lists members of this group. Obviously the group includes MHC, Ig, and TCR genes, but a surprising collection of related molecules also have been identified, including those related to T-cell structure and function, growth factor receptors, and proteins specific to the nervous system.

These relationships have been defined primarily on the basis of gene *structure*. Conserved domains frequently are identified through homology search algorithms applied to new DNA sequences and based on protein and DNA databases. These relationships must reflect the

Table 11.2 ▶ **Immunoglobulin gene superfamily members**

Immunoglobulin gene superfamily members
T-cell receptor components (TCR, CD3 α and β)
T-cell adhesion and related proteins (CD1, CD2, LFA3)
T-subset antigens (CD4, CD8, CTLA4)
Brain and lymphoid antigens (Thy1, MRC, Ox2)
Immunoglobulin receptors (PolyIgR, Fcγ2b/γ1R)
Neural adhesion molecule (NCAM)
Myelin protein (Po)
Myelin-associated protein (MAG)
Carcinoembryonic antigen (CEA)
Platelet-derived growth factor receptor (PDGFR)
Colony-stimulating factor 1 receptor (CSF1R)
Basement membrane link protein (LINK)

* Adapted from Williams and Barclay, *Annu Rev Immunol* 6:381, 1988.

evolutionary relatedness of the genes and the preservation of topologic and possibly functional features of the encoded proteins that have become adapted to specialized functions through subsequent changes. Only the immunoglobulin genes themselves undergo somatic mutation, however, clearly distinguishing them from other members of this family. The recombination events occurring with TCR and MHC genes help increase the multiplicity of proteins produced, but the overall three-dimensional structure of the resulting proteins is remarkably conserved. Of course, local differences are the principal determinants of antigenic diversity, but these are superimposed on a larger, conserved framework.

F. SINGLE-GENE DISORDERS OF IMMUNE FUNCTION

In addition to the more generalized aspects of immune function described above, several important disorders reflect specific mutations of individual parts of host defense. These are important because of their often striking clinical presentations as well as for the important physiologic insights that their elucidation has provided. As we will discuss in Chapter 14, they also have been targets for specific treatments. Several of these disorders affect individual parts of the immune system; others are more generalized. Table 11.3 presents a listing of some of the disorders in this general category, grouped according to physiologic dysfunction; this grouping often provides a clue to distinguishing among them.

Table 11.3

Single-gene disorders of immune function

Physiologic Category	Location	OMIM #
Combined Immunodeficiency		
Severe combined immunodeficiency (SCID)		
SCIDX1 (Swiss)	Xq13.1	300400
Adenosine deaminase (ADA) deficiency	20q12	102700
Purine-nucleoside phosphorylase deficiency	14q11.2	164050
Wiskott-Aldrich syndrome	Xp11.2	301000
Ataxia-telangiectasia	11q22.3	208900
T-cell dysfunction		
DiGeorge syndrome	22q11	188400
Mucocutaneous candidiasis		212050
B-cell dysfunction		
X-linked (Bruton) hypogammaglobulinemia	Xq21.3	300300
X-linked immunoproliferative syndrome	Xq25	308240
Hyper-IgM-associated immunodeficiency	Xq26	308230
Phagocyte dysfunction		
Chediak-Higashi syndrome	1q42	214500
Chronic granulomatous disease		
Cytochrome b α-subunit	16q24	233690
Cytochrome b β-chain	Xp21	306400
Myeloperoxidase deficiency	17q23.1	254600
Glucose-6-phosphate dehydrogenase deficiency	Xq28	305900

1. COMBINED IMMUNODEFICIENCY

Infants with severe combined immunodeficiency (SCID) are severely affected, usually presenting with inanition and diarrhea, along with multiple infections. These "SCID infants" lack a thymus gland, tonsils, and lymph nodes, as well as B and T cells. They have recurrent infections and may show systemic responses to live-virus vaccines. In addition, the Wiskott-Aldrich syndrome (OMIM #301000) is notable for thrombocytopenia. Individuals with ataxia-telangiectasia (OMIM #208900) have cerebellar ataxia and telangiectases on the skin and in the eyes. They have more variable immune deficiency, usually with anergy and reduced levels of IgA and IgE.

Note that two of the disorders in this category on Table 11.3 are X-linked; this can assist in diagnosis. Individuals in these categories are remarkable for showing non-random X-chromosome inactivation; there is preferential inactivation of the X-chromosome carrying the abnormal gene. Because methylation patterns and polymorphisms can distinguish the inactive X-chromosome, prenatal diagnosis can be based on a determination of which maternal X-chromosome has been inherited by a male fetus at risk.

In general, SCID infants are severely ill and their prognosis is poor. Nevertheless, important progress has been made in treatment of several of these children with bone marrow transplantation as well as enzyme replacement (through either intravenous infusion or gene therapy; see also Chapter 14).

2. T-CELL DYSFUNCTION

Infants with DiGeorge syndrome (OMIM #188400) generally have congenital malformations in addition to their immune problems (see Chapter 9). In addition to facial dysmorphism, hypoparathyroidism, and cardiac anomalies, they have an absent or aplastic thymus and lymphopenia. Their infections are generally more chronic and can respond to treatment. Transplantation with early fetal thymus tissue also has been successful. Mucocutaneous candidiasis (OMIM #212050), as its name implies, is notable for chronic fungal infections that respond poorly to treatment.

3. B-CELL DYSFUNCTION

Disorders in this category are, as expected, deficient in functions related to immunoglobulins. Infants with "Bruton" hypogammaglobulinemia (OMIM #300300) often do well until maternal immunoglobulins disappear at about 6 months of age, at which time severe bacterial infections develop, often due to *H. influenzae* and *S. pneumoniae*. Such infants are deficient in all immunoglobulin classes as well as in B lymphocytes. Immunoglobulin replacement can promote some clinical improvement.

Finding elevated levels of IgM suggests another syndrome, hyper-IgM-associated immunodeficiency (OMIM #308230), in which the presentations and treatments are similar to those of Bruton hypogammaglobulinemia. The X-linked lymphoproliferative syndrome (OMIM #308240) is clinically different, involving particular problems with viral infections, especially with Epstein-Barr virus (EBV), that can be fatal. Burkitt lymphoma also is seen, presumably reflecting EBV infection.

4. PHAGOCYTE DYSFUNCTION

Children with chronic granulomatous disease (OMIM #233690, #306400) generally have recurrent bacterial infections in infancy. Problems develop because defective phagocytosis often prevents killing of the bacteria despite the use of antibiotics, leading to chronic involvement of and damage to many organs. Obviously, the first step in treatment is aggressive antibiotic administration, followed by careful observation to be certain that the infection has been controlled.

Individuals with the Chediak-Higashi syndrome (OMIM #214500) often show at least partial albinism, as well as hepatosplenomegaly. They also have recurrent bacterial infections due to decreased natural killer cell function. Control of infections can be difficult, and these children also are at risk for neurodegeneration and lymphoma.

An exception to the rule of poor prognoses is seen for individuals with myeloperoxidase deficiency (OMIM #254600). Although bacterial killing is slowed, it is still detectable. In general, these individuals do well clinically, without complications, and often are undetected unless purposely studied.

An interesting subset of individuals with G6PD deficiency (OMIM #305900, see also Chapter 10) show recurrent bacterial infections. Such individuals usually have less than 5% of normal G6PD levels. In addition to the problems with drug-induced erythrocyte instability described earlier, these patients lack adequate levels of G6PD in neutrophils to produce NADPH for the respiratory burst involved in oxidation after phagocytosis.

5. COMPLEMENT DEFICIENCIES

The multiple roles of the complement system can be affected by mutations as well. Specific defects of most of the factors have been identified; most are autosomal recessive traits. Exceptions to this transmission pattern are seen in properdin deficiency (OMIM #312060), which is X-linked, and C1-inhibitor deficiency (OMIM #106100), which is autosomal dominant.

Three general categories of clinical presentations are seen; the relevant complement components are given:

- Increased susceptibility to infections:
  encapsulated bacteria (C3)
  Neisseria spp (C5-C9)
- Rheumatic diseases (C1-C4)
- Angioedema (C1 inhibitor)

The interactions of immunology and genetics are complex summations of discrete, heritable gene changes and variable somatic events. Their varied clinical effects reflect the central importance of both host defense and cell surface recognition processes. As discussed above, changes in details of this system can lead to many different, and at times unexpected, consequences. Appreciating these possibilities is important to appropriate diagnosis and care.

CASE STUDY

Paul appeared healthy at birth but began having problems with diarrhea by the middle of his third month of life. He had positive stool cultures for *Salmonella* and *E. coli,* and control of these infections proved very difficult so that he had poor weight gain. Thrush was a persistent problem. At age 5 months he developed pneumonia with a *Pseudomonas* infection that responded only poorly to antibiotics; his x-ray showed no obvious thymic shadow. Laboratory studies showed leukopenia with consistently low lymphocyte counts after the first pulmonary infection. Serum immunoglobulin levels were very low. He died at 8 months of age when, after exposure to chickenpox, he developed a disseminated varicella infection.

Comment

This unfortunate boy had adenosine deaminase deficiency and the clinical picture of severe combined immunodeficiency (SCID, OMIM #102700). Such children are rare but SCID infants have a generally similar clinical course despite multiple genetic forms as discussed in the text and Table 11.3.

- Such children usually begin to show problems several months after birth when the effectiveness of maternal antibodies has diminished.
- In the absence of effective treatment, these children show an inevitable downhill course with disseminated infections that often are never completely eradicated.
- The lethality of varicella is well recognized.
- Individuals with this condition have been the target for early attempts at gene replacement with adenosine deaminase and various delivery systems. Some results have been promising, and it is clear that relentless deterioration is the underlying natural pattern in the absence of effective intervention.

STUDY QUESTIONS

1 While studying individuals on an island in the North Atlantic, you are impressed with the high frequency of Q disease. In an effort to evaluate this further, you propose to evaluate the association of Q disease with HLA

markers. What important considerations do you need to take into account as this study progresses?

2 You are taking care of a healthy 43-year-old man who may become a kidney donor. As part of the evaluation, you discover that he is HLA-DR3 positive. You think:

a He is at risk for maturity-onset diabetes.
b He never had insulin-dependent diabetes, so there may be a laboratory error.
c Other family members may be at increased risk for diabetes.
d The absence of diabetes in your patient provides evidence that DR3 and diabetes are not associated in this family.

3 While working with animal cell cultures as hosts for virus studies, you have isolated a line that appears resistant to the dangerous encephalopathic Z virus. This virus has been associated with disease in a small population in northern China. You conclude:

a Your cells lack a surface receptor for the Z virus.
b You need to perform an HLA susceptibility study in the Chinese population.
c You may have a mutant of the virus.
d You should find out if the host animal (the source of the tissue culture cells) is susceptible to the Z virus.

FURTHER READING

Barrett DJ et al. Antibody deficiency diseases, in Scriver CR, Beaudet AR, Sly WS, Valle D (eds.). *The Metabolic and Molecular Bases of Inherited Disease,* 7th ed. New York, McGraw-Hill, 1995, p 3879.

Blaese RM. Genetic immunodeficiency syndromes with defects in both T and B lymphocyte function, in Scriver CR Beaudet AR, Sly WS, Valle D (eds.). *The Metabolic and Molecular Bases of Inherited Disease,* 7th ed. New York, McGraw-Hill, 1995, p 3895.

Bodmer JG et al. Nomenclature for factors of the HLA system. *Hum Immunol* 43:149, 1995.

DeVries RRP and van Rood JJ. Immunogenetics and disease, in King RA, Rotter JI, Motulsky AG (eds.). *The Genetic Basis of Common Disease.* New York, Oxford University Press, 1992, p 92.

Forehand JR et al. Inherited disorders of phagocyte killing, in Scriver CR, Beaudet AR, Sly WS, Valle D (eds.). *The Metabolic and Molecular Bases of Inherited Disease,* 7th ed. New York, McGraw-Hill, 1995, p 3995.

Hammer RE et al. Spontaneous inflammatory disease in transgenic rats expressing HLA-B27 and human β_2m: an animal model of HLA-B27 associated human disorders. *Cell* 63:1099, 1990.

McDevitt HO. The HLA system and its relations to disease. *Hosp Pract* 20:57, 1985.

Ochs et al (eds). *Genetics of Primary Immunodeficiency Diseases.* London, Oxford University Press, 1998.

Robson KJH. Haemochromatosis: A gene at last? *J Med Genet* 34:148, 1997.

Schwartz BD. Infectious agents, immunity, and rheumatic diseases. *Arth Rheum* 33:457, 1990.

Schwartz BD. Structure, function and genetics of the HLA complex in rheumatic disease, in Koopman WJ (ed.). *Arthritis and Allied Conditions,* 13th ed. Baltimore, Williams & Wilkins, 1997, p 545.

Taurog JD et al. The germ-free state prevents development of gut and joint inflammatory disease in HLA-B27 transgenic rats. *J Exp Med* 180:2359, 1994.

Tiwari JL and Terasaki PI. *HLA and Disease Associations.* New York, Springer, 1985.

Williams AF and Barclay AN. The immunoglobulin superfamily-domains for cell surface recognition. *Annu Rev Immunol* 6:381, 1988.

Winkelstein JA et al. Genetically determined disorders of the complement system, in Scriver CR, Beaudet AR, Sly WS, Valle D (eds.). *The Metabolic and Molecular Bases of Inherited Disease,* 7th ed. New York, McGraw-Hill, 1995, p 3911.

USEFUL WEB SITES

GENECARDS

http://bioinformatics.weizmann.ac.il/cards/index.html
Presentation of integrated data from large genetics databases in a concise screen of data. Data include gene name, synonyms, gene locus, protein product(s), associated diseases. Searching can be extended from basic cards to other sites.

HELIX (INCLUDES GENLINE)

http://www.hslib.washington.edu/helix
Medical Genetics Knowledge Base. Electronic textbook under development relates testing to diagnosis, management, and counseling. Directory of laboratories providing testing for genetic disorders.

OMIM (ONLINE MENDELIAN INHERITANCE IN MAN)

http://www.ncbi.nlm.nih.gov/omim
Basic catalogue listing mendelian genes and traits. Clinical descriptions, references, and links to other databases.

NATIONAL ASSOCIATION FOR RARE DISORDERS (NORD)

http://www.NORD-rdb.com/~orphan
Primary nongovernmental clearinghouse for information about rare disorders. Over 1100 disease reports. Broad resource guide. Includes Rare Disease Database, a wide range of information for families and professionals. Listings by symptoms, causes, affected populations, treatments. Listing of support groups. Includes orphan drug designation database.

OFFICE OF RARE DISEASES (NIH)

http://cancernet.nci.nih.gov/ord/index
Broad-based NIH index including support groups, clinical research database, investigators and extensive glossary.

CHAPTER 12 *Genetics and Common Diseases*

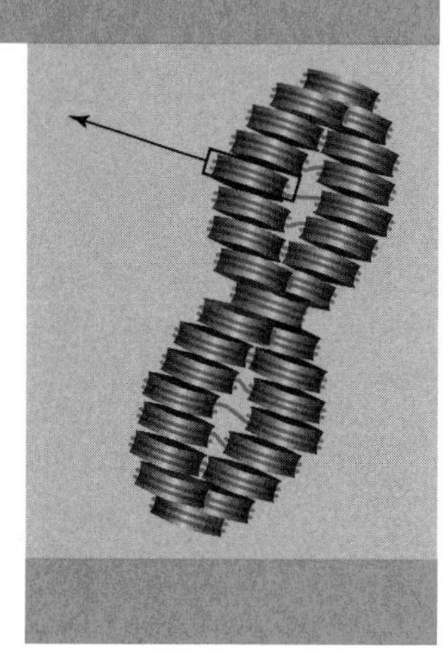

A. BACKGROUND

One of the main reasons for the historical underemphasis placed on genetics in medicine has been the relative rarity of clear-cut mendelian conditions in general clinical practice. Although it is true that mendelian conditions are individually rare, they are not collectively rare. Nevertheless, the likelihood that many affected individuals will be encountered in the average medical practice is low. Despite the fact that single-gene mendelian disorders are *not* common, there is a long history of recognizing inherited contributions to conditions that *are* common. Although we do not yet have a complete molecular underpinning for this recognition in terms of having identified all of the contributing loci and alleles, empiric estimates of the contributions of inheritance to common clinical problems have been made by thoughtful clinicians for many years. Some consequences of this recognition have been the aggressive use of screening protocols for those considered to be at high risk and observations that certain conditions in certain population groups respond better or worse to a specific treatment. Such rather vague statements are slowly yielding to detailed analysis, however, and future research will inevitably establish the molecular basis for at least some of these observations by identifying alleles to guide identification and management. Also, it is fair to say that individual gene contributions to complex but common illnesses will be the focus of considerable epidemiologic, genetic, and therapeutic study in the future.

Table 12.1 ▶ Kindred relationships and proportions of genes in common with a proband*

Relationship	Proportion of genes in common
Monozygotic twin	1/1
1st degree relative (parent, dizygotic twin, other sib, child)	1/2
2nd degree relative (uncle, aunt, grandparent, nephew, niece, grand-child, half-sib)	1/4
3rd degree relative (first cousin, great-grandparent, great-grand-child)	1/8
4th degree relative (second cousin)	1/16

* See also Figure 1.27.

It is important to recall that specific family relationships imply certain proportions of genes in common. Several of these are presented in Table 12.1. Recall that the number of autosomal genes held in common between kindred members equals $1/2^n$, where n is the degree relationship. The term "genes in common" means gene alleles derived from a common ancestor (also referred to as "identical by descent," IBD). The relationships were introduced in Figure 1.27.

Several lines of data support the notion of genetic contributions to common diseases.

1. TWIN STUDIES

Comparisons between fraternal (dizygotic) and identical (monozygotic) twins have provided strong support for the idea that there are heritable features underlying common traits. Obviously, monozygotic twins are genetically identical, with all alleles shared. Dizygotic twins represent individual conceptions and thus have one-half of their genes in common (the same as all siblings). Table 12.2 presents concordance rates between members of twin pairs for several conditions. Such data are

Table 12.2 ▶ Twin concordance data*

Condition	Percent Concordance Monozygotic	Dizygotic
Hyperthyroidism	47	6.5
Clubfoot	23	2.3
Cleft lip with or without cleft palate	30	4.7
Congenital hip dislocation	41	2.8
Insulin-dependent diabetes	56	11
Schizophrenia	60	10
Pyloric stenosis	22	2
Coronary artery disease	46	12

* Data taken from Carter, *Br Med Bull* 25:52, 1969.

Table 12.3

Correlation of finger ridge counts*

Relationship	Correlation coefficient	Correlation expected
Monozygotic twins	0.95 ± 0.07	1.00
Dizygotic twins	0.49 ± 0.08	0.50
Sibling–sibling	0.50 ± 0.04	0.50
Parent–child	0.48 ± 0.04	0.50
Parent–parent	0.05 ± 0.07	0

* Adapted from Carter, *Br Med Bull* 25:52, 1969.

impressive and help justify current efforts to identify loci whose variations contribute to conditions such as schizophrenia that have eluded clarification. Note that although some traits show relatively low concordance, they are still more frequent in monozygotic than in dizygotic twins.

2. FAMILY CLUSTERS

The appearance of similar features in family members at a rate higher than expected by chance and with some relation to the degree of genetic concordance is additional evidence for underlying genetic contributions to these features. This can vary from the striking similarity of numbers of finger ridges (a close parallel to the number of genes shared; see Table 12.3) to less prominent but still notable similarities in frequencies of traits such as death from coronary artery disease (see Table 12.4). The number of finger ridges is determined before birth and thus should be relatively independent of the environment. By contrast, the environment clearly can affect the development of coronary artery disease. Figure 12.1 summarizes data from a study comparing the distribution of blood pressure scores in relatives of individuals with hypertension

Table 12.4

Ratios of death from coronary artery disease related to age of death of twin*

Age at death (years)	Relative risk	
	Men	Women
Monozygotic		
36–55	13.4	
56–65	8.1	14.9
66–75	4.3	3.9
76–85	1.9	2.2
≥86	0.9	1.1
Dizygotic		
36–55	4.3	
56–65	2.6	2.2
66–75	1.7	1.9
76–85	1.4	1.4
≥86	0.7	1.0

* Adapted from Marenberg et al. *N Engl J Med* 330:1041, 1994.

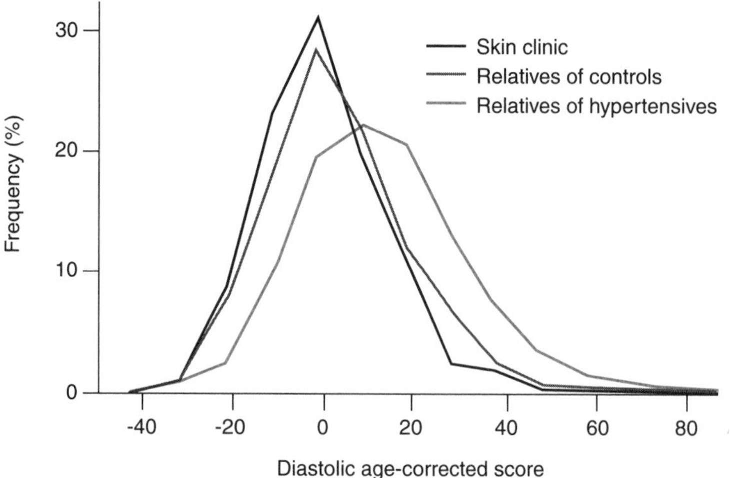

Figure 12.1 *Distribution of blood pressure in three clinic populations. Note the displacement of the data for relatives of individuals with hypertension. (Adapted from Weiss, Genetic Variation and Human Disease, 1993, p 106.)*

to that in both a control group and a group of patients in another clinic. The shift of the phenotype distribution is apparent.

3. POPULATION CONCENTRATIONS

Our earlier discussions of specific single-gene changes emphasized that some are concentrated in specific ethnic or other population groups. As discussed in Chapter 6, certain examples show strong associations—for example, sickle cell disease in populations of African origin. Others are less clear cut, including cystic fibrosis in Caucasians, PKU in Northern Europeans, and adult Gaucher disease in Jews.

More frequently encountered traits and diseases also show biased population distributions. For example, in Hawaii, the Chinese population has the lowest rate of clubfoot deformity and Polynesians have the highest rate; Caucasians have an intermediate rate. There are many examples of this sort of association, not only with traits themselves but also with their clinical characteristics. For example, hypertension in the young black male differs from its counterpart in the elderly Caucasian female in severity, age of onset, response to medications, and complications. Thus, in a search for responsible (or at least contributing) gene alleles, it is essential to define the condition carefully, because it is likely that groups separated by such factors as geography, ethnicity, religion, etc., will have different concentrations of individual alleles that contribute to common phenotypes.

4. MARKER ASSOCIATIONS

Chapter 11 covered the prominence of HLA haplotypes in certain conditions (see Table 11.1). Although the cell-surface antigens themselves may not specifically cause the disease (as noted in Chapter 11, all people with HLA-B27 do not get ankylosing spondylitis, for example), these relationships are statistically significant. They also can be helpful in terms of diagnosis, prognosis, and treatment.

B. MODELS

There are different perspectives on explaining the potentially very complex relationships between different alleles of multiple genes and environmental factors to produce a range of phenotypes. As alluded to above, however, many of the common phenotypes—hypertension, diabetes, longevity, height, obesity—are pathophysiologically vague. There are undoubtedly multiple ways to affect these poorly understood clinical endpoints. We will consider two systems to help organize thoughts in this area, but other approaches also can be envisioned.

1. THE MULTIPLE ADDITIVE LOCUS MODEL

First formulated by Sewall Wright, this approach views phenotypes as the sum of the values of the contributing alleles. For any locus with two alleles x and y there exist three possible situations—xx, xy, and yy—and for n loci there are 3^n genotypes. Interestingly, at even relatively small values of n, the distribution of genotype values resembles the sort of continuous distribution encountered in studies of quantitative traits (see Figure 12.2).

This model provides considerable comfort to those who try to reconcile a broad distribution of phenotypes (in terms of severity, age of onset, etc.) with multiple alleles. Clearly, however, once several loci are involved, distinguishing individual contributions can be difficult. Furthermore, distributions of similar shape can be found for different population groups even though very different allelic (and even genetic) contributions may be involved. Thus, identifying alleles with relatively large effects in an Asian population may say little about the critical loci (or alleles) affecting the phenotype distribution in Finland. The idea persists, however, that alleles of sufficient phenotypic consequence can be identified to explain disease predispositions in important segments of the population.

2. THE THRESHOLD MODEL

This model (also called the "discontinuous model") can be represented as a continuous curve, with affected individuals appearing only at one end of the distribution (see Figure 12.3). The point at which a combination of genetic and environmental factors causes the appearance of the trait can be considered a threshold. Because relatives of affected

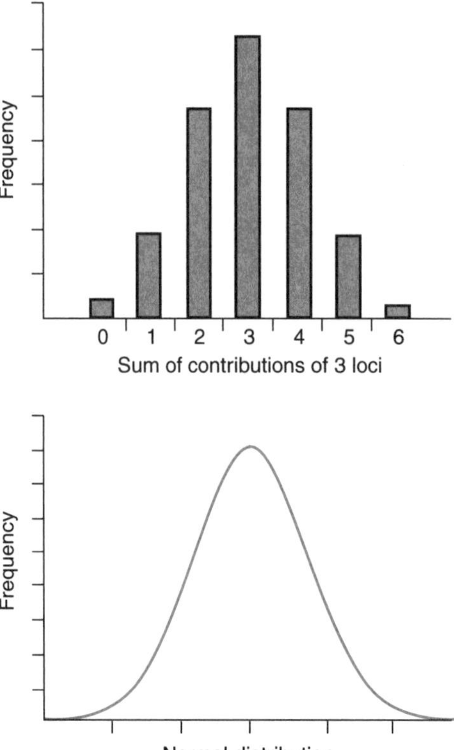

Figure 12.2 *Distributions for a multifactorial trait consistent with the multiple additive locus model. The additive interaction of only three independent diallelic loci gives the distribution of phenotypes shown in the upper panel. The lower panel shows a normal distribution with the same mean and variance. Contributions of additional loci can make the distributions appear even more similar. (Adapted from Weiss, Genetic Variation and Human Disease, 1993, p 102.)*

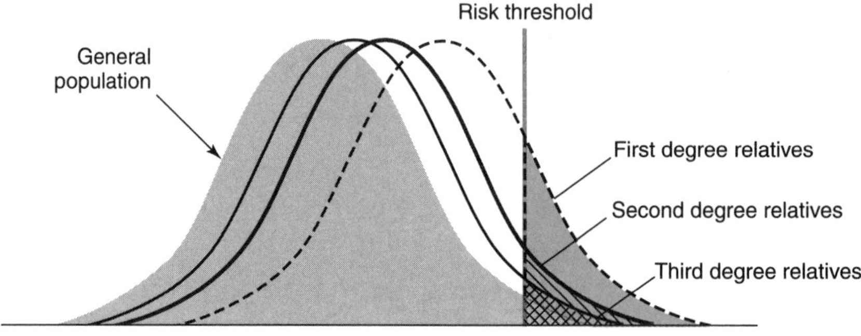

Figure 12.3 *Distributions for a multifactorial trait consistent with the threshold model. The dotted line indicates the threshold, and the distributions for different groups of relatives are shown. The darker shaded areas indicate the expected risk for clinical manifestations. (Adapted from Carter, Hosp Pract 5:45, 1970.)*

individuals share alleles (recall Table 12.1), one can anticipate that the distribution of their liability for the same trait will be shifted (to the right, in the conventions of Figure 12.3) as shown. The degree of displacement of the distribution should be related to the proportion of genes held in common with the proband. As implied by Figure 12.3 and shown in Figure 12.4, the risk falls off sharply for third-degree and higher relatives. Similar models have been applied to traits such as cleft lip and/or palate and pyloric stenosis. The prevalence equals the fraction of individuals to the right of the threshold.

This type of model leads to important predictions, some of which are not immediately obvious:

- **Different families can be expected to have different risks of recurrence.** Because families differ in their genetic complements, their combinations of susceptibilities likely will differ for different conditions. Thus, although the recurrence risks for congenital heart disease are known for the general population (see Table 9.7), individual family groups may have different risks.
- **Having more than one relative affected raises individual risk**. The notion underlying this is that increased risk will have been identified by the presence of multiple manifesting related individuals. Such increases have been estimated based on population studies; several are shown in Table 12.5.
- **Increased risk with parental consanguinity.** Even though a defined Mendelian pattern of transmission is not present, consanguinity

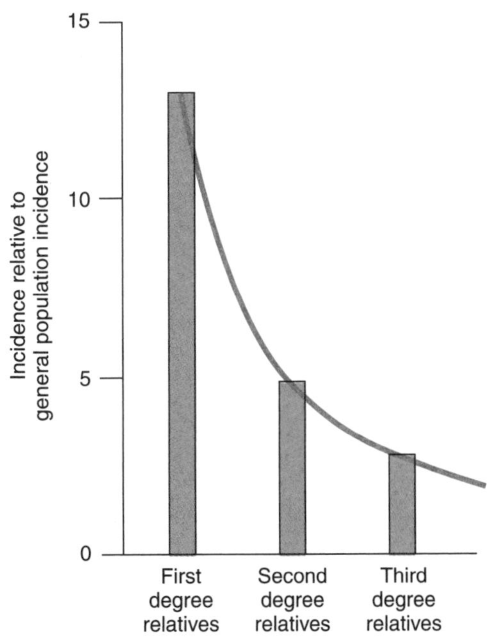

Figure 12.4 *This graph shows the relative darkened areas for the different degrees of relatedness shown in Figure 12.3. Note the rapid fall in risk according to this model. (Adapted from Carter, Hosp Pract 5:45, 1970.)*

Table 12.5 ▶ Congenital malformations in families*

Congenital finding	General population incidence	Incidence relative to general population Monozygotic twins	1st-degree relatives	2d-degree relatives	3d-degree relatives
Cleft lip, with or without cleft palate	~1/1000	400X	40X	7X	3X
Clubfoot	~1/1000	300X	25X	5X	2X
Neural tube defects	~1/500		8X		2X
Hip dislocation (♀)	~1/500	200X	25X	3X	2X
Pyloric stenosis (♂)	~1/200	80X	10X	5X	1.5X

* Adapted from Carter, *Br Med Bull* 25:52, 1969; Smith & Aase, *J Pediatr* 76:653, 1970.

increases the number of gene alleles in common, raising the likelihood of their contributing to a phenotype.

- **More severe changes imply a greater risk.** Again, this implies an increase in underlying accumulations of gene alleles contributing to the phenotype.
- **Relatives of an individual with a very rare problem have a higher risk than they would if the problem were relatively common.** A common problem is assumed, with this model, to require fewer unique genetic contributions for its appearance. Hence, multiple gene alleles may predispose a person to the same clinical end point. Rarer problems can reflect the additive effects of relatively rare alleles that might be concentrated in families, raising the risk for close relatives.
- **If the sex distribution of a trait is not uniform, relatives of an affected individual of the *less*** frequently manifesting sex are ***more*** likely to have problems. The model predicts that the presence of a larger contribution of risk-imparting alleles leads to manifestations in an individual of the less frequently involved sex; thus, relatives would have a greater chance of receiving alleles that would increase their own risk (see Table 12.5).

C. CANCER

Important notions of the biology of neoplastic changes were introduced in Chapter 4. In addition, we already have considered conditions such as neurofibromatosis (see Chapter 5), for which neoplasia is a known complication. In such cases, however, the underlying diagnosis of VRNF is usually so obvious that it justifies suspicion and preemptive care for tumors. Furthermore, other situations involving characteristic recurrent tumor phenotypes are well recognized by specialists. These include such conditions as familial adenomatous polyposis (OMIM #175100), Peutz-Jeghers syndrome (OMIM #175200), multiple endocrine neoplasia type I (MEN I) (OMIM #131100) and type II (MEN II) (OMIM #171400), acoustic neuroma (NF2) (OMIM #101000), and basal cell nevus (OMIM #109400) syndromes, as well as xeroderma pigmentosa (OMIM #278700), immune deficiencies (see Chapter 11), and certain inherited anemias.

More frequently encountered, however, are familial aggregations that often do not meet obvious mendelian criteria but nevertheless are difficult to ignore. At the outset, it is important to emphasize two things. First, "cancer" is not a single diagnostic entity. Second, cancer is frequent in the population. Thus, simply on the basis of chance, even without genetic predisposition, one can expect to encounter families in which multiple individuals have had cancer. This complicates the job of both the epidemiologist and the physician, because it is necessary to establish whether there is a predisposition in a given kindred or whether one might be seeing an example of uncomplicated "bad luck."

Some assistance in interpreting the cancer data has been provided by analysis of large numbers of extended families, in which cancer has affected many individuals. Such so-called cancer families often have proved enlightening. In many such kindreds, the diagnosis of a specific type of cancer cannot in itself explain the entire picture. However, adding the complete data from all family members can increase the likelihood that a mendelian contribution will be found.

The left side of Figure 12.5 shows a family reporting only colon cancer on a questionnaire. The presence of more than one affected individual, while interesting, is not particularly unusual. In contrast,

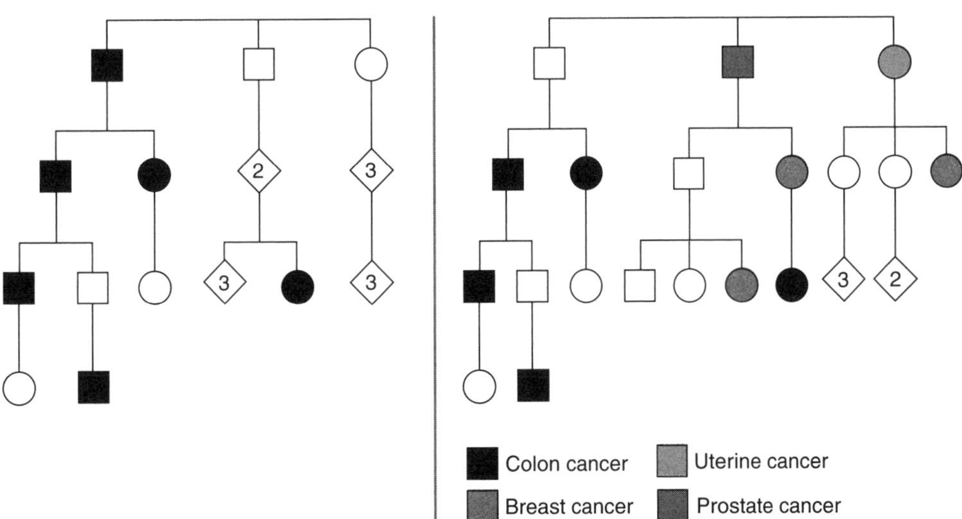

Figure 12.5 *The left side of this figure shows information on a kindred in which colon cancer is frequent. Although this is impressive, further ascertainment showed the pattern on the right. Considering the entire kindred adds a great deal to the original information and leads to the suspicion of an underlying general predisposition.*

the right side of Figure 12.5 shows the same family with complete ascertainment. Here, a predisposition to various types of neoplasia must be more strongly suspected.

An example of this sort of kindred for which a molecular explanation exists can be shown by examples of the rare Li-Fraumeni cancer syndrome (OMIM #151623) (see Figure 12.6). A mutant allele of the p53 gene can be traced through this kindred. The central gatekeeping function of the p53 protein in controlling cell division provides a pathophysiologic explanation for at least some of the predisposition to cancer in family members with the mutant allele. The Li-Fraumeni cancer syndrome is quite rare, however, and inherited mutations in the p53 gene are uncommon. *Acquired* mutations in this gene are relatively frequent, however; they have been found in half of the tumors examined. Such acquired mutations are important in the biology of the tumors (see Chapter 4) but obviously reflect a different situation from the perspective of mendelian inheritance.

Breast cancer is the most common form of cancer in women; it will develop in roughly one in nine women during their lifetimes. The risk of developing breast cancer is not uniform, however, and an increased likelihood in women in certain families has been recognized for many years. This recognition led to guidelines for mammographic screening that considered women in such kindreds to be those most likely to benefit from early and regular mammography. The intensity of work in this area and the remarkable frequency of the problem has led to the identification of two genetic loci: BRCA1 (OMIM #113705) and BRCA2 (OMIM #600185). These were detected by finding mutant

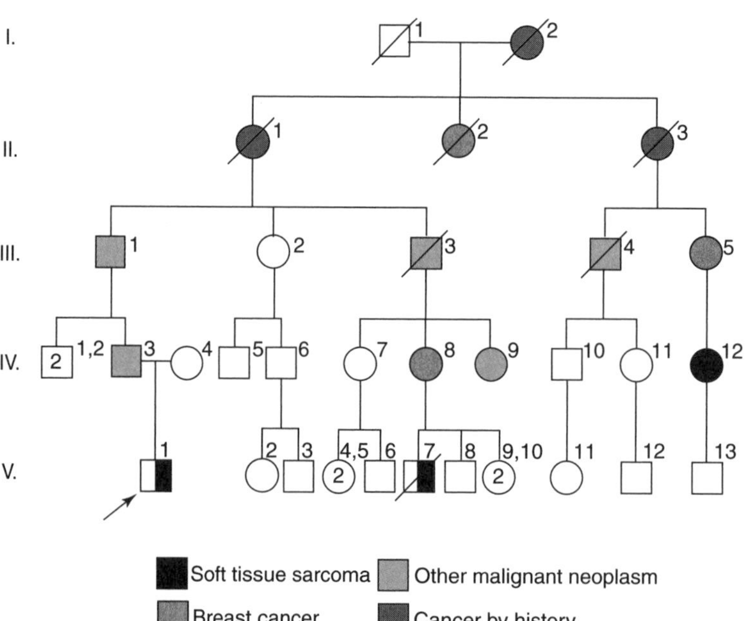

Figure 12.6 *Li-Fraumeni cancer syndrome kindred. A mutation in the gene for p53 was found to segregate with the predisposition to neoplasia. (Modified from Li and Fraumeni, Ann Intern Med 71:747, 1969.)*

alleles in families with dominant transmission of susceptibility to breast cancer. Such kindreds are rare. While this does not mean that the data are not useful, it is too early to generalize about all of the implications of alleles of BRCA1 and BRCA2.

The fact that alleles and mutations of BRCA1 and BRCA2 can be detected must be viewed in light of the impressive frequency of breast cancer in general and of the threat that it presents to women at risk. This has led to an unusual but not unprecedented clinical conundrum. The urgency, prominence, and high frequency of breast cancer have led to the desire to perform large population studies to identify carriers of mutant alleles of BRCA1 or BRCA2. Such screening, however, necessarily involves a heterogeneous population. While some individuals, most likely a small fraction, come from kindreds in which mendelian transmission is suggested, others may fall into the category of bad luck. Because of the complex nature of the mutations that accumulate during the development of neoplastic cells (see Chapter 4), it is hardly surprising that a complex mix of genetic predispositions and environmental factors can lead to cancer in a given patient and that either of the models presented above might be supported. The relative contribution(s) and consequences of a mutation or simply a polymorphism in the BRCA1 and BRCA2 genes in a given individual must be interpreted against this background.

Unfortunately, not all supporting family data are consistently available. Furthermore, the clinical decisions to be made on the basis of detected gene changes are not always clear. What is the appropriate management for an individual who is found to have a given mutation but who lacks an incriminating family history? What is an appropriate management recommendation for an individual with a positive family history? Are there ways of determining which families or which associated alleles more strongly determine or are associated with the development of malignancy?

An example of the confusion regarding these implications can be appreciated by considering a recent report showing that 6% of nurse practitioners, 28% of nonsurgeon physicians, and 50% of surgeons would recommend prophylactic double mastectomy to women identified as having a genetic predisposition to breast cancer. One-third of the members of the same groups would test a 13-year-old girl for the gene mutations, even though no proven preventive measures exist. Four percent of women said they would abort a female fetus with a breast cancer gene mutation, and 9% of physicians said that they would discuss the possibility with a patient (*Gene Lett* 2:1, 1997).

Observations such as this indicate that a complex set of questions must be considered in offering testing, counseling, and clinical advice. Current data do not support indiscriminate use of such tests in the absence of careful counseling and clinical follow-up. Only as the relative values of these and related tests are established will management be adjusted to the appropriate risks. This is the current position of the American Society of Human Genetics.

D. HYPERTENSION AND VASCULAR DISEASE

Elevated blood pressure is one of the most common findings in medical practice. Clearly, it is a condition with many contributing causes. Some of these can be eliminated with standard management, which includes reducing dietary salt and alcohol and increasing exercise. Even after these exogenous factors are controlled, however, there remains a range of blood pressures and evidence for familial aggregations (recall Figure 12.1). Before the latter can be understood, however, it is important to remember that hypertension is common and widely distributed. In addition, it is not unusual to have more than one affected person in a family. This is similar to the situation with cancer, as described above. Nevertheless, family clustering has important implications, and, because of the broad spectrum of hypertension and its medical importance, efforts have been made to distinguish different types.

An important group of individuals with hypertension is distinguished by low plasma renin activity. This is seen more frequently in black individuals. Another 15% of people who otherwise meet the criteria for "essential hypertension" (OMIM #145500) have high plasma renin activity. Members of these two groups respond differently to treatment. Despite these relatively distinguishable features, no clear basis for the renin levels is known, and the general category of "essential hypertension" has been used to describe both groups.

Any condition that has known variation correlating with age, race, sex, smoking, alcohol use, weight, cholesterol levels, and diabetic status (to consider only the most prominent) clearly represents an end point of multiple interacting factors. In addition, it has been particularly difficult to identify unequivocal genetic contributions in humans. However, some interesting and instructive studies have been performed in rats. Specific rat strains are consistently hypertensive; others are not. Breeding experiments can involve a sufficient number of animals to permit genetic linkage analysis. Thus, it has been shown that the interaction of as few as two different genetic loci, one on rat chromosome 10, can account for most of the variance in blood pressure between members of the two strains (recall the multiple additive locus model).

The variations among animals (and, presumably, humans) are generally not "all or none" phenomena. Thus, relatively minor differences in the contributions of individual gene products that affect salt metabolism, transport across membranes, vascular tone, or blood vessel structure (to suggest only a few) can be expected to affect blood pressure. At this time, there are few absolute indications for using different clinical management on the basis of these population variations. It is likely, however, as will be discussed below, that additional definition of responsible gene variations will be able to identify populations particularly likely to respond to selected agents, either currently available or to be developed.

The more general problem of vascular disease also has had considerable study. Several references have been provided for additional details, but two areas have been particularly instructive. In the first, inherited variations in lipoproteins and cholesterol transport and metabolism have been associated with specific forms of atherosclerosis. Perhaps the most dramatic example of this has come from the valuable studies of familial hypercholesterolemia (OMIM #143890). Multiple genetic defects have been associated with this phenotype. Another example is homocystinuria (OMIM #236200), a rare recessive disorder associated with premature vascular disease. Although individuals homozygous for mutant alleles in both of these conditions are rare and usually obvious, it is again important to recall the Hardy-Weinberg relationship and to realize that heterozygotes for these alleles may have much more subtle signs. This has led to identifying elevated blood levels of homocysteine and (even more obviously) cholesterol as independent risk factors for vascular disease; such measurements can now be considered a valuable part of screening for cardiac risk reduction.

E. DIABETES

Diabetes is frequent, affecting as many as 2% of the population. The broad distribution of diabetes and its important clinical consequences have led to extensive efforts to understand genetic contributions. The first significant aspect of diabetes in this regard is that it exists in several different forms. There is clearly a difference, clinically and physiologically, between insulin-dependent diabetes mellitus (IDDM) (OMIM #222100) and non-insulin-dependent diabetes mellitus (NIDDM) (OMIM #125853). There also exists another category, referred to as maturity-onset diabetes in the young (MODY). Within this category, two types have been associated with specific molecular defects. MODY I (OMIM #125850) is related to defects in hepatocyte nuclear factor 4α, on chromosome 20. MODY II (OMIM #125851) is associated with alleles of the glucokinase gene, on chromosome 7. Many other genes involved in glucose metabolism, including transport and uptake, have been studied with respect to NIDDM. This is a particularly prominent area of study, because concordance rates for NIDDM in identical twins approach 100%.

By contrast, IDDM generally develops in younger patients and has a susceptibility locus on chromosome 6 (recall the HLA-DR3 and HLA-DR4 association discussed in Chapter 11). This is clearly not the entire explanation for this condition, however, and an allele in the 5′ region of the insulin gene also shows linkage with IDDM susceptibility.

The destruction of pancreatic beta cells in IDDM develops rather rapidly, and at least some of its features are consistent with a viral infection and specific target organ destruction. Genetic variations affecting susceptibility to such an infection are likely to be important determinants of the phenotype.

It is important to remember that the glucose intolerance characteristic of diabetes also can be a sequela to other conditions. For example, hemochromatosis (OMIM #235200), a Mendelian condition involving chronic iron storage, leads to pancreatic damage and secondary diabetes in addition to other problems. An important distinction here is that using phlebotomy to remove erythrocytes and the iron they contain is a very different form of treatment than using insulin or oral hypoglycemic agents.

The remarkable clinical differences among IDDM, NIDDM, MODY, and diabetes secondary to hemochromatosis have led to entirely different treatment strategies, based on the very different underlying physiologic problems. Additional genetic dissection likely will identify individuals who have a higher or lower likelihood of responding to particular treatments and for whom appropriate prophylactic and therapeutic measures may differ.

F. INFECTIOUS DISEASES

Infectious diseases represent the interaction of the host with exogenous agents—viral, bacterial, fungal, or parasitic. It is not difficult to imagine that many differences would affect not only the initial infection but the clinical outcome. Furthermore, in some populations where endemic infectious agents (for example, those for schistosomiasis and malaria) are plentiful, genetic resistance may have arisen. Studies of large populations have begun to elucidate the basis for variations in infectious disease susceptibility.

In some African populations, 25% of individuals have died from falciparum malaria in each generation. Invasion of the erythrocyte by the malaria parasite is central to the infection. The Duffy blood group antigen is essential to this process. Vivax malaria cannot infect individuals who do not express the Duffy antigen; such individuals are thus genetically resistant (OMIM #110700).

The importance of the sickle-cell trait in malaria resistance already has been noted in Chapter 6. It is the most prominent example of population variation in infectious disease susceptibility. In individuals with the sickle-cell trait, the variation is not at the level of binding of the parasite to the erythrocyte, as with the Duffy blood type. Rather, carriers for the sickle-cell trait present an inhospitable intraerythrocytic environment for parasite survival.

Another example of a common endemic illness is schistosomiasis. It is possible to evaluate entire village populations in which the exposure to the agent is uniform. Studies of such populations have shown that of 100 individuals challenged with schistosomes, 20 to 30 will develop a heavy infection on the basis of egg counts; another 15 will develop severe disease, but under 10 will have a lethal outcome. This variance cannot be explained by age or gender. Furthermore, high or low susceptibility to schistosomiasis infection represents a stable family phenotype. Segregation analysis has shown that this variation can be explained by two alleles of a single major gene in the 5q31-33 region (OMIM #181460).

Tuberculosis is another common infectious disease. It has a complex pathophysiology, and latent tuberculosis develops in only some hosts. Family studies have shown important variations in susceptibility. In an isolated family from Malta, four children who shared a common ancestor (and, hence, were consanguineous) developed related but different mycobacterial infections. These children were not immunodeficient. Linkage analysis showed a relationship of this susceptibility to an allele on the long arm of chromosome 6. A mutation in the stop codon for the IFN-γ receptor, part of the Class III immune response group described in Chapter 11, appears to explain this susceptibility (OMIM #209950).

Another example of tuberculosis and mycobacterial susceptibility has been found in European populations in places where vaccination with *Bacille Calmette-Guérin* (BCG) is used for children. A very low incidence (fewer than 1 in 10^6) of disseminated BCG infection can be expected after large-scale vaccination programs are carried out. It is important to note that over 30% of individuals who develop the disseminated infection show consanguinity. In three different kindreds with this susceptibility, a single base pair deletion in the IFN-γ receptor gene also was identified. Several different mutant alleles have been identified in these kindreds. These two different sets of observations and alleles indicate that the IFN-γ receptor helps determine susceptibility to mycobacterial infection in humans.

Studies of genetic susceptibility to infections in humans are counterparts of studies performed in animals. Mice, in particular, have been very valuable study organisms because of distinct genetic variations among inbred strains. Studies in mice also have permitted evaluation of the contributions of multiple susceptibility genes and have shown that the effects can be additive in some cases.

Humans with inherited immunodeficiencies often present with profound infection susceptibilities early and throughout their lives (see Chapter 11 and Table 11.3). Such individuals are very rare. However, when considering such rare homozygotes, it is important to recall the principles of Hardy-Weinberg equilibrium as described in Chapter 6. These indicate that the rarity of individuals homozygous for certain alleles can mask a far more common distribution of heterozygotes in the population. Because not all of these genes and their alleles have been identified and certainly not all are tested for in routine clinical practice, it is not known how they can contribute to individual susceptibilities to a wide range of human pathogens. Studies on isolated populations and the search for parallels to animal systems already have identified underlying susceptibility genes, however, and certainly will continue to do so.

Evaluating individual susceptibilities in outbred populations is obviously more complicated. Here it is difficult to begin with a set of population-limiting assumptions and the notion of screening for susceptibility markers has potential relevance, as will be discussed in Chapter 13. When the potential for different individual susceptibilities to different infectious agents is added to the considerations of Chapter 10 with respect to individual drug sensitivity differences, it is obvious that the care of individuals with infectious diseases from the perspectives of both prevention and treatment needs to take genetic factors into consideration.

STUDY QUESTIONS

1 As a member of the Public Health Service, you have been assigned to the care of an isolated tribe of native Americans. You are impressed with the number of cases of pneumonia you have encountered in the children in the area. You suggest:

- a Sanitary conditions need improvement.
- b Children should take prophylactic penicillin.
- c A dominant trait may underlie these findings.
- d A recessive trait may underlie these findings.
- e Tests should be performed to detect consanguinity.

2 You are working in the health clinic of a large university. In the first semester, two students have presented with symptoms of stroke. Although the manifestations resolved, you are concerned at this atypical situation. You suggest:

- a Measurement of plasma renin levels in a group of students.
- b Conferring with the cafeteria to assess the salt content of the food.
- c Closing the two fast food restaurants across from the main entrance to the campus.
- d Examining the histories of the two individuals.
- e Treating everyone with aspirin.

3 A 20-year-old woman has come to you for advice. She is an Ashkenazi Jew and talked a graduate student friend into performing tests for BRCA1 and BRCA2 mutations. She was found to be heterozygous for a recognized mutation in BRCA2. Her family history is negative for cancer over three generations. What do you tell her?

4 A 24-year-old man comes to you and tells you that he has hemochromatosis and already is receiving treatment and is doing well. His particular concern

is that his father also has hemochromatosis (and has responded similarly well to treatment) and also has ankylosing spondylitis. What do you tell your patient?

a These conditions are not pathophysiologically related, so he should be at low risk for having both.
b His mother must be at least a carrier for hemochromatosis.
c He should have spinal x-rays.
d He should have a test for HLA-B27.

FURTHER READING

Breast Cancer Linkage Consortium. Pathology of familial breast cancer: Differences between breast cancer in carriers of BRCA1 and BRCA2 mutations and sporadic cases. *Lancet* 349:1505, 1997.

Breslow JL. Lipoprotein transport gene abnormalities underlying coronary heart disease susceptibility. *Annu Rev Genet* 42:357, 1991.

Carter CO. Genetics of common disorders. *Br Med Bull* 25:52, 1969.

Carter CO. Multifactorial genetic disease. *Hosp Pract* 5:45, 1970.

Hadley TJ and Peiper SC. From malaria to chemokine receptor: The emerging physiologic role of the Duffy blood group antigen. *Blood* 89:3077, 1997.

Hamilton M et al. The etiology of essential hypertension. 4. The role of inheritance. *Clin Sci* 13:273, 1954.

Hobbs JL et al. The LDL receptor locus in familial hypercholesterolemia: Mutation analysis of a membrane protein. *Rev Genet* 24:133, 1990.

Jacob HJ et al. Genetic mapping of a gene causing hypertension in the stroke-prone spontaneously hypertensive rat. *Cell* 67:213, 1991.

Jouanguy WE et al. Partial interferon-gamma receptor defect in a child with tuberculoid bacillus Calmette-Guérin infection and a sibling with clinical tuberculosis. *J Clin Invest* 100:2658, 1997.

Kluijtmans LAJ et al. Molecular genetic analysis in mild hyperhomocysteinenia: A common mutation in the methylenetetrahydrofolate reductase gene is a genetic risk factor for cardiovascular disease. *Am J Hum Genet* 58:35, 1996.

Krainer M et al. Differential contributions of BRCA1 and BRCA2 to early onset breast cancer. *N Engl J Med* 336:1416, 1997.

Marenberg ME et al. Genetic susceptibility to death from coronary heart disease in a study of twins. *N Engl J Med* 330:1041, 1994.

Marquet S et al. Genetic localization of a locus controlling the intensity of infection by *Schistosoma mansoni* on chromosome 5q31-q33. *Nat Genet* 14:181, 1996.

Risch N. Linkage strategies for genetically complex traits. 1. Multilocus models. *Am J Hum Genet* 46:222, 1990.

Smith DW and Aase JM. Polygenic inheritance of certain common malformations. *J Pediatr* 76:653, 1970.

Struewing JP et al. The risk of cancer associated with specific mutations of BRCA1 and BRCA2 among Ashkenasi Jews. *N Engl J Med* 336:1401, 1997.

Weiss KM. *Genetic Variation and Human Disease.* Cambridge, England, Cambridge University Press, 1993.

Welch GN and Loscalzo J. Mechanisms of disease: Homocysteine and atherothrombosis. *N Engl J Med* 338:1042, 1998.

Wright S. An analysis of variability in number of digits in an inbred strain of guinea pigs. *Genetics* 19:506, 1934.

USEFUL WEB SITES

THE GENE LETTER

http://www.geneletter.org
Newsletter about genetics and potential policy questions. Useful and timely summary articles. Chat opportunities. Links to other sites.

NATIONAL HUMAN GENOME RESEARCH INSTITUTE HOME PAGE

http://www.nhgri.nih.gov
Review of basic genetics. Genetic maps. Discussion of "How to conquer a genetic disease." Background on Human Genome Project. Discussion of ethical and social issues.

HOWARD HUGHES MEDICAL INSTITUTE

http://www.hhmi.org
Summaries of sponsored research in genetics. Timely news updates. Useful section on structural biology.

CHAPTER 13

Integrating Genetic Knowledge into Medical Practice

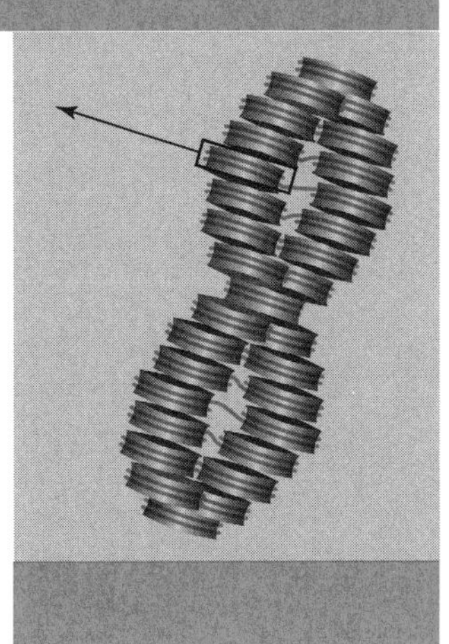

The enormous amount of information currently available and being developed regarding mendelian inheritance and the influence of genetics on clinical medicine must be connected to the care of individuals and their health. While the most important consideration in attempting to integrate genetic information into medical practice is thinking about it in the first place, it also is essential to be able to have efficient access to this growing body of information. Historically, this has been difficult; the broad distribution of disciplines contributing to knowledge about specific genetic conditions, as well as more general aspects of genetic influences, has not been integrated. Individual specialties and sometimes even single laboratories have concentrated on limited areas. Furthermore, large and cumbersome compendia always present formidable barriers to easy access.

It is important to remember that genetic information is intrinsically complex. With the potential ramifications of changes at one or more positions in 3×10^9 nucleotides, simplicity cannot be anticipated. Nevertheless, emerging information management strategies hold considerable promise for increasing both the access to and the usefulness of genetic information. As such strategies develop, they also will permit wider access to relevant literature in order to assist detailed study and decision making.

A. THINKING ABOUT A GENETIC EXPLANATION

A family history always is an important part of a complete medical evaluation, and any evidence for disorders in relatives is valuable. As discussed earlier, however, Mendelian problems may present *without* obvious antecedents. In such situations, it is important to maintain suspicion. Suspicion should be heightened in the presence of particular aspects of the presentation or background. Several of these features are summarized in Table 13.1.

Table 13.1 provides only a basic list. It also is important to recall the principal features of specific types of inheritance patterns presented in Chapters 5–8. In disorders with these patterns, recognizing variable expressivity can be particularly valuable. For many dominant disorders, obtaining an accurate family history can be challenging because different members with the *same* basic mutation can have different manifestations (recall Marfan syndrome). Mitochondrial inheritance patterns are different from those related to nuclear genes; in addition, they are complicated by different ages of onset and variable severity and progression in different individuals.

When suspicion regarding a genetic contribution(s) to a particular problem is present, the next step involves knowing where to turn for additional information and help. Suggestions are presented below.

Table 13.1 ▶ **Bases for heightened suspicion about genetic disease**

Family History
- Ethnic group
- Consanguinity
- Individuals with illness similar to those of patient
- Health problems in multiple relatives, particularly if of early or atypical onset
- Reproductive difficulties, habitual abortion, neonatal deaths, infertility
- Delayed puberty, mental retardation, congenital malformation(s), neurologic or muscular disease(s)

Personal History
- Age of parents (particularly if older)
- Problems with wound healing or bleeding
- Drug reactions
- Poor health in childhood
- Special diet(s)
- Fertility

Physical Examination
- Major malformation(s)
- Unusual stature
- Developmental delay
- Retardation
- Dysmorphic features
- Abnormal sexual development

Present Illness
- Atypical presentation (age of onset, severity, management difficulties) of a common disorder
- Exposure to teratogen(s)

B. GENETIC DATABASES AND THEIR USE

The notion of a listing for all conditions with recognized mendelian inheritance is appealing. Based on the broad range of information that now must be integrated into such a listing, however, it is clear that complexity is a concern. It is still possible to approach many conditions through books and collected articles addressing those conditions specifically; some of these have been introduced as references in earlier chapters. Many of these collections, monographs, etc., already exist and are particularly useful for health practitioners, as well as for patients and their families. Nevertheless, such collections (see Further Reading) generally emphasize the most common inherited and congenital problems and, sometimes, those peculiar to or concentrated in particular populations. Less frequently encountered conditions often do not evoke sufficient general interest to justify independent volumes.

Fortunately, the assembling of information regarding inherited conditions can build on a useful catalog established over 30 years ago. With the first publication of *Mendelian Inheritance in Man* (MIM), a framework was established for organizing information about genetic traits and conditions. This catalog has had many printed revisions, but recent years have clearly shown the inadequacy and untimeliness of such an approach. Accordingly, the National Center for Biotechnology Information (NCBI) has developed an online version of this catalog, known as "Online Mendelian Inheritance in Man" (OMIM). This database provides a continuously updated collection of much information relevant to the diagnosis, management, and investigation of inherited diseases. Placing this material online also has permitted the use of an integrating database approach in order to link information regarding clinical features, diagnostic approaches, gene structure, mutation analysis, protein sequences, reference material from Medline, etc. OMIM identification numbers have been used throughout this book.

Access to OMIM is available through its Web site (see Useful Web Sites for the URLs of this and other relevant Web sites, with comments). One can search this database using several different approaches. Several representative screens are shown in Figure 13.1. The broad base of information now available for a given trait or condition can be accessed through appropriate database links as indicated. Descriptions of individual problems are available, as are comments on clinical features, medical management, biochemical pathways, and relevant molecular biologic and chromosomal details. Reference lists have links to Medline, enabling users to obtain abstracts directly. Because this catalog database undergoes constant editorial revision, and because its image presentation has improved, it is becoming possible to obtain images based on clinical photographs for comparison with findings in problems of interest. Such images are available on other sites as well (see Useful Web Sites).

This database has made available a remarkable amount of timely

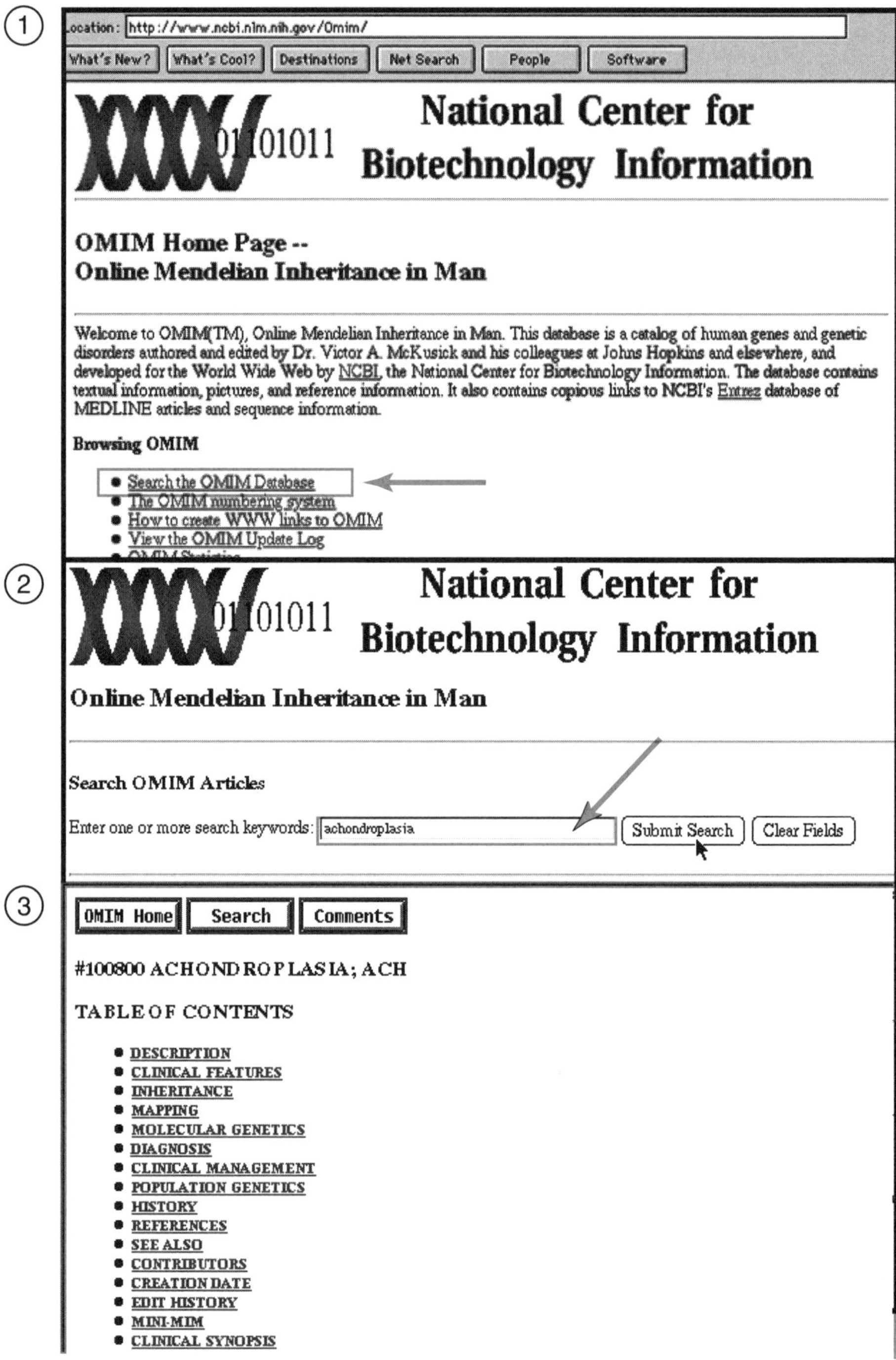

Figure 13.1 *Several pages from the (OMIM) Web site. Its organization permits rapid transfer between clinical information, references, DNA database entries, and cross-references. Beginning at the home page, a search for information about achondroplasia (OMIM #100800) (see Chapter 5) is illustrated.*

information to anyone with online access. The alternative to online searching of this database is to use a CD-ROM, updated quarterly. The CD-ROM version has the advantages of portability, convenience, and unlimited access without delay or the need for Internet connections.

Other databases can be reached independently, without using OMIM. However, many of these are oriented toward more specialized interests rather than clinical applications. These include the genome database (GDB), the gene map, and protein databases. Fortunately, many can be reached through OMIM as well, so that it is possible to use OMIM as the basic access site.

While the immediate supply of detailed information about various mendelian conditions is a valuable resource, it does not always address the concerns of the clinician faced with an unusual situation. Here, one often must be more resourceful. It is particularly important when confronting an individual with multiple changes and anomalies or a perplexing biochemical or developmental abnormality to know where to begin.

Often the first approach is made through specialized texts. For instance, cardiology textbooks have sections with detailed information regarding cardiac malformations and their relationships to other syndromes. A textbook of neurology will have extensive information regarding neurologic and myopathic syndromes, along with clinical and laboratory tests that are helpful in establishing a specific diagnosis. Large compendia of information regarding birth defects are available as catalogues and references, as well as online (see Useful Web Sites). The three-volume book *The Metabolic and Molecular Bases of Inherited Disease* is useful in itself; it also is available as a CD-ROM. Such a large collection offers access to detailed genetic information and remarkably thorough discussions of individual metabolic problems. OMIM includes lists of comparative clinical features for many of these disorders.

Still another useful set of databases is based on the cumulative experience of genetic and metabolic diagnostic testing laboratories. Here, information derived from past testing experience often can be integrated rapidly into the laboratories' reference bases to provide a reasonable list of diagnostic possibilities to explain a specific result, as well as suggestions for further metabolic or genetic testing. It is important to identify institutional, regional, and/or commercial laboratories for testing and advice.

Useful information regarding human chromosome anomalies is provided by both texts and databases of chromosome changes (see Useful Web Sites). Some of these are accessible through OMIM. These are usually most helpful when used in conjunction with karyotype reports so that references can be generated quickly relating a specific chromosome finding to appropriate literature or other diagnostic studies. It should be obvious that the analysis of physical chromosome changes is a large-scale parallel to the analysis of multiple individual gene variations. Chromosomal changes, as discussed earlier, frequently involve complex clinical presentations, because of the likelihood that multiple genes are involved in the region of a single visible change. Nevertheless, the convergence of large-scale chromosomal and more localized molecular information is inevitable and obviously will lead to a more useful and integrated picture of genetic changes and their consequences, as discussed in Chapters 1–3. For example, it is difficult to discuss the visible chromosome changes underlying Prader-Willi and Angelman syndromes without a knowledge of the detailed molecular biology of imprinting in the affected gene loci. As more information becomes available, such relationships will become more obvious and more useful.

C. IMPLICATIONS OF INDIVIDUAL GENETIC VARIATIONS

Chapter 1 discussed mutations and their frequency. Even after the valuable activity of correction mechanisms that are part of DNA replication and repair, the frequency of genetic changes is on the order of 10^{-10} per replication event. Although this is a low number, the total number of cells in a mature human is on the order of 10^{15}. Thus, the opportunity to have multiple genetic differences among cells (and their lineages) in mature individuals is substantial. Chapter 1 also indicated that a number of mechanisms, in addition to repair enzymes, minimize the consequences of these differences. Such mechanisms include the use of redundant codons that accommodate base changes without causing amino acid changes, the presence of noncoding regions of DNA (both introns and inter-genic regions), and the ability to tolerate many amino acid replacements in proteins without their causing notable biologic effects (called "conservative changes"). Some acquired differences may be obvious, however, even though they are rare. For instance, the development of café-au-lait spots and neurofibromas in a body segment, such as an arm or leg, or in a single dermatome, presumably derived from a postconception mutation in the neurofibromin gene during growth and development, can be striking. Another area of particular interest is the acquisition of a mutation that confers growth instability leading to cancer cells. Such a change can occur in a pivotal caretaker or gatekeeper gene during the course of random mutation and erroneous repair. The development of some local tumors reflects mutation events, as discussed in Chapters 4 and 12. Accumulated mutations hasten the neoplastic process and enhance the aggressiveness of the tumor.

While these sorts of events are all genetic, in the sense that they involve changes in genes that can be transmitted to progeny cells, they are not necessarily heritable unless they involve cells in the germ line. Of growing clinical interest is the interaction of inherited mutations and the notion of determining the status of carriers for multiple gene changes. As discussed in Chapter 6 in regard to the Hardy-Weinberg formulation, the frequency of alleles for many recessive mutations is surprisingly high (recall Figure 6.11). It is not difficult to conclude that *any* individual is heterozygous for multiple recessive mutations that have no obvious clinical consequence. Considering such differences makes it clear that, at least to some extent, the distinction between dominant and recessive inheritance is somewhat arbitrary. The major difference is that in a dominantly inherited condition the heterozygote manifests overt difficulties. The fact that we are confronted with no obvious clinical signs reflecting heterozygosity for a recessive mutation does not mean that the mutation itself has no clinical consequences in heterozygotes, however.

An example of predisposition to risks for heterozygotes is shown in homocystinuria (OMIM #236200), discussed in Chapter 12. Homozygotes for homocystinuria are known to have difficulties with thrombotic disease. They also usually have other striking abnormalities, including growth and skeletal changes and mental retardation. The fact that those obvious clinical manifestations are absent in heterozygotes does not mean that there are no consequences for them, however. Evidence now has accumulated that elevated blood levels of homocysteine (such as are found in heterozygotes) increase an individual's risk of thrombotic disease. Thus, many years after the description of the homozygous phenotype, it is clear that carriers (previously considered asymptomatic) also are at risk for problems.

The other extreme can be encountered in individuals with achondroplasia (OMIM # 100800). As discussed in Chapter 5, heterozygotes for this autosomal dominant condition have a reasonably predictable phenotype. In contrast, homozygotes have profound skeletal and developmental changes that are usually lethal; such individuals are encountered very rarely and do not survive to reproduce.

Considering this important spectrum, which is continually being broadened by detecting the consequences of heterozygosity, we cannot assume that carrier status is without biologic liability. Currently, there is simply insufficient information to predict the extent of this liability for most heterozygotes.

If the data are hard to collect and assess for carriers of single heterozygous mutations, they currently are not even conceivable for carriers of multiple independent heterozygous mutations. As heterozygotes continue to be detected through screening programs (see below), and as specific susceptibilities or clinical characteristics become associated with individual heterozygosities, the breadth and consequences of inherited genetic variability may become more obvious. Until such information becomes available, however, the clinician, in whatever role, must remain alert to the possibility of significant individual biologic variation. The fact that such variation may never have been documented in published reports or may not produce a single identifiable metabolic or structural abnormality does not mean that a set of underlying predispositions may not have led to (or at least contributed to) a particular clinical presentation (recall the multiple additive locus model for common diseases introduced in Chapter 12). This possibility is further complicated by considering that the effects of acquired mutations may be very different in the presence of different genetic backgrounds. The opportunity to detect such events has been recognized and exploited by students of genetically isolated animals such as inbred strains of mice, in which new mutations may be detected more readily because of the underlying genetic uniformity. It is far more difficult to detect such changes in outbred populations.

D. CLINICAL USE OF BIOLOGIC PROFILES

The previous section presented the rationale for considering individual genetic variations and their consequences for human medical biology. With rare exceptions, this has not been a clinically relevant consideration in the past, because the test systems were either not available or not widely used. Thus, with relatively rare exceptions, there exists no large body of data to guide clinical decision making. The gathering of data regarding individual variations must be the basis for making informed decisions regarding potential differences in clinical biologic reactivity or susceptibility to environmental factors, pathogens, or drugs, as well as in natural history. Because such data ultimately must reflect changes at the level of the DNA sequence, several basic requirements must be met:

- There must be a consensus about the basic structure of the human genome. Ultimately, this must be resolved at the nucleotide level with a complete sequence of the human genome. This is now a foreseeable development, based on the efforts of the Human Genome Project; it will build on the framework of the linkage map of genetic markers already available. Much of this currently is coordinated through the genome data base (GDB) and may be accessed online directly or through OMIM, as discussed above (see Useful Web Sites).
- Methods must exist for detecting individual variations quickly, reliably, and inexpensively. It simply is not feasible to propose sequencing the entire genome (3×10^9 nucleotides) for everyone. Even if the sequencing could be done, the data management problems would be overwhelming. A more reasonable approach is to develop methods to study certain critical DNA regions whose variations appear to have prognostic significance. Several proposals are now being investigated to accomplish this. For example, a proprietary method for attaching many artificial polynucleotides to a matrix chip can provide an array of sequences varying at specific positions. With appropriate redundancy in the sequences on the chip, hybridization with an individual's DNA could then determine microsequence information for selected regions based on the pattern of hybridization with the fixed oligonucleotide array (see Figure 13.2).

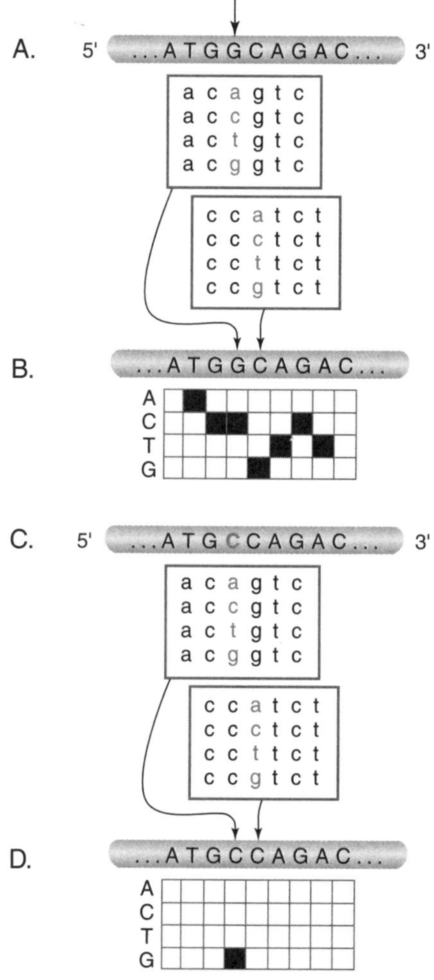

Figure 13.2 *DNA chip technology allows the attachment of multiple short polynucleotides (oligonucleotides), differing at single bases along their length, to a solid-phase support. (A) A given base sequence (shown at the top in capital letters) is tested by a panel of four complementary oligonucleotides—each varying at the same position (lower-case letters). The next base to the right in the sequence is tested by another panel of four oligonucleotides, staggered one position to the right, as shown. (B) Hybridizing the DNA sequence for which the sampling arrays (now attached to a chip matrix) were designed should give the pattern of densities shown. (C) The test DNA shown has a single base change (G→C). (D) When the test DNA is hybridized with the same array of test sequences on the matrix, the resulting pattern is remarkably different.*

Such hybridization patterns can be read optically, and computer-generated interpretation can identify all variations present. By choosing the oligonucleotide arrays to represent mutations and polymorphisms of interest, such an approach permits screening for thousands of changes in a single test. Only short oligonucleotides have been shown for simplicity; in practice, longer sequences are used. (Modified from Chee et al., Science 274:610, 1996.)

When automated, such a matrix analysis system could provide information about thousands of sites of genetic variation for an individual, establishing a profile of changes (mutations). Similarly, population variation profiles could form the basis for defining study populations and comparing clinical outcomes.

- A base of information must be developed relating thousands of individual genomic changes to each other and to their individual and collective clinical consequences.

The last requirement will be by far the most difficult to achieve. Several important questions will need to be addressed. Some of these will be considered in more detail below:

- How many individuals will need to be studied in order to provide reasonable predictive power for the information?
- What is the duration and extent of clinical information needed to arrive at reasonable predictions? Do we need 75-year follow-up studies for degenerative disorders?
- How will individuals be followed for a sufficient time to generate predictive data? (Individuals' decisions, such as moving, changing physicians, or changing health care patterns, will complicate this.) It should not be surprising that relatively small societies and those with well-integrated systems of medical care (e.g., Finland and Sweden, respectively) have provided some of the best opportunities for longitudinal population studies.
- How generalizable will the conclusions be if they are based on a limited ethnic or geographic group? Differences among populations already are recognized (see Chapter 12).
- What effects will *different* treatments have on outcomes? For example, as discussed in Chapter 12, different genetic backgrounds in individuals with "hypertension" may lead to different responses to beta blockers, ACE inhibitors, and calcium channel blockers; there already is some evidence for this in larger ethnic group studies. Will we need cohorts followed for long periods on identical regimens? Can any of these data be generated from retrospective studies?

1. SCREENING

The idea behind screening is to identify individuals at risk for a particular problem. Having identified these people, some reasonable intervention must be available in order to minimize predictable consequences. As discussed above, screening already has been useful in human genetics. The Guthrie test for PKU (OMIM # 261600) is particularly instructive. As discussed in Chapter 6, this test is designed to detect elevated levels of phenylalanine in the blood of newborn infants (see Figure 6.10). Once elevated levels are found, further detailed testing is necessary to establish the specific diagnosis. If PKU is present, dietary and medical management can minimize the potential neurologic consequences.

The simple paradigm of the Guthrie test conceals a complex set of assumptions and requirements for any effective screening program—not all of them unique to genetic problems. An ideal screening program should accomplish several things:

- As many affected individuals as possible should be identified (i.e., there should be few false negatives).
- Few unaffected individuals should be identified as affected (i.e., there should be few false positives).
- The test must be broadly applicable, simple to administer, and inexpensive.
- There must be a system for prompt and effective investigation of presumptively identified individuals.
- Long-term follow-up and treatment must be available.

Often, the idea of screening for a particular condition is considered but not instituted for one or more important reasons:

- There is no treatment for the condition if it is found.
- There is no simple, inexpensive test system.
- The population has segments differing in risk level. For example, screening for sickle-cell anemia in Caucasians or for Tay-Sachs disease (OMIM #272800) in non-Jews would each be misleading, inappropriate, and wasteful.
- The problem is so rare that the "noise level" of the test might identify more nonaffected than affected individuals (too many false positives). This also can happen when the end point of the test is difficult to establish because of a broad range of possible results. This is, of course, not the case if specific nucleotide changes are sought but such testing is not yet practical for one or more reasons, as discussed earlier.

2. COMPLEX TRAITS

As discussed in Chapter 12, complex traits present an important challenge for the clinical application of genetic principles. There are several aspects to this challenge. First, the diagnosis is often too vague. "Hypertension" and "cancer" are not single entities. In those relatively uncommon areas in which *specific* types of hypertension or cancer can be identified, genetic studies are justifiable. Colon cancer presents a particularly important example of this. If any one of several distinct familial forms is identified, specific gene-based testing can be used. For example, the gene for adenomatous polyposis coli (APC) (OMIM #175100) can be studied in individuals at risk on the basis of simple pedigree analysis. However, this is reasonable only in the setting of an *identified* affected family member. The testing can be performed considerably before the onset of symptoms, so that genetically affected persons can be offered close follow-up with regular colonoscopy and, if necessary, colectomy. The recent identification of an A → T 3902 mutation in the APC gene predisposing to colon cancer specifically in Ashkenazi Jews (OMIM #175100.0029) is an example of the success of this approach.

Identification and treatment of an individual are potential benefits of this approach. On the other hand, if the diagnosis is too vague, testing for distinct gene changes subjects the individual to anxiety, frustration, and the possibility of both false negative and false positive results. As mentioned in Chapter 12, studies of hypertensive rats have shown that discrete variations at individual (ultimately) identifiable loci can contribute substantially to the final blood pressure level. For humans, the more complicated goal is for an individual to have testing for variations at multiple markers and then have the pattern of changes compared with population patterns to assign a risk or likelihood of a certain clinical outcome.

In the past this was a formidable problem, because of the large number of possible gene changes to assay. However, as noted above, test approaches such as DNA chip techniques offer simple testing for as many as 1000 changes in a single hybridization step. Such a test can identify groups of variations in single genes as well as in chromosomal regions implicated by linkage studies. When such testing becomes generally available, the task will be to assemble the test results and generate clinically useful conclusions. The report "Assessing Genetic Risks" (1994) from the Institute of Medicine recommended distinguishing disorders for which early treatment made a difference to the health of the individual from those for which it did not. Screening for the former would be required under state law, while screening for the latter would be voluntary and not accessible unless authorized.

This approach is likely to help in the care of individuals with complex traits, but it will be particularly helpful when Mendelian features can be identified and individual clinical details can be integrated into both testing and interpretation (with prognosis and treatment). As discussed above, the lack of follow-up data will reduce the prognostic usefulness of some of these techniques for some time. Also, if anonymous polymorphic DNA markers (which will be developed first from linkage studies of determinants of complex traits—see Chapters 1 and 12) are used for testing, they will not necessarily provide information about the physiologic change(s) underlying the clinical picture. Specifically, it is unlikely that many individuals with complex traits will be homozygous and, hence, unable to synthesize an essential structural protein or perform a critical metabolic reaction (since such an inability might well be lethal). Rather, multiple subtle alterations of enzymatic activity, control of metabolic reactions, or slight changes in protein-protein interactions may join to give a final clinical picture that is unpredictable from first principles.

As emphasized earlier, the more precise the definition of a disorder, the more reliable will be the testing, prognosis, and treatment. This is true even for single-gene conditions, as has been emphasized for PKU (OMIM # 261600). Figure 13.3 shows several levels of clinical distinction for this disorder. Children with PKU generally have lighter hair color; this might be a way of distinguishing them. Unfortunately, as the bottom panel of Figure 13.3 shows, this is not very discriminating. A similar problem arises when head size is measured. Although microcephaly is recognized in PKU, smaller head size provides little diagnostic help. The general presence of mental retardation begins to offer a distinction but is not diagnostic because many conditions lead to it. A better test—more closely related to the defect—is to measure the blood level

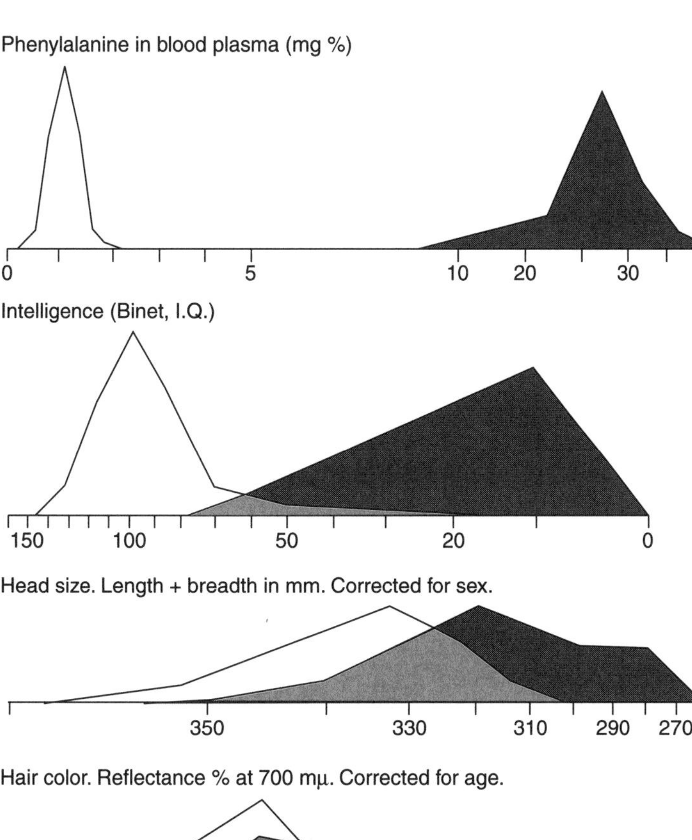

Figure 13.3 *Frequency distributions of individuals with phenylketonuria (PKU) (shaded, right side) and control populations (left side) with regard to other characteristics associated with PKU. Note the increasing clinical distinction with measurements biologically closer to the basic defect. (Adapted and modified from Penrose, Ann Eugenics 16:134, 1951.)*

of phenylalanine (the basis for the Guthrie test). As the top panel of Figure 13.3 shows, this offers considerably better discrimination, although there still are some false positives and false negatives, as discussed in Chapter 6. In contrast to these approaches, recall that testing for a distinct PKU gene mutation (when this can be performed) offers unequivocal discrimination for both homozygotes and heterozygotes. Based on this evidence of improving discrimination for a single-gene disorder, it is clear that the closer the measurements can be to the variation(s) in the critical determinant(s) of complex conditions, the better the diagnostic and prognostic accuracy will be. This same set of considerations is a reminder that the better the molecular or pathophysiologic difference(s) between clinical phenotypes can be appreciated, the better the testing, counseling, and, optimally, treatment will be.

The potential usefulness of genetic/biologic profiles has not yet been realized. Nevertheless, their implications are great, and they should contribute to improvement in both immediate and long-term patient care. For example, multiplex testing using arrays of as many as 20,000 DNA chips could screen for all genetic conditions that are diagnosable prenatally. Criteria for such multiplex screening would combine data about the frequency of each disorder in the population, the cost of each test, the ability of each test to detect all or most of those affected, and the degree of disability associated with each disorder. Another problem confronting those wishing to use such valuable information is how it can be transmitted reliably and confidentially (see also next section). One appealing notion is to prepare a data card containing profile results in stored form. Such a card could be carried by the individual and, upon decoding, regenerate the basic biologic information as a printed or projected display. At least in theory, such an approach could be universal, providing simple and reliable access to these data in any clinical setting. Although not yet available, this has considerable promise for making these valuable findings available.

3. CONFIDENTIALITY

The development of genetic profiles will change the management of complex, common conditions even as it specifically identifies individuals at risk for rarer problems. Thus, this information will be of fundamental importance to an individual's health. As discussed above, individuals might carry a small card encoded with information regarding medications, drug sensitivities, and medical diagnoses. This sort of basic, current medical data could allow quick access to sometimes life-saving information in emergency and travel situations. The longer-term plan for such a card is far more extensive and informative, however. Variations at thousands of genetic markers and loci also could be stored on such a card. Reviewing this information, most likely with the help of a computer-based interpretation system, could immediately disclose a broad array of predispositions and long-term medical considerations. It is important to remember that although such data are essential to the care of the health of an *individual,* they should not be public knowledge. Just as current medical records are securely maintained, it is essential that access to these far more broad-ranging individual data sets be protected. This is further complicated by the realization that access to this information is essential for developing prognostic statements and appropriate management strategies that are, necessarily, based on population data.

There are no immediate answers to this problem. The important thing to recall is the potential impact of the release of what could become a very large amount of information, affecting many aspects of an individual's life. Encryption is becoming more sophisticated, but probably no system is fully protectable. Some may choose to limit the information available through such a system or even refuse to use it entirely. Unfortunately, the fewer individuals who participate, the less broadly useful the information will be. Furthermore, having such information offers the possibility of obtaining and comparing long-term natural history data for clinical studies. Such analysis will permit physicians to make improved prognostic statements, as well as modify past predictions in light of follow-up information.

4. INSURANCE

An important concern about the availability of genetic information has been its potential for affecting insurability. The basic notions of insurance involve sharing risks and anticipating costs on the basis of population-based, actuarial predictions. Based on data collected earlier, insurers already consider such factors such as age, sex, weight, the presence or absence of diabetes or hypertension, and the use of cigarettes and alcohol. Each of these affects the statistical likelihood of incurring a problem for whose coverage the insurer may be responsible. To ignore such factors would be irresponsible from the perspective of the insurer, an organization with responsibilities to its policyholders and its owners, as well as to the larger community that relies on its solvency for stability in difficult times. From the perspective of the individual, however, such considerations may be interpreted differently. It is unlikely that anyone could claim that it is unfair to charge an octagenarian more for life insurance than the same coverage would cost for a 20-year-old (because such a policyholder has a much greater likelihood of "collecting" and a shorter period to pay for the policy's protection) or to have higher insurance rates for property damage coverage for an explosives factory than for a mattress warehouse (again, reflecting the potential for catastrophe). Such actions clearly are "discriminatory," but the entire insurance enterprise is based on recognizing and rating risk (as well as on minimizing the need for claims by prophylactic measures whenever possible).

In this context, "genetic risk" has unusual features. First, in many cases it is not recognizable by conventional clinical criteria. Particularly for later-onset, multifactorial conditions, such as hypertension, maturity-onset diabetes, breast or colon cancer, and atypical presentations (mutations) of cystic fibrosis, there may be no detectable features on physical examination and no tips to glean from the history. Yet such conditions clearly affect risk, as considered in earlier chapters. The objection often is raised that individuals identified as being at risk will not be able to purchase insurance because of their identified risk (alternatively, their cost of a "rated" policy will be very high) and that this is "discriminatory" or "unfair." The other perspective is that failure to detect and appropriately charge high-risk individuals unfairly raises the cost of insurance for those at lower risk (a situation commonly encountered in drivers' insurance rates).

There is no simple solution to this dilemma. However, earlier chapters have emphasized the pervasiveness of genetic variation in the population. Considering recessive conditions in particular, the health implications of carrier status are largely unknown, but the possible additive effects of being a carrier for multiple gene changes have never been assessed; Chapter 12 emphasized the potential for such effects. Carrier females for X-linked conditions were previously assumed to be unaffected (Queen Victoria had no known manifestations of hemophilia A), but subtle changes in some heterozygous women have been found for X-linked conditions such as adrenoleukodystrophy (OMIM #300100).

The question can be reduced to "How far should you look for consequential differences among individuals?" Because sequencing the entire genome (3×10^9 bp) for everyone is clearly impractical and because we do not have the long-term follow-up data needed to interpret the results even if the sequencing were performed, this is not the answer. The course chosen in legislative decisions in several states in the United States has been to prevent genetic data from being considered as criteria for insurability. This is certainly the most "fair" approach, in that everyone is assumed to have similar risks, but it should be obvious from the information above that this is equivalent to the "don't ask, don't tell" approach considered in other areas.

Although this may be the simplest solution now, it could have the negative effect of failing to identify those at special increased risk for important, preventable, or presymptomatically treatable conditions. As the breadth of testing and its predictive power increase, the last group will include individuals who are particularly likely to benefit from being identified but who may fear or refuse to have testing. We may be forced to consider these new perspectives carefully as population-based testing and screening are used more broadly, particularly as they affect the detection of liabilities for problems of later onset.

5. EMPLOYMENT

Another concern raised by testing for genetic traits and conditions has been the possibility of workplace discrimination based on the results. A 1989 study by Northwestern National Life Insurance Company found that by the year 2000, 15% of 400 employers surveyed planned to check the genetic status of prospective employees and their dependents prior to making employment offers. A 1995 Harris poll of the general public revealed that over 85% of those surveyed indicated that they were "very" or "somewhat" concerned that insurers or employers might have access to genetic information.

Fourteen states have enacted laws to protect against genetic discrimination in the workplace. While some of these laws specify conditions—such as sickle-cell carrier status in Florida and Louisiana—others—e.g., Texas and Oregon—ban worker discrimination on the basis of *any* genetic information.

The other side of the question relates to the possibility that identifying certain traits might improve worker safety. For example, variations in metabolic pathway X might directly relate to susceptibility to the potential toxicity of certain industrial chemicals; thus, despite "appropriate" environmental standards (meeting or exceeding OSHA criteria), certain individuals could still be at risk. In a January 1998 statement, Vice President Gore said that genetic testing "should be permitted in some situations to ensure workplace safety and health and to preserve research opportunities" *if* the employee has provided consent and confidentiality is maintained.

STUDY QUESTIONS

1 You are a consultant to the medical hygiene office of a well-known chemical manufacturer. This company has a scrupulous record of adherence to environmental and OSHA guidelines. It has recently found a novel method of recycling bottles that is efficient and remarkably inexpensive compared with older approaches; unfortunately, it involves several solvents that, individual and collectively, have known hepatotoxicity. This toxicity may be more pronounced in certain individuals, for whom a gene test has been developed. The company has asked for your advice. What do you suggest?

2 You have just completed a comprehensive medical evaluation, including a stress test and flexible sigmoidoscopy, on a senior industry executive. He is 51 years old, and his examination came out well. Ten days after his visit, he calls to ask if you can complete a 6-page form regarding his suitability for a $5-million insurance policy. He also relates the disturbing news that his brother, age 47, recently has been found to have localized colon cancer after suspicion was raised by "some gene test." The family is of Ashkenazi

Jewish ancestry, and the test is available to you. You tell him:

a He should have the test.
b He is unlikely to develop the problem because he is older than his brother.
c The brother's health had been reported as "normal" on your earlier report (which has already been sent to the insurance company, because you have an efficient secretary) and will have to be revised.
d He should have a full colonoscopy.
e If he is unwilling to consider colectomy, there is no use in having the test.

3 A 26-year-old woman has come to you to have "genetic cancer tests" done. She is particularly interested in testing for breast and colon cancer. She has no affected family members and comes from an Irish family with several generations notable for longevity. What do you tell her?

FURTHER READING

Andrews LB (eds). *Assessing Genetic Risks: Implications for Health and Social Policy.* Washington, DC, National Academy Press, 1994.

Chee M et al. Accessing genetic information with high-density DNA arrays. *Science* 274:610, 1996.

Collins FS. BRCA1—Lots of mutations, lots of dilemmas. *N Engl J Med* 334:186, 1996.

FitzGerald MG et al. Germ-line BRCA1 mutations in Jewish and non-Jewish women with early onset breast cancer. *N Engl J Med* 334:143, 1996.

Hudson KL et al. Genetic discrimination and health insurance: An urgent need for reform. *Science* 270:391, 1995.

Jones KL. *Smith's Recognizable Patterns of Human Malformation,* 5th ed. Philadelphia, Saunders, 1997.

Laken SJ et al. Familial colorectal cancer in Ashkenazim due to a hypermutable tract in APC. *Nat Genet* 17:79, 1997.

Langston AA et al. BRCA1 mutations in a population-based sample of young women with breast cancer. *N Engl J Med* 334:137, 1996.

Lapham EV et al. Genetic discrimination: Perspectives of consumers. *Science* 274:621, 1996.

Marshall E. The genome program's conscience. *Science* 274:488, 1996.

Pokorski RJ. Genetic information and life insurance. *Nature* 376:13, 1995.

Rimoin DL et al (eds). *Emery and Rimoin's Principles and Practice of Medical Genetics,* 3d ed. New York, Churchill Livingstone, 1996.

Rothenberg K et al. Genetic information and the workplace: Legislative approaches and policy challenges. *Science* 275:1755, 1997.

Scriver CR et al (eds). *The Metabolic and Molecular Bases of Inherited Disease,* 7th ed. New York, McGraw-Hill, 1995.

Seashore MR, Wappner S. *Genetics in Primary Care and Clinical Medicine.* Englewood Cliffs, NJ, Appleton & Lange, 1996.

Welch GN, Loscalzo J. Mechanisms of disease: Homocysteine and atherothrombosis. *N Engl J Med* 338:1042, 1998.

@ USEFUL WEB SITES

• CLINICAL INFORMATION

HELIX (INCLUDES GENLINE)

http://www.hslib.washington.edu/helix
Medical Genetics Knowledge Base. Electronic textbook under development relates testing to diagnosis, management, and counseling. Directory of laboratories providing testing for genetic disorders.

OMIM (ONLINE MENDELIAN INHERITANCE IN MAN)

http://www.ncbi.nlm.nih.gov/omim
Basic catalogue listing mendelian genes and traits. Clinical descriptions, references, and links to other databases.

HEALTHLINKS

http://healthlinks.washington.edu/
Broad access to many related areas of basic and clinical genetics. Includes "Genome Machine II," database of genes and diseases mapped to a particular chromosome location. Also chromosome information.

NATIONAL ASSOCIATION FOR RARE DISORDERS (NORD)

http://www.NORD-rdb.com/~orphan
Primary nongovernmental clearinghouse for information about rare disorders. Over 1100 disease reports. Broad resource guide. Includes Rare Disease Database, a wide range of information for families and professionals. Listings by symptoms, causes, affected populations, and treatments. Listing of support groups. Includes orphan drug designation database.

• TESTING

CENTER FOR BIOETHICS, UNIVERSITY OF PENNSYLVANIA

http://www.med.upenn.edu/~bioethic
Articles on cloning and bioethics primer. Articles with comment options. Access to virtual library. Literature summaries about insurance, discrimination, privacy, screening.

COMMUNITY OF SCIENCE WEB SERVER

http://cos.gdb.org/best.html
Links to Medline, Federal Register, and U.S. patent citation database. Access to selected journals.

EUBIOS ETHICS INSTITUTE

http://www.biol.tsukuba.ac.jp/~macer/index.html
Newsletter for bioethics and biotechnology; based in Japan and New Zealand. Access to articles in Japanese. Meeting contents and abstracts.

GENETIC PRIVACY ACT	http://www.ornl.gov/hgmis/resource/elsi.html Ethical, legal, and social issues pages for Human Genome Project. Includes access to genetics support groups. Useful cross-links to books, journals, and programs. Useful glossary. Model legislation supported by Human Genome Program.
HUMAN GENOME PROJECT INFORMATION	http://www.ornl.gov/hgmis Includes section on ethical, legal, and social issues; chat opportunities.
NATIONAL BIOETHICS ADVISORY COMMITTEE	http://www.bioethics.gov General source of advice to National Science and Technology Council. Notices of meetings and transcripts. Access to publications.
NATIONAL CENTER FOR GENOMIC RESOURCES	http://www.ncgr.org Genetics and public issues pages including legislative proposals. Section on continuing medical education about genetics. News updates. Access to Genome Sequence Database.
PROMOTING SAFE AND EFFECTIVE GENETIC TESTING IN THE U.S. (1997)	http://www.med.jhu.edu/tfgtelsi Report of Task Force on Genetic Testing including discussions of safety, quality control, communication, and rare disease studies. Recommendations for providing gene tests in United States.
THE GENE LETTER	http://www.geneletter.org Newsletter about genetics and potential policy questions. Useful and timely summary articles. Chat opportunities. Links to other sites.
UNDERSTANDING GENE TESTING	http://www.gene.com/ae/AE/AEPC/NIH/index.html Illustrated brochure from National Cancer Institute with useful information for lay public.

● CHROMOSOME STUDIES

CYTOGENETICS	http://www.waisman.wisc.edu/cytogenetics Useful review of chromosome changes in tumors
CYTOGENETICS GALLERY	http://www.pathology.washington.edu:80/Cytogallery Includes acquired and constitutional chromosome anomalies and karyotypes. Useful comparison of banding patterns. FISH examples in color. Access to idiograms for generating custom figures.

CYTOGENETIC RESOURCES

http://www.kumc.edu/gec/geneinfo.html
A particularly valuable entrance to multiple sites including Cytogenetics Images Index, Cytogenetic images and animations, Gene Map of human chromosomes, Human Cytogenetics Database, Chromosome Databases, Chromosome empiric risk calculations, and Karyotypes of normal and abnormal chromosomes. Excellent cross-references to other sites.

MOLECULAR CYTOGENETICS AND POSITIONAL CLONING CENTER (BERLIN)

http://www.mpimg-berlin-dahlem.mpg.de/~cytogen/
Describes a family of yeast artificial chromosomes (YACs) cytogenetically and genetically anchored, spread evenly over the entire human genome. A resource for FISH mapping. There also is a collection of YAC probes for all human chromosomal ends, averaging several cM from actual ends. This is a reference panel for studies of telomere rearrangements.

RESOURCES FOR HUMAN MOLECULAR CYTOGENETICS

http://bioserver.uniba.it/fish/Cytogenetics/welcome.html
Libraries of partial chromosome paints, recognizing a definite region of a chromosome. >900 fragments characterized from both normal and radiation-induced somatic cell hybrids.

CYTOGENETIC IMAGES AND ANIMATIONS

http://www.tokyo~med.ac.jp
Animations of meiotic interchanges with normal and anomalous chromosome structures.

CHAPTER 14 Treatment

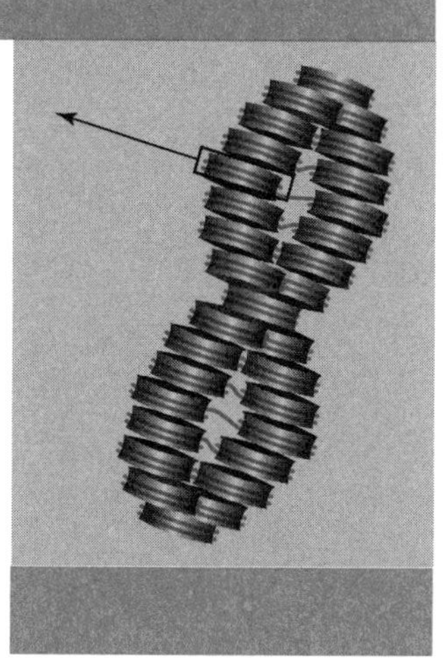

The goals of medicine historically have been relieving suffering and improving the quality of life. Many past efforts in developing the nosology of genetic diseases have aimed at clarifying patterns of illness and identifying their causes. Therapeutics has not been a prime consideration. In the absence of treatment options, much past work has emphasized prenatal diagnosis, a discipline that has made much progress. Unfortunately, options for management after diagnosis have been limited, and prevention rather than treatment has necessarily dominated much thinking and practice. Molecular biology and biotechnology are slowly beginning to alter the focus of genetics as applied to clinical medicine, however. Treatment, with restoration, maintenance, or control of growth, development, and function, is emerging as the critical element whose absence often has kept genetics on the sidelines of clinical practice. Although some aspects of this important topic have been introduced in earlier chapters, this section will examine several areas in which treatment considerations are prominent.

A. NATURAL HISTORY DATA

As testing and diagnostic methods are improved, individuals with genetic (and some congenital) disorders are being identified earlier, often before symptoms develop. In the past (and all too often currently), establishing a particular diagnosis has not always been reassuring, because no treatment options existed. Nevertheless, establishing reasonable expectations and prognosis must underlie effective care. Fortunately, the identification of many individual inherited conditions over the last several decades has permitted some longitudinal observations. Those observations have led to a reasonable set of expectations regarding clinical expectations and natural history for individuals affected with any of a sizable group of disorders.

Some of the implications are rather obvious. For instance, the susceptibility to hemorrhage and hemarthrosis in hemophilia A (OMIM #306700) (see Chapter 7) are clear. Drug sensitivities in G6PD deficiency (OMIM #305900) (see Chapters 7 and 10) are also well-established. The value of aggressive management of crises in sickle-cell disease (OMIM #141900.0243) (see Chapter 6) are well known, and the importance of carefully following neurofibromas or other unusual growths in individuals with neurofibromatosis (OMIM #162200) (see Chapter 5) also is evident.

Less obvious, however, have been conditions that develop considerably later in life. This problem also is complicated by the pleiotropism discussed in Chapter 5. For example, the possibility of spinal stenosis and its complications in individuals with achondroplasia (OMIM #100800) (see Chapter 5) is now recognized. Affected individuals can thus be observed for early signs before irreversible damage occurs. Similarly, recognizing the possibility for orthopedic intervention for leg straightening and hip positioning in these individuals has improved later function. The ability to detect aortic dilatation in individuals with Marfan syndrome (OMIM #154700) (see Chapter 5) through early noninvasive testing such as echocardiography has established a reasonable set of expectations for long-term management. Furthermore, the advent of effective and reliable surgical repair techniques now permits presymptomatic treatment of this problem in individuals who show early changes.

The entire rationale for developing longitudinal data is to anticipate problems that will occur later and, when possible, to minimize their consequences or even to prevent their development. Recognizing the problems that are likely to occur by having some general understanding of the pathophysiology of the condition (this does not always presuppose precise molecular understanding) at least has the possibility of suggesting treatments. Treatment plans should be consonant with the physiologic changes anticipated. This approach has been the basis for introducing the use of beta blockade to reduce the rate of aortic dilatation in individuals with Marfan syndrome (OMIM #154700) (see Chapter 5). Having observed the development of a set of predictable problems in similarly affected individuals, the physician should be able to make prognostic suggestions and provide intervention options to minimize later difficulties.

Although effective prevention of problems may be the ideal role of the physician in dealing with specific genetic and congenital conditions, it is not always possible. As discussed earlier, a major limitation to intervention in any disorder involves inadequate understanding of its underlying pathophysiology. Furthermore, as indicated in Chapter 12, identifying a specific gene abnormality may be only part of the picture of complex interactions with other gene changes. Particularly in situations in which interactions between the gene and the environment and/or other genes are prominent, prediction becomes more difficult.

As detailed genetic testing becomes available, physicians, other health professionals, and individuals and families will be confronted with the identification of very specific changes that do not necessarily have very specific and predictable clinical consequences. For example, as discussed earlier, finding mutations in the *BRCA1* and *BRCA2* genes (OMIM #113705, #600185) (see Chapters 12 and 13) identifies individuals at altered risks for developing breast cancer. Even from family studies, however, it is clear that identifying such a gene change does not automatically indicate that cancer will develop or, if it does, how it will present or behave.

Another frustration with current practice procedures is that, despite the fact that some diagnoses have predictable consequences, the approach to their management is changing. For example, identifying an individual at risk for APC (OMIM #175100) (see Chapter 13) is a strong predictor of the development of colon cancer. When this is combined with endoscopic discovery of multiple colonic polyps, the development of cancer is almost inevitable. Nevertheless, the only definitive treatment at the time of this writing is preemptive colectomy. This obviously precludes the possibility of using any later treatment that might have a more specific rationale based on newly developed knowledge and that might permit salvage of the colon. The rapid progress in diagnostic, experimental management and treatment approaches assures us that this will remain an area of difficulty in terms of decisions for individuals and their physicians.

Under any circumstances it is important to assure parents and affected individuals, as well as sibs and others at potential risk, that a specific diagnosis has been established (or eliminated). Often the challenge of a counseling situation involves trying to assess the implications of an insecure diagnosis for other members of a family or for the affected individual. Establishing the diagnosis unequivocally (whether on molecular, chromosomal, or clinical grounds) is the first step. The remarkable value of making a specific diagnosis for every person at risk cannot be minimized. The difficulties, frustrations, and disappointments that arise from having to go from clinic to clinic and still not getting an answer are considerable. Nevertheless, this cannot be the end of counseling and interactions, particularly as new treatments are developed.

B. GENE OR METABOLIC CONTROL

Treatment for individuals with inherited and congenital disorders must be based on expectations of their natural histories, as discussed above. Also, to be maximally effective, treatment should reflect the underlying biochemical and pathophysiologic changes. Although these are not yet understood for all inherited conditions, there have been encouraging successes, some of which have been discussed earlier (see Chapters 5–7, 10). Presymptomatic management is the ideal approach. In such a situation, an affected individual, identified early, can have some appropriate regimen begun that either circumvents or minimizes the consequences of the genetic change. The goal for these approaches is to use diet and/or some exogenous molecule to alter or control gene expression or protein/metabolic function. Such effects will be pharmacologic—transient and repeatable; for example:

- Using restricted diets in PKU (OMIM #261600) or galactosemia (OMIM #230400) to minimize levels of potentially toxic metabolites yet permit growth and development (see Chapter 6).
- Treating congenital hypothyroidism with exogenous thyroid hormone to eliminate the threat of cretinism.
- Using drugs such as hydroxyurea to increase the expression of fetal hemoglobin in individuals with sickle cell disease (OMIM #141900.0243), thereby lowering the relative concentration of sickle hemoglobin in erythrocytes and reducing the likelihood of a sickle cell crisis (see Chapter 6).
- Using hormonal cycling to induce and maintain adolescent development in individuals with Turner syndrome, as discussed in Chapter 3.
- Administering vitamin B_6 to some individuals homozygous for homocystinuria (OMIM #236200) to (at least partially) compensate for the change(s) in the target enzyme—cystathionine β-synthase. Supraphysiologic levels of this cofactor can minimize the dysfunction of at least some of the mutant alleles, lowering blood homocysteine levels and reducing vascular damage.
- Using chelation for copper in Wilson disease (OMIM #277900) or a combination of phlebotomy and chelation for iron in hemochromatosis (OMIM #235200) to reduce metal accumulation and attendant tissue damage.
- Administering sodium phenylbutyrate to provide an alternative route for nitrogen excretion in the urea cycle disease ornithine transcarbamylase deficiency (OMIM #311250). Glutamine becomes conjugated with phenylacetyl coenzyme A to yield phenyl acetylglutamine, which can be excreted. As shown in Figure 14.1, this leads to resynthesis of glutamine by the amidation of glutamate, helping to eliminate nitrogen.
- Using heme derivatives as feedback inhibitors to reduce metabolic

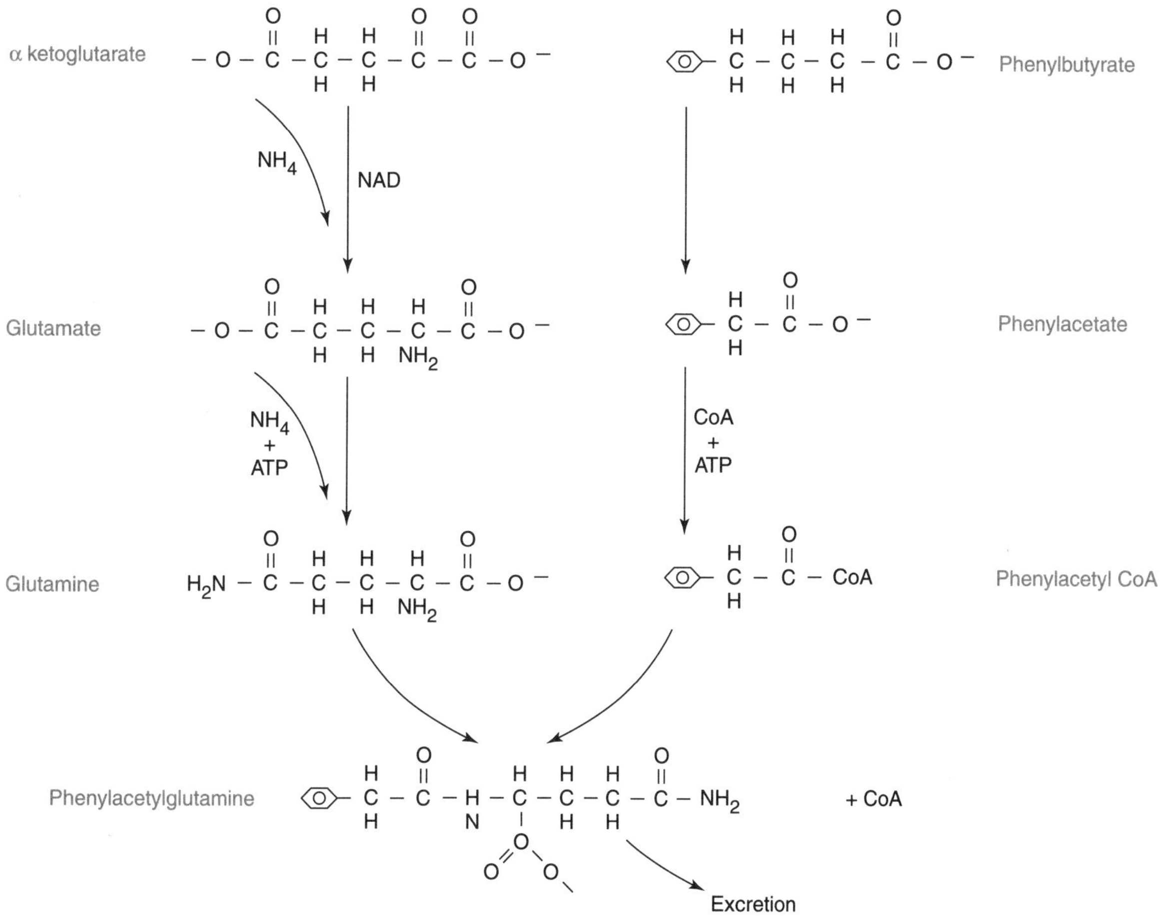

α ketoglutarate → (NH4, NAD) → Glutamate → (NH4 + ATP) → Glutamine

Phenylbutyrate → Phenylacetate → (CoA + ATP) → Phenylacetyl CoA

Phenylacetylglutamine + CoA → Excretion

Figure 14.1 *Use of sodium phenylbutyrate to assist with nitrogen excretion in ornithine transcarbamylase deficiency. Phenylacetyl coenzyme A is joined to glutamine to form phenylacetylglutamine, which can be excreted.*

flux through the heme synthesis pathway, thus minimizing the consequences of an attack of acute intermittent porphyria (OMIM #176000), as discussed in Chapter 10.

Each of these examples emphasizes exploiting knowledge of the defective metabolic pathway, proteins, or gene(s) involved. Although these treatments are pharmacologic and inherently transient, their effects on the quality of life of treated individuals can be remarkable. These approaches also can eliminate the potential difficulties and current unavailability of formal gene replacement.

C. SURGERY

The repair or replacement options presented by surgery can be particularly valuable for individuals with congenital changes, as discussed in Chapter 9. Surgical approaches to the care of individuals with clubfoot deformities, congenital hand malformations, or heart defects often are quite effective. Hemangiomas also can be approached through a combination of laser treatment, embolization, and resection, as discussed.

As described in Chapter 5, surgery also has changed expectations for individuals with achondroplasia (OMIM #100800) and Marfan syndrome (OMIM #154700), with intervention to control spinal stenosis and leg and hip alignment in the former and scoliosis, eye problems, and aortic and heart valve disease in the latter. Individuals with Type I Gaucher disease (OMIM #230800) (see Chapter 6) also have had considerable benefit from surgical repair of bone problems, particularly in the hips and knees. Many of these interventions are one-time events and do not need later revision. For instance, repair of hand malformations, closure of septal or valvular cardiac defects, and control of scoliosis are generally done as part of a single (although perhaps protracted) episode of management. Often, surgical procedures are not considered until well after the diagnosis has been established. This means that they may be only part of the unfolding, long-term picture of the care of affected individuals. Thus, it is important that individuals who may have been treated early in life remain under careful observation. Furthermore, documentation and clear delineation of the problems addressed (and presumably solved) early in life are important in terms of genetic prognosis and counseling for offspring of the treated individual. This is particularly clear for individuals with congenital heart disease, for example (see Chapter 9).

D. PROTEIN REPLACEMENT

Numerous genetic disorders are due to failure to produce an appropriate protein. This protein may be an enzyme or a structural protein, but the clinical consequences often are predictable on the basis of the known function of that protein. Thus, the option of administering an exogenous protein(s) to treat the clinical problems has always had rational appeal. We will consider several instructive examples.

- Administering Factor VIII protein to individuals with hemophilia A (OMIM #306700) has been remarkably successful. The history of the improving quality of Factor VIII preparations and the unfortunate incidents in the history of these efforts have underscored the fact that this has not always been a trivial undertaking. However, the option of developing recombinant, pathogen-free products through biotechnology has greatly altered the safety and reliability of the care of these individuals.
- Exogenous immune globulins have been helpful in the care of

individuals with defects in specific immunoglobulin production. Because the beneficial effects affect multiple aspects of host defense and because (in general) entire families of proteins are administered, a broad level of protection against a range of pathogens is afforded to the recipient. This approach has been useful in both children and adults (see Chapter 11).

- Type I Gaucher disease (OMIM #230800) also has been the focus of considerable efforts in protein replacement. As discussed in Chapter 6, the use of exogenous preparations of β-D-glucosidase (trade name Ceredase) has been remarkably effective in selected individuals. The original product was derived from placental tissue, but recombinant material should improve dosage and reliability and may reduce cost. About 800 individuals worldwide are receiving this treatment.
- Adenosine deaminase can be complexed with polyethylene glycol (PEG). The mixture survives longer in blood and has proved useful for treating individuals with adenosine deaminase deficiency (OMIM #102700).

Protein replacement is not without its complications, however. One of the most difficult has been the development of an antibody response against the administered molecule. In many (but not all) recipients, the administered protein is recognized as foreign and the development of antibodies against the administered protein can be a barrier to successful treatment. An additional complication is based on the underlying genetics of the disorders. This is related to the fact, mentioned earlier, that many individuals who fall into the same clinical disease category do not, in fact, have the same molecular mutation. Thus, the broad spectrum of individuals with hemophilia A includes those with partially functioning endogenously produced Factor VIII protein as well as those who make none at all. The natural histories of individuals in different categories within the same basic diagnostic classification can justifiably be expected to differ.

Another aspect of protein replacement that can be confusing, particularly early after the product has been developed, is the design of appropriate administration protocols. For example, many individuals with hemophilia A are most effectively managed by presymptomatic, preemptive Factor VIII administration; others need the protein only in emergencies. A similar problem arises with protein replacement treatment in Type I Gaucher disease. Here, dose and treatment schedules are still being optimized; these affect cost, the potential for antibody development, and, ultimately, the quality of life for those treated. Because each of these situations has no immediate precedent, a great deal of clinical observation and protocol evaluation is necessary. Many treatment protocols must be individualized.

E. GENE REPLACEMENT

One of the most highly publicized areas related to genetic conditions has been the notion of gene replacement. The rationale is simple: If a defective gene leads to either inadequate or aberrant function of its product, then replacing it with a functional copy should cure or reverse the problem. Although this obviously is an oversimplification, it has gained appeal, particularly because such replacements can be performed in experimental animals. In the fruit fly Drosophila (*Drosophila melanogaster*), for example, replacement of chromosomal DNA segments with exogenous counterparts is a reliable and useful technique. Such relatively simple protocols have not been successful in mammals, however.

Because the areas of gene replacement and "gene therapy" have been so extensively evaluated in many publications, they will not be presented in detail here. Rather, an outline of several approaches will be considered. This is an area in which the success of the procedures will depend critically on detailed protocols and recombinant techniques that are under development currently. For this discussion, we will emphasize efforts to achieve *permanent* expression of the replacement gene(s) in the recipient cell(s). Transient expression will have less therapeutic value.

Several aspects of gene replacement in humans are of particular clinical concern:

- The desired product must be produced in the recipient cells, ideally with repression or elimination of the aberrant product.
- The product must be produced in the appropriate cells. For example, it is not useful to produce normal globin in lung cells of patients with sickle cell disease; the protein must appear in appropriate erythrocyte precursors.
- Ideally, the recipient cells should persist in the individual, so that repeated treatments are not required. This has been a consistent challenge with treatments for cystic fibrosis (OMIM #219700) and α_1-antitrypsin deficiency (OMIM #107400), both of which have prominent pulmonary symptoms. The rationale of delivering replacement genes by aerosol is appealing because the replacements can reach the dysfunctional respiratory epithelial cells. Nevertheless, these cells turn over relatively rapidly, and thus the treatment protocols necessarily involve repetition to treat the new pulmonary cells. Such repeated treatment is not only cumbersome and expensive but increases the possibility that side effects will develop.
- The treatment must be reliable, with consistent consequences. The potential for disappointing outcomes, leading to frustration or even litigation, requires consistently high quality.
- In conditions involving the central nervous system, it is essential

to be able to manage neurologic as well as somatic gene replacement. This is particularly true for neurodegenerative disorders that also have somatic components. One of the aspects of greatest concern in developing treatments for such individuals is that they may have improvement in terms of somatic difficulties while their neurodegeneration continues.

Assuming that the gene of interest has been isolated and characterized, there are two essential elements in the design of gene transfer procedures. The first relates to the group of sequences surrounding the gene. These sequences should help to direct the insertion of the exogenous gene into the appropriate position within the chromosomal DNA of the recipient. This is important for several reasons. As has been well established in bacterial systems, random insertion of a piece of exogenous DNA can act as a mutagenic event—it can disrupt a host gene and change control of local genes. Putting the exogenous DNA into a region of the chromosome where transcription is repressed (recall X-inactivation and imprinting) may reduce the amount of product produced and, hence, the therapeutic effect.

Another function of the surrounding sequences is to control the level and tissue specificity of the gene's expression. A great deal of information is known about the DNA sequences (mostly 5′ to the structural gene) responsible for controlling transcription and translation of genes. The goal is to have the exogenous gene expressed in the same manner as the original gene.

The second essential element in planning gene replacement is the delivery vehicle responsible for getting the exogenous gene and its control sequences into the target tissue(s). One group of vehicles uses derivatives of viruses to act as vectors. The rationale for using viruses is clear:

- The natural function of viruses is to introduce foreign genetic material into cells.
- The genetic material of some viruses becomes a permanent part of the cellular DNA, so methods for inserting viral gene sequences into chromosomes already exist.
- Viruses, as part of their natural host range, have specific cell targets, and thus an appropriate match between the recipient cell and the virus vector can be envisioned.
- Viruses can induce the synthesis of new proteins in the cells that they infect, thus potentially permitting the expression of desired new genes.
- Many details of the molecular biology of viruses are known.
- Recombinant DNA techniques permit construction of virtually any desired virus-nucleic acid hybrid.

Table 14.1 ▶ **Gene delivery systems**

Vehicle	Nucleic acid	Insert size	Comment
Retrovirus (Maloney)	ss RNA	~8 kb	Integrates into genome of dividing cells
Adenovirus	ds DNA	~7 kb	Epichromosomal; inflammatory
Herpesvirus type 1	ds DNA	>10 kb	Epichromosomal
Adeno-associated virus	ss DNA	4.7 kb	Integrates; needs helper
Liposome–plasmid complex	ds DNA	Unlimited	Epichromosomal; inefficient
DNA–ligand complex	ds DNA	Unlimited	Epichromosomal; degradation in endosomes
Direct DNA injection	ds DNA	Unlimited	Epichromosomal; used in muscle; inefficient

Several delivery systems have been emphasized in gene delivery protocols. These are summarized in Table 14.1 and Figure 14.2.

- Retroviruses (single-stranded RNA; based on the Maloney murine leukemia virus) (see Figure 14.2A). The genes within the virus—*gag, pol, env*—are removed and replaced by ≤8 kb of exogenous RNA. The long terminal repeat (LTR) domains of viral sequences are left. The RNA is packaged and used to infect recipient cells. On entering the cell, the RNA is copied into DNA (producing a "provirus") that is transported to the nucleus and inserted into chromosomal DNA. The recipient cell must be dividing to permit integration.
- Adenoviruses (double-stranded DNA) (see Figure 14.2B). A segment of viral genes is removed, permitting insertion of ~7 kb of exogenous DNA. The virus invades the cell, where its DNA remains

Figure 14.2 *Viral gene delivery systems. (A) Retrovirus: Three genes—gag, pol, env—are removed and replaced with the desired sequence. After entering the cell, viral reverse transcriptase forms a copy of double-stranded DNA that can be integrated into nuclear DNA. (B) DNA viruses (adenovirus, HSV-1): A group of viral genes is deleted and replaced by the desired sequence. The virus enters the cytoplasm, and the double-stranded DNA is delivered to the nucleus, where it persists epichromosomally. Although they are very different in other ways, the main difference between these two vectors in this context is the size of their genomes (adenovirus: ~36 kb; HSV-1: ~150 kb). (C) Adeno-associated virus: This small, single-stranded DNA virus forms a double-stranded molecule after infection. It can then enter the nucleus and integrate.*

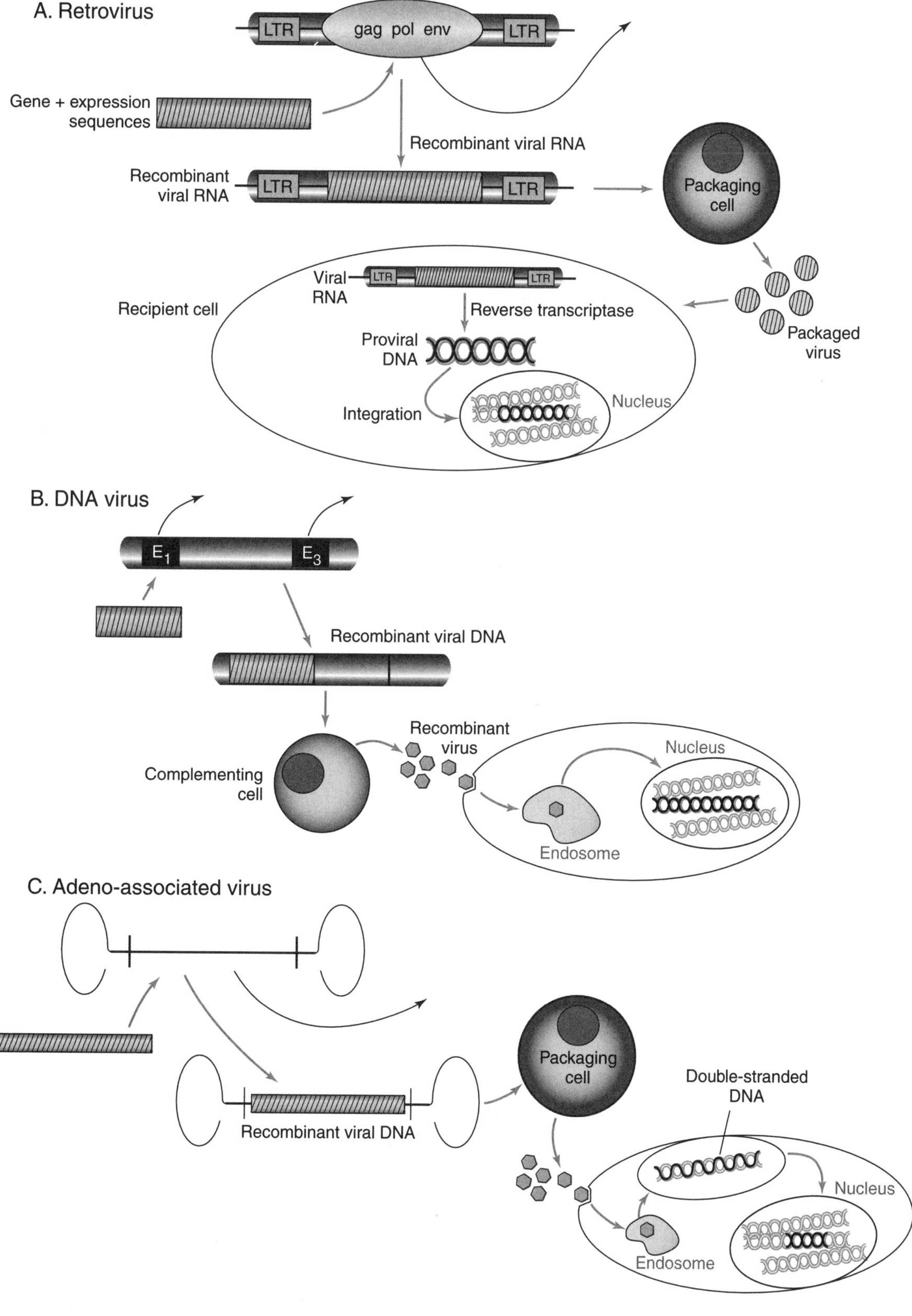
A. Retrovirus
LTR
gag pol env
LTR
Gene + expression sequences
Recombinant viral RNA
Recombinant viral RNA
LTR
LTR
Packaging cell
Viral RNA
LTR
LTR
Recipient cell
Reverse transcriptase
Proviral DNA
Packaged virus
Integration
Nucleus
B. DNA virus
E_1
E_3
Recombinant viral DNA
Recombinant virus
Nucleus
Complementing cell
Endosome
C. Adeno-associated virus
Packaging cell
Double-stranded DNA
Recombinant viral DNA
Nucleus
Endosome

epichromosomal. It does not permanently change the genotype of the host cell. Unfortunately, adenovirus vectors generally evoke an inflammatory response, limiting the duration of their effectiveness.

- Herpesvirus type 1 (HSV-1; double-stranded DNA) (see Figure 14.2B). This virus has a large genome (~150 kb). It may have a tropism for neurons, although it also infects other cells. It remains epichromosomal. It can carry a longer DNA sequence than adenoviruses.
- Adeno-associated viruses (AAV; single-stranded DNA) (see Figure 14.2C). A relatively short length of DNA can be inserted into this virus. The virus undergoes stable chromosomal integration with high rates of infectivity. It has no known human pathogenicity but does require adenovirus or herpesvirus as a helper for replication.
- Liposome-plasmid complexes (double-stranded DNA). These complexes can be positively charged, permitting them to associate readily with DNA. They appear to be inefficient as vectors but are not inflammatory. Their DNA likely remains epichromosomal.
- DNA-ligand complex. Asialoglycoproteins can target complexes to the appropriate hepatocyte receptor. Transferrin-DNA complexes can go to the transferrin receptor. Such complexes are not targeted to the nucleus, so the DNA often ends up degraded in endosomes.
- Direct DNA injection (double-stranded DNA). This method has been used to deliver the dystrophin gene to muscles of mice. The DNA remains epichromosomal. The process is inefficient, but this approach may be useful in situations where local concentrations of a product are needed.

Gene replacement is an area of intense investigation. The emphasis has not been limited to genetic disorders; cancer treatment has been an important consideration as well. While the entire field has moved slowly, it has shown progress. Clearly, the notion of transferring genes into humans *is* feasible. This has been shown in several examples:

1 Adenosine deaminase (ADA) cDNA has been introduced into T lymphocytes of children with ADA deficiency (OMIM #102700).
2 LDL receptor cDNA has been transferred to isolated hepatocytes of individuals with LDL receptor mutations (OMIM #143890).
3 CFTR cDNA has been introduced into the nasal epithelium of individuals with cystic fibrosis (OMIM #219700).

Several important barriers remain in this challenging field: (1) The inconsistency of the results, one of the most frustrating problems, likely reflects lack of knowledge about all of the critical variables. (2) Long-term toxicity has not been assessed and cannot reliably be extrapolated from animal studies. (3) The ideal delivery system has not been identified (recall Table 14.1). (4) Scaling up of vector production for broad-based clinical trials has been difficult.

Efforts to develop reliable gene replacement undoubtedly will exploit many still undiscovered aspects of cell biology and gene expression and will require careful short- and long-term monitoring. It is important to realize that effective therapy must work for the *lifetime* of the patient.

F. BONE MARROW TRANSPLANTATION

Although not specifically considered gene replacement, the use of allogeneic bone marrow transplantation (AGMT) is another approach to the care of individuals with various genetic conditions. Again the rationale is simple: when defective stem cells are replaced with normal cells, one can generate healthy, mature blood cells. Alternatively, bone marrow replacement may be used to introduce cells that produce a needed serum protein product, like Factor VIII in hemophilia A or β-glucosidase in Gaucher disease. Bone marrow transplantation obviously is not an option for individuals with central nervous system difficulties, because the positive effects cannot be expected to distribute themselves within the central nervous system. It has been used successfully for certain hematologic conditions and lysosomal storage disorders, however.

In a study of 63 European patients treated with AGMT for lysosomal disorders, visceral improvement was common but neurologic deterioration was not reliably arrested, especially if neurologic signs were present before the procedure. Individuals with Type I Gaucher disease had particularly good clinical results. AGMT also has been used successfully for some children with refractory anemias.

AGMT remains a complex approach with its own significant morbidity and mortality. Nevertheless, it has improved enormously in effectiveness and now can be considered as an approach to the care of selected individuals with genetic disease.

The use of bone marrow cells also offers the possibility of using cells from an individual's bone marrow as recipients for exogenous genes ("ex vivo" gene replacement). If such recipient marrow cells, extracted from the patient's bones, can stably accept replacement DNA sequences, they can then be returned to the host without the difficulties associated with AGMT and proliferate appropriately to express the desired gene product.

G. LIVER TRANSPLANTATION

This approach has been particularly effective in individuals with conditions associated with the synthesis of abnormal proteins by the liver. Individuals with one group of disorders—dominant amyloid neuropathies due to prealbumin gene mutations (OMIM

#176300)—have shown encouraging results. Iron-induced liver disease in individuals with homozygous β-thalassemia (OMIM #141900) also can respond. Liver transplantation also has been successful in individuals with copper accumulation in Wilson disease (OMIM #277900).

H. ORPHAN DISEASES

One of the difficulties with developing specific management techniques for individuals with inherited conditions is the relative rarity of such conditions. Developing new protocols and recombinant and pharmacologic agents is time-consuming and expensive. It is difficult to justify the costs involved in such ventures on the basis of the anticipated financial recovery from the investment. Thus, these conditions have become designated "orphan diseases." Such a designation (based on the Orphan Disease Act of 1983) implies that the potential recipient population for the treatment is small and the organization or company developing the treatment (with its attendant development, regulatory, and monitoring costs) will be given the exclusive right to sell their product for 7 years. Using this important classification assists companies in justifying their necessary investments. It also limits the number of competitors likely to develop a similar product. Despite the presence of orphan disease designations, many start-up and investigational costs are borne by private and academic organizations dedicated to the study of particular disorders. Only after there is a reasonable expectation of success can commercial groups be expected to become involved.

I. POPULATION QUESTIONS

One of the most important aspects of considering genetics in medicine and the specific treatment of affected individuals arises from the fact that genetic treatments change expectations regarding natural history. As stated earlier in this chapter, an appreciation of the natural history of inherited conditions is essential. It establishes expectations and is useful for counseling. It can be reassuring and sometimes provide some comfort to parents and affected individuals to know what to anticipate on the basis of clinical experience with similarly affected people.

The nature of many (but not all) inherited conditions is that their consequences develop over a long period of observation. For instance, spinal stenosis in individuals with achondroplasia (OMIM #100800) is rare before age 20. Aortic dissection in Marfan syndrome (OMIM #154700) is also a relatively late problem. Thinning of cortical bone occurs late in individuals with Type I Gaucher disease (OMIM #230800). Thus, in terms of the clinician's interaction with affected individuals who have received a novel treatment, an entirely new set of questions arises. Examples are:

- What are the clinical expectations for individuals treated with restricted diets for PKU (OMIM #261600) and galactosemia (OMIM #230400)?
- What is the situation on 30-year follow-up for these individuals? (These follow-up data are beginning to become available.)
- Where are the 30-year follow-up data for individuals with achondroplasia who have had laminectomies for spinal stenosis?
- What is the likelihood of continued vascular integrity in individuals who have had aortic valve and ascending aorta replacement in Marfan syndrome? Can aneurysms be expected to develop elsewhere?
- What will be the effect of Ceredase in the long-term management of individuals with Gaucher disease?

The answers to these questions are simply not known, and far more questions will be raised as additional individuals are treated and other treatments are devised.

One of the obvious effects of controlling complications that are likely to be lethal in childhood is that a group of treated individuals will reach adulthood and reproductive age. This may never have occurred for similarly affected individuals in the past. For example, the entire picture of Alzheimer-like cognitive loss, now recognized as common in older patients with Down syndrome, was not even imagined when affected individuals either were institutionalized or did not survive. Thus, questions regarding the management of pregnancy in affected women will need to be addressed, as will more general aspects of care for older individuals. An additional important consideration is that current treatment protocols are strictly limited to somatic cells; no effect on the genetic constitution of the germ cells is anticipated. Thus, all of the offspring of an individual homozygous for a recessive condition must of necessity be at least heterozygous for the condition. While in most cases such heterozygous individuals are asymptomatic (see Chapter 6), they themselves may have questions regarding their carrier status and their likelihood of encountering a heterozygous partner.

Such considerations recall earlier unsavory aspects of a field known as "eugenics." Prominent and widely discussed early in the twentieth century, the notion of increasing population "fitness" by minimizing transmission of "undesirable" traits from "inferior" individuals led to major social and political disasters, including laws regarding immigration and sterilization in the United States. Unfortunately, not all of the ideas that served as the basis for such decisions have been eliminated. It is important to recall, however, that anyone in an outbred society is heterozygous for multiple mutant alleles, homozygosity for any of which could lead to identifiable genetic conditions or even prove lethal. Attempts to selectively eliminate any of these are doomed to failure simply on the basis of the number of existing gene variations to be addressed, not to mention the likelihood of new mutations. Rather than viewing genetic traits as problems to be eliminated (which is clearly impossible), a better perspective on the care of individuals with inherited gene abnormalities is that identifying these mutations will justify developing entirely new and challenging treatment efforts in the future. Although such a statement may appear simplistic, the pace of current research clearly justifies this point of view. This perspective also is justified because, as indicated in Chapter 12, the essential contributions of single-gene changes to complex traits and common diseases are becoming appreciated. Although such conditions are not yet fully understood, more detailed study will help identify high-risk individuals within larger categories of common disease, as well as those likely to benefit from specific treatment protocols. All of these advances will permit increased appreciation of the value of detailed investigations of molecular pathophysiology for which the mutations themselves will serve as critical starting points.

J. SUPPORT ORGANIZATIONS

The care of and for individuals and their families in the context of inherited and congenital problems is challenging. An important aspect is maintaining appropriate support for long periods of time. Numerous questions arise that cannot all be resolved at initial encounters and often are not even considered until considerable time has passed. Furthermore, personnel changes in clinical settings and movement by patients and their families often create additional barriers to consistent care. A particular challenge under such circumstances is to provide opportunities for personal support and an alternative to the sense of isolation that often is created by the rarity of some of these conditions.

Fortunately, a remarkable number of support organizations have developed, often beginning with affected individuals and/or their parents and relatives. Some are based around research groups whose efforts have attracted referrals. Other groups are organized around specific disorders, for example, the Neurofibromatosis Society and the Marfan Society. Still other groups are centered around common clinical features, such as short stature, for which the Little People of America has been a remarkably effective organization.

Such groups are valuable adjuncts to medical care. They can provide helpful ideas, contacts with research and treatment opportunities, and personal emotional support. Fortunately, these can be contacted through several agencies, as shown in Table 14.2. They also can be reached through other Web sites (see below) and OMIM, making referrals convenient. These groups can supply literature, referrals to physicians located close to affected persons, and a wide array of ideas and social opportunities for patients and their families.

◄ Table 14.2

Genetic support groups

Group	E-mail address	Street address
Alliance of Genetic Support Groups	alliance@capaccess.org	35 Wisconsin Circle, Suite 440 Chevy Chase, MD 20815 1-800-336-4363
Office of Rare Diseases	http://www.rarediseases.info.nih.gov/ord	
MedHelp International	staff@medhelp.org	

STUDY QUESTIONS

1 You have recently seen a new patient at risk for myotonic dystrophy (OMIM# 160900). She is 20 years old and very concerned about her status. What can you offer her?
It turns out that she has a very wealthy relative who is eager to use "gene treatment" for her. What would you propose?

2 You have a 9-year-old patient with Duchenne muscular dystrophy (OMIM #310200) who has not been doing well. His parents have heard about "injection treatments in mice" and have asked you whether this might be an option for their son. What is the rationale for these treatments? Could you recommend anything like them for this boy?

3 The gene for hemophilia A (OMIM #306700) covers over 200 kb in the Xq28 region. You have been given the assignment of presenting proposals for gene replacement therapy. What suggestions would you make? What vector(s) would be appropriate? Would bone marrow transplantation be a useful option?

FURTHER READING

Crystal RG. Transfer of genes to humans: Early lessons and obstacles to success. *Science* 270:404, 1995.

Flotte TR and Carter BJ. Adeno-associated virus vectors for gene therapy. *Gene Ther* 2:357, 1995.

Holmgren G et al. Clinical improvement and amyloid regression after liver transplantation in hereditary transthyretin amyloidosis. *Lancet* 341:1113, 1993.

Hoogerbrugge PM et al. Allogeneic bone marrow transplantation for lysosomal storage disease. *Lancet* 345:1398, 1995.

Levine F and Friedmann T. Gene therapy. *Am J Dis Child* 147:1167, 1993.

Ludmerer KL. *Genetics and American Society.* Baltimore, Johns Hopkins University Press, 1972.

Monahan PE et al. Direct intramuscular injection with recombinant adeno-associated virus vectors results in sustained expression in a dog model of hemophilia. *Gene Ther* 5:40, 1998.

Olivieri NF et al. Brief report: Combined liver and heart transplantation for end-stage iron-induced organ failure in an adult with homozygous beta-thalassemia. *N Engl J Med* 330:1125, 1994.

Yáñēz RJ and Porter ACG. Therapeutic gene targeting. *Gene Ther* 5:149, 1998.

USEFUL WEB SITES

GENERAL

OMIM (ONLINE MENDELIAN INHERITANCE IN MAN)

http://www.ncbi.nlm.nih.gov/omim
Basic catalogue listing mendelian genes and traits. Clinical descriptions, references, and links to other databases.

OFFICE OF RARE DISEASES (NIH)

http://cancernet.nci.cih.gov/ord/index
Broad-based NIH index including support groups, clinical research database, investigators, and extensive glossary.

GENE THERAPY

http://www.wiley.co.uk/genetherapy/
Summary of aspects of gene therapy including genes, vectors, and diseases. Summary of clinical trials. Bimonthly update. Bibliography.

MARCH OF DIMES

http://modimes.org/
Resources for families and professionals about birth defects. Professional education section.

NATIONAL HUMAN GENOME RESEARCH INSTITUTE HOME PAGE

http://www.nhgri.nih.gov
Review of basic genetics. Genetic maps. Discussion of "How to conquer a genetic disease." Background on Human Genome Project. Discussion of ethical and social issues.

NATIONAL ASSOCIATION FOR RARE DISORDERS (NORD)

http://www.NORD-rdb.com/~orphan
Primary nongovernmental clearinghouse for information about rare disorders. Over 1100 disease reports. Broad resource guide. Includes Rare Disease Database, a wide range of information for families and professionals. Listings by symptoms, causes, affected populations, treatments. Listing of support groups. Includes orphan drug designation database.

ALLIANCE OF GENETIC SUPPORT GROUPS

http://www.medhelp.org
Access through Med Help International. Basic listing of support organizations.

BIOETHICS

CENTER FOR BIOETHICS, UNIVERSITY OF PENNSYLVANIA

http://www.med.upenn.edu/~bioethic
Articles on cloning and bioethics primer. Articles with comment options. Access to virtual library. Literature summaries about insurance, discrimination, privacy, screening.

EUBIOS ETHICS INSTITUTE

http://www.biol.tsukuba.ac.jp/~macer/index.html
Newsletter for bioethics and biotechnology; based in Japan and New Zealand. Access to articles in Japanese. Meeting contents and abstracts.

GENETIC PRIVACY ACT

http://www.ornl.gov/hgmis/resource/elsi.html
Ethical, legal, and social issues page for Human Genome Project. Includes access to genetics support groups. Useful cross-links to books, journals, and programs. Useful glossary. Model legislation supported by Human Genome Program.

HUMAN GENOME PROJECT INFORMATION

http://www.ornl.gov/hgmis
Includes section on ethical, legal, and social issues; chat opportunities.

NATIONAL BIOETHICS ADVISORY COMMITTEE

http://www.bioethics.gov
General source of advice to National Science and Technology Council. Notices of meetings and transcripts. Access to publications.

NATIONAL CENTER FOR GENOMIC RESOURCES

http://www.ncgr.org
Genetics and public issues pages including legislative proposals. Section on continuing medical education about genetics. News updates. Access to Genome Sequence Database.

THE GENE LETTER

http://www.geneletter.org
Newsletter about genetics and potential policy questions. Useful and timely summary articles. Chat opportunities. Links to other sites.

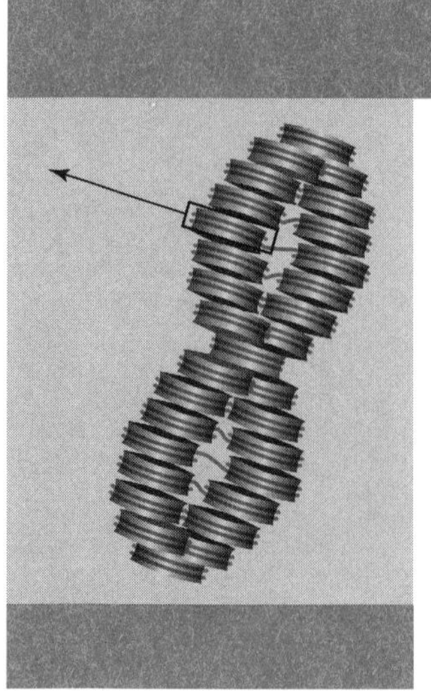

Epilogue

Genetic contributions to all aspects of medicine are becoming more apparent for several reasons.

- The molecular bases of inheritance and its variations are being clarified.
- It is becoming possible to measure gene changes that affect clinical practice.
- Database management is growing to accommodate the enormous body of genetic information in an accessible format (frequently on the Internet).
- There is a growing appreciation of the fact that population heterogeneity implies different predispositions and biologic responses.
- The emergence of biotechnology in both academic and commercial venues is revealing details about pathophysiology and its permutations and presenting new treatment options.

It remains impossible to escape the fact that genetic information will remain intrinsically complex. However, this should be seen as a challenge to further investigation and understanding of exciting new aspects of human biology, as well as to a lifelong effort to integrate a genetic perspective into whatever aspect of medicine one pursues.

As set forth at the beginning, the purpose of this book is to emphasize the broad range of involvement of genetics in human and medical biology. The text is only a place to begin. The references are only as timely as their publication dates. The databases are under continuing revision. Appreciating the excitement of the basics and watching their implications develop should add challenge and satisfaction to a lifetime of medical practice.

Answers to Study Questions

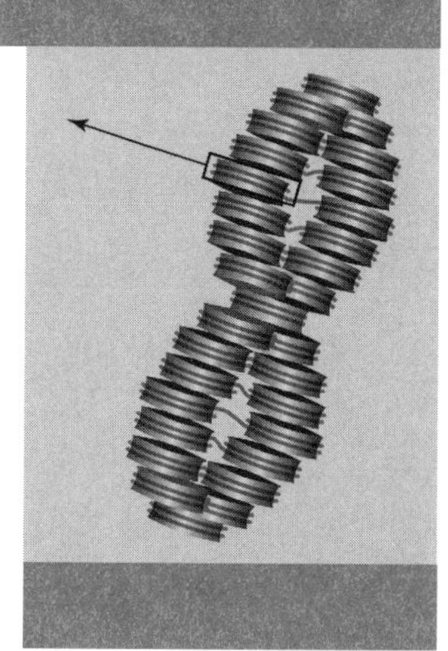

CHAPTER 1

1 a Deleting this dinucleotide in the middle of the intron should have NO physiologic effect. It will not interrupt splicing (it does not involve the splicing sites) and, because it will be eliminated BY splicing, it will not affect the encoded gene. Such mutations often can serve as useful position markers for linkage studies.

b There are two possibilities. TTA, encoding Leu, could become TCA, encoding Ser. Alternatively, TTA could become CTA, also encoding Leu. The first "point mutation" causes an amino acid change while the second is biologically silent due to codon redundancy. Either nucleotide transition might be useful as a linkage polymorphism.

c As described, there are numerous $(CA)_n$ clusters in noncoding regions. They are generally genetically and clinically silent but can serve as very useful polymorphic position markers for linkage analysis.

2 There are several possible explanations, based on codon redundancies.

```
        Native:        Arg  Phe  Phe  Ser  Pro  ...
                       CGC  TTT  TTC  TCA  CCA  G..
i.  T deletion                   Δ
                       CGC  TTT  TCT  CAC  CAG  ...
                       Arg  Phe  Ser  His  GluN
```

ii. C→G transversion ↓
CGC TTT TTC TGA CCA ...
Arg Phe Phe term

iii. TCA deletion Δ
CGC TTT TTC CCA ...
Arg Phe Phe Pro

iv. T insertion ↓
CTG CTT TTT CTC ACC
Leu Leu Phe Leu Thr

3 a,b,c,d
4 a,b,c
5 a,b,c,d
6 III-3 was not the father of IV-4; a case of nonpaternity in which the biologic father had what appeared to be "allele 3" but did not have YoYo disease.

CHAPTER 2

1 The best answer is "c" because one cannot know the mutations to seek based on this information alone. Health records for their deceased child ("a") may suggest a mutation, but the phenotype is not generally sufficiently distinguishable for different mutations. Reply "b" is technically correct, but the mutation(s) sought must be known before testing. Reply "d" is irrelevant.

2 This is not an unusual clinical situation. This couple's risk for a Down syndrome pregnancy is low ("a") but not zero. Triple screening, although suggestive, does not "establish" the diagnosis. The second part of the question relates to a follow-up visit and certainly "a," "b," and "c" are correct, with "c" followed by "b" being most reasonable. The likelihood of recurring depression cannot be established by the data presented because it is not sufficiently defined; for example, dominant transmission of bipolar illness is recognized. Adoption "e" is always an option.

3 Part One: "b" Prenatal diagnosis cannot be certain because the SSCP pattern shows one pair of single-strands of DNA migrating at normal positions and the other pair at anomalous positions. Thus, although the anomalous pair can be identified easily, the status of the Q syndrome gene on the other strand cannot be determined.

Part Two: Because of the explanation for Part One, the answer must be "c." The pair of single-strands with anomalous migration is present; presumably these account for the mutation on one of the two parental chromosomes. The other pair of strands could represent either the mutant or the normal chromosome from the other parent, indistinguishable by this method. Thus, the likelihood that the fetus is affected is "iv" 1 in 2.

4 This woman is below the age for which further prenatal diagnostic studies are usually recommended. However, the elevated MSAFP clearly changes the risk. Referring to Figure 2.6 gives some perspective on the risk present. Two approaches are useful. First, ultrasound can be helpful in searching for abdominal wall or neural tube defects. Second, amniocentesis can also be helpful for both Down syndrome and further AFP testing.

CHAPTER 3

1 a The best answer is "a." This patient has many features suggesting Down syndrome: a. He is much younger than his sisters, indicating that his mother was older when he was born. b. He has a simian crease. c. He is short and pleasant (possibly retarded because he speaks little). d. He has epicanthal folds. e. He has a heart murmur. Why is he pale? Why does he have the thigh bruise? A stress test is not the best study for him as you discover in:

1 b An atrial septal defect is present, strengthening your suspicion of Down syndrome. Answer "b" could be helpful because catheterization will define the lesion better and assess heart function and the possible value of surgical correction. He is still not doing well, however.

1 c He has leukocytosis and anemia. Is he infected? You cannot confirm a diagnosis with the test possibilities listed, so you suggest:

1 d a, b, c are reasonable but:

1 e all cultures are negative. The next step should be a bone marrow examination. This shows that your patient with Down syndrome has one of the known complications of his disorder—acute lymphocytic leukemia. He needs a referral to Oncology.

Comment: As noted earlier in Chapter 3, the diagnosis of Down syndrome should be reasonably established on clinical grounds for an individual from any population group. The suspicious features present here serve to strengthen the impression even without chromosome studies. All of his presenting features can be explained as aspects of this single diagnosis. His breathlessness likely reflects reduced cardiac efficiency (due to the septal defect) unmasked by the anemia developing as part of his leukemia.

2 a both "a" and "b" are appropriate

2 b Nearly 3 years with a lack of success in conceiving is unusual for a healthy young couple. Perhaps the three episodes of late cycles and heavy flow were early spontaneous abortions. What about the husband's cousin with fatal heart disease at 2 years of age? Could this child have had Down syndrome as well? Could the father be a carrier for a balanced translocation? The best answer is "d."

3 The best answer is "d." Breast cancer is very rare in males and certainly at this age. The diagnosis of XXY (Klinefelter) syndrome would be a unifying diagnosis and would offer the best help with subsequent management.

4 This girl has a characteristic presentation for Turner syndrome and she shows mosaicism. As noted earlier in the chapter, this is not unusual. A different ratio of XX to XO cells might have been found if a different tissue had been sampled (although this is not necessary). There is no need to repeat the study or to study the parents. They should be encouraged that their daughter is healthy and otherwise doing well.

5 You are correct in thinking that these numbers of abnormal karyotypes are high. However, they were derived from CVS. Many of these (possibly all) would have been spontaneously aborted and not seen in amniocentesis studies that are necessarily performed later in pregnancy (recall Figure 3.1).

6 These three children have different genetic bases for their phenotype.
#1 shows uniparental disomy for a paternal chromosome.
#2 shows absence of the middle band from the maternal chromosome shown on the left of the parental side of the figure.
#3 shows loss of sequences from both maternal bands shown on the right of the parental side of the figure.

ALL individuals show abnormalities of maternal sequences—either 1. total absence in uniparental disomy, 2. deletion, or 3. substantial reorganization. This raises the suspicion of imprinting as an important contributing factor in the disorder.

CHAPTER 4

1 The notion of defective DNA repair as the basis for malignancy is consistent with the model proposed in Chapter 4. Here, the final derangement of gene changes, culminating in uncontrolled cell proliferation, represents an accumulation of changes—mutations, deletions, rearrangements. At least some of these events are eliminated in healthy cells by repair mechanisms; the absence of effective repair can speed the development and accumulation of these defects.

2 This is not a rare situation. The death of her father at an early age complicates the interpretation but the pedigree information that is available suggests an autosomal dominant trait. Of course, having unaffected individuals in such a sibship is not unusual because of the expected 50% transmission rate (see also Chapter 5). This eliminates possibility "b." Possibility "a" is a radical move in the absence of any clinical indications. On an immediate basis, she could benefit from a colonoscopy to clarify her clinical status (option "d"). Her family also should be investigated to find out whether a recognized gene change is present; if one is identified, option "g" will be helpful for her and her sibs. Options "c" and "f" are reasonable for almost anyone.

3 Wilms tumor is one of the conditions that shows clinical patterns consistent with the Knudson hypothesis as presented in Chapter 4. Inherited forms are generally multicentric and bilateral; hence sibs would be expected to show this pattern. Isolated, unilateral Wilms tumors are generally not inherited, and while LOH studies may show changes in the region of the Wilms tumor gene locus, they would not be diagnostically useful.

4 The chemotherapy did not eliminate the Ph^1 chromosome from the blood cells but reduced the proportion of Ph^1-positive cells substantially. The persistence of this important chromosomal marker indicates that florid leukemia likely will recur, presumably because the malignant progenitor cells were not eliminated. Thus, the patient likely will do better for some time, although the rate of redevelopment of substantial numbers of tumor cells is difficult to predict.

5 The best answers are "b" and "c." The prediction of metastasis patterns ("a") is difficult; "d" may become true in the future; "e" is true from a trivial perspective.

CHAPTER 5

1 "c" is the best answer. Although catheterization will show many details, ECHO cardiography is likely to give more valuable information. There are several forms of this condition with different ages of onset and progression. Study of an affected relative may show that he or she falls into one of the subsets of the disorder for which the gene (and, possibly the mutation) has been identified. Obviously, this latter situation will change the diagnostic approach to direct gene analysis.
2 There are many genetic contributions to early atherosclerotic disease; not all are related to cholesterol levels (see also Chapter 12). Thus, option "d" is not necessarily going to be helpful. The best option is "e"—defining the father's status as carefully as possible may reveal the underlying problem.
3 The answer is "d." His ECHO studies have only evaluated the heart valves and aortic root. This man may have an abdominal aortic aneurysm, and these symptoms could indicate dissection.
4 This is not an unusual situation. Despite the fact that her brother has symptoms, at age 43 she is not too old to manifest problems if she has the amplified gene as well. The most direct recommendation is "e" so that her own situation can be determined. At her age, she also needs "d," and the possibility of a neural tube defect ("c") is something that should be clarified by regular prenatal and obstetric care.
5 This situation indicates a problem of testing without plans for handling the outcome or informing the subjects of the risks. The best response is "d." Prognosis is still unclear (recall Figure 5.15).
6 Neither has enough to establish the diagnosis of neurofibromatosis (recall Figure 5.5). The second student may have a mutation in the neurofibromin gene that developed after conception and is confined to a developmental segment. Assuming that no neurofibromas are present, the second student will be appropriately followed for any changes; there should be no concerns about the first.

CHAPTER 6

1 Many factors contribute to the clinical severity of sickle cell disease. Currently, most cannot be identified. Specialized laboratories can measure the amount of fetal hemoglobin (HbF) present in erythrocytes; the higher the amount, the less likely the cells will be to undergo the sickling process. Another globin mutation might be present that stabilizes the hemoglobin tetramer. Changes in the structure or stability of the erythrocyte membrane might help resist sickling. These modifying factors are important to identify through future studies.
2 A priori the obvious answer must be "e" because the child must inherit one sickle globin gene from the father. Appropriate advice should include "c" because knowing Tanya's genotype will clarify the risk for homozygous sickle cell offspring. Prenatal diagnosis ("d") is not appropriate unless the mother is a carrier and the parents then request the test.
3 As described in Chapter 6, Type II Gaucher disease is clinically distinct

from the adult form, Type I. However, both involve mutations in the same gene. Thus, choices "d" and "e" are the best, but further evaluation is indicated.

4 The primary skin pigment is melanin. As shown in Figure 6.9 melanin is derived from tyrosine. Failure to synthesize tyrosine due to phenylalanine hydroxylase deficiency can lead to a "downstream" deficit of melanin.

CHAPTER 7

1 Although the setting may appear bizarre, the problem is not irrelevant. Obviously, you cannot perform the DNA studies, so you must rely on indirect analysis. Knowing the location of the FraX site (recall Figure 7.2) you might consider linkage studies. If you were able to find color blindness (using the Ishihara plates) or G6PD deficiency (both of these are highly polymorphic as discussed), you could study both affected and unaffected males to find out if either the absence *or* the presence of either of these traits is found. If such a relationship could be established, you could use the principles of linkage analysis (recall Chapter 1) to arrive at simple predictions. The major limitation to the study would be the genetic distance (and, hence, likelihood or crossing-over) between FraX and either G6PD or CB. The intelligence test also could be helpful in clarifying whether individuals thought to be affected are in fact retarded or whether some other problem might be present (brain damage, chronic illness, etc.).

2 The problem of vague diagnosis is not rare. This boy has had other studies that were not diagnostic, and the parents would like to have healthy children. Your goal is to define the problem as well as possible. A karyotype (answer "d") is the most useful of the listed options. The percentages ("a" through "c") cannot be used without some notion of the nature of the problem. Similarly, prenatal sex determination would be irrelevant if the problem is not X-linked.

It is important to note that answer "a" is unlikely in any instance: The problem is not inherited as a dominant trait (parents are unaffected) and, even if it is an X-linked trait, passed from an asymptomatic carrier mother, the chance of recurrence would only be 25% (50% chance of a male conception X 50% chance of receiving the X-chromosome with the mutation).

3 This kindred shows many of the useful aspects of X-linkage for predictions and family studies. Having affected individuals in multiple generations permits reconstruction of allele patterns.

- a Affected individual IV-1 will have the allele linked to the trait. Thus the question is whether IV-2 has that allele as well; if so she is a carrier. (The allele on the single X-chromosome of III-1 should be determined so that you can minimize potential confusion.) A similar approach can be used for individual IV-4.
- b Yes, based on the answer to part "a" and assuming minimal recombination between the marker and the trait for the gene.
- c IV-5 should be easily clarified because he has only a single X-chromosome. Does he have the same marker pattern as III-3, for example?

III-6 should similarly be identifiable; does he have the allele pattern of II-3? You also could determine the carrier status of II-4. Does she have the same pattern of X-chromosome alleles as II-2?

CHAPTER 8

1 This question confronts the difficulties of diagnosis in mitochrondrial diseases, especially LHON. You can easily exclude answers "c" (she is clinically unaffected) and "f" (her own status is unknown). Examining her mother ("a") is unlikely to be very helpful; "b" and "e" are actually related; the facts that "e"is true and that you do not know her level of heteroplasmy eliminates "b." The best answer is "d," but recall that different tissues can have different degrees of heteroplasmy and symptom development may be difficult to predict.

2 This is a complicated situation. It is difficult to accept the diagnosis of "muscular dystrophy" made on his sister 35 years ago. He could, in fact, have Becker muscular dystrophy ("a"). The most reasonable approach includes "b" and "e"; it is premature to consider problems in his children.

3 Answer "c" is the best; not all disorders due to mitochondrial dysfunction show structurally abnormal mitochondria.

CHAPTER 9

1 You have no diagnosis based on the information available. He, of course, needs more detailed study, so "d" is appropriate. The picture may be clarified by examining the parents ("c"). You may not need an ECHO study but at least auscultation would be valuable, further evaluation being guided by the findings in "c" and "d."

2 This important observation developed after study of multiple kindreds. The answer is unknown, but the temporal parallel of cardiac and upper arm development as well as the fact that the heart is a left-sided structure suggest that a developmental defect is affecting both processes.

3 The problem is that treatment has been requested, and a diagnosis has not been established. With multiple spots, an important consideration is neurofibromatosis even though no nodules have been noted (recall Figure 5.5). An ophthalmologist should be able to identify Lisch nodules (recall Figure 5.3); finding them would put the diagnosis and management into a different category. Thus, the best answer is "c."

CHAPTER 10

1 As described in the text, isoniazid inactivation occurs by acetylation of the drug. Thus, the rapidity of acetylation affects the ratio of acetylated to nonacetylated forms of the compound. If one of these species was more likely to be recognized by the immune defense system as "foreign," any

change in the ratio could affect the immune response to the agent; such a response could possibly become generalized into a "lupus-like" reaction.

To address this possibility, it will be necessary to assemble groups of individuals matched for as many variables as possible (age, sex, socioeconomic group, etc.) but differing in their rate of isoniazid acetylation. Such groups will necessarily need to be followed for at least several years. The problem then becomes how the administration of isoniazid can be justified. This has been a difficulty for other similar studies. Should patients with tuberculosis be followed? Should one choose individuals with a skin test conversion and who could be candidates for isoniazid treatment? Does the presence of tuberculosis (or at least the conversion of a skin test) add complicating immunologic features?

As an alternative to such a prospective study, could individuals with past treatment be examined? For instance, could you study clinically stable patients who are being treated for resolving (but historically active) tuberculosis? What is the incidence of lupus in members of such a group? How does the incidence of lupus compare in individuals within the group who differ in isoniazid inactivation rates? How do such incidence figures compare with the expected frequency of developing lupus in a population similar except for the presence of tuberculosis (a population that may be difficult to identify)?

If longitudinal studies of large groups of patients are impractical, what about laboratory investigations? Could you measure the rate of isoniazid acetylation in a population with lupus and compare it with a similar population without lupus? What about examining individuals with a history of developing a lupus-like syndrome after exposure to other drugs—hydralazine, procainamide, or quinidine? Could these comparisons help indicate specific features of individuals at risk?

The complexities of this problem indicate why a definitive test of the hypothesis has not yet been performed. The important question is based on determining the effect(s) that altered metabolism of drugs (and other exogenous agents) might have on the immunologic changes that lead to lupus. Lack of full clarification of the pathophysiology of lupus makes this difficult to answer.

2 As the individual responsible for intra- and postoperative care, your goal is to limit the complications, preferably through a preventative strategy. This really is a question about screening for rare defects (see also Chapter 13). You already ask about past surgical experiences of the individual and his or her family members—but is this enough? You have to know several things:

a What is the distribution of these defects in the population you and your staff are likely to encounter? Can you identify those at higher risk?

b What is the reliability of the test: How many individuals will have abnormal results but no defect ("false positives")? How many affected individuals will you miss ("false negative")? The more rare the defect(s) the more important are these considerations.

c What is the cost of the test(s)? How will these be covered?

d What will you do with the results? Having identified an individual with

one of these defects, will you deny anesthesia? Will you refer him or her elsewhere? Will you look at this as an opportunity to show that your team can handle anything?
(N.B.: Currently, no screening programs are used for these conditions.)

3 This problem addresses the same sort of consideration as #2, but here we are concerned with a single person. The "worst case" is that he and his sister have the defect for acute, intermittent porphyria (recall that it is an autosomal dominant as discussed in the chapter). Being male, he has not had the cyclic hormone changes that contribute to the increased frequency of attacks in women. Furthermore, he has generally been healthy and has used very few drugs. A simple blood test can establish or eliminate the diagnosis for him by measuring uroporphyrinogen-1 synthase activity in his erythrocytes. If his level of enzyme activity is low, you need to be careful with drugs. For instance, barbiturate use in anesthesia would be dangerous. You also would need to check on the list of drugs proposed by Oncology and identify those dangerous for porphyria.

Finding a normal level of erythrocyte uroporphyrinogen-1 synthase does not eliminate the presence of OTHER types of porphyria, however; several of these can be tested by blood analysis in specialized laboratories.

CHAPTER 11

1 In general, the process of establishing the relationship between HLA markers and some trait (in this case, "Q" disease) is straightforward using the "Relative Risk" notions introduced in Chapter 11. However, this situation is complicated by finding the disorder in a geographic isolate. As with any sort of isolate, one cannot assume that the "normal" population distributions of traits and markers are present. The most important consideration here is to establish the distribution and frequency of HLA haplotypes on *this* island. Only when this distribution is known can you begin to relate a specific haplotype to the trait of interest. Without knowing this information, you might find an association that is invalid due to a distorted underlying distribution of haplotypes. This provides another example of the "founder effect" in which a limited population shares more gene changes in common than anticipated and where these peculiar gene changes can be traced to the island's founders. This also is related to the more frequent appearance of homozygotes for otherwise rare genotypes in such populations.

2 Recall that this important association (diabetes and DR3) is *not* causal. Certainly, "a" and "c" are true—he and his family remain at risk. Option "d" is clearly incorrect; the absence of diabetes in this family does not "prove" anything. Answer "b" is trivial but always possible.

3 This is a problem of considerable potential significance. Note that individuals "resistant" to HIV infection *have* been identified and changes in a cell surface molecule have been found. You cannot *conclude* "a," although you may suspect it. Answer "c" is possible, although you should have established that before doing your screening. Answer "d" is a possible extension of

this observation (recall the HIV findings). Answer "b" might be interesting but would not help with understanding your findings about the viral resistance.

CHAPTER 12

1 This situation is not unknown and is an example of the sort of potentially valuable insight that can be made by simple epidemiologic observations. *If* you suspect an inherited predisposition you obviously need more data. Answer "c" is possible but you cannot tell without kindred data. Answer "d" also is possible and could be strengthened by finding consanguinity in ancestors of affected individuals ("e"). Answer "a" may be valuable but will miss the point of the underlying susceptibility. Whether prophylactic penicillin will be useful ("b") will depend on whether you think that a real susceptibility is present.

2 This is a potential finding in any group. The critical thing is to find out more details about the affected students ("d") to determine if any features suggest some sort of predisposition. Measuring plasma renin levels ("a") might help but likely will not prove valuable. Options "b," "c," and "e" are likely to be failures, and the frustration they generate may undermine your credibility and impede your chances of resolving the question.

3 As discussed in Chapter 12, testing this young woman was not recommended by genetic reference panels, partly because the optimum management recommendations are unknown. Nevertheless, she now *has* been identified as a heterozygote for a BRCA2 mutation. You need to tell her several things:

1. The natural history of BRCA2 heterozygotes is unknown.
2. Her negative family history for cancer is important information, but we do not know the BRCA2 gene status for any of them. (For example, showing multigenerational segregation of the mutation in people who lived long lives without cancer would be useful information.)
3. She should remain under careful observation, possibly in a center specializing in women's health.

4 This information is a useful example of several associations. The association between ankylosing spondylitis and HLA-B27 is well known, as discussed in Chapter 12. Moreover, the gene for hemochromatosis falls within the MHC (recall Figure 11.1) and is therefore very closely linked to HLA markers. Thus, while it is true that ankylosing spondylitis and hemochromatosis are not linked pathophysiologically, there is a reasonable expectation that they should cosegregate. (This would eliminate answer "a.") Recall that hemochromatosis is a dominant trait, so he could have inherited it from his father. (This eliminates answer "b.") *If* the trait came from his father and *if* the father's ankylosing spondylitis was HLA B27-associated, the patient should have a test for HLA-B27 ("d"). If he has received *both* the HLA-B27 marker and the hemochromatosis gene from his father, he certainly is at increased risk for ankylosing spondylitis and baseline spinal x-rays would be valuable ("c").

CHAPTER 13

1 This is a typical example of the sort of difficulties raised by workplace screening. If you run the gene test, you likely will identify current and prospective employees who might wisely avoid the plant where this work is done. What will you do if:

1. There is a concentration of potential hepatotoxicity in a specific ethnic or racial group.
2. Working at the new plant is a path to more rapid advancement.
3. Advanced worker age turns out to increase risk.

Obviously, any answer will be controversial and potentially viewed as discriminatory.

2 This is another example of a problem with testing. If you try to revise the report ("c"), you will be revealing confidential information to the insurance company. Age of onset of these tumors is variable and answer "b" does not apply. To take care of *this patient* you should advise him to have the test "a" and, pending that result, he could have a colonoscopy ("d"). Even if he is unwilling to consider the colectomy option, you and he can care for him better if you know his potential for future colon cancer.

3 You need to investigate her family history in more detail including ages at death and general health of other family members. The current recommendation for screening in individuals who have a *positive* history of colon cancer is to perform colonoscopy as a baseline but no gene tests are known for her ethnic group. As noted earlier, the current test guidelines for breast cancer do not include gene testing, particularly for women without a family history of affected individuals. Basically, this woman has no indications for screening.

CHAPTER 14

1 Presumably, the "risk" for myotonic dystrophy is the presence of an affected family member. Depending on who was (is) affected, you can consider a study of the length of the trinucleotide repeat in the gene (recall Figure 5.15). At least such a study will give you a reasonable idea about risk. You should make it clear that no current treatment is appropriate for myotonic dystrophy. If her trinucleotide repeat length places her at high risk for disease, she needs attentive follow-up; if her repeat length is normal, she is unlikely to have problems.

2 Studies in mice have used intramuscular injection of the gene for dystrophin. Some muscle cells then show the presence of the protein encoded by the introduced gene. Physiologic data about the treated muscles are not yet available. Muscles have been considered possible targets for directed gene introduction because of their syncytial nature—a broad distribution of gene product(s) can spread through many cells. This approach has not yet been developed with sufficient reliability or efficacy to be recommended for use in humans.

3 This question integrates much current knowledge about strategies for gene introduction. The very large size of the gene virtually precludes using any viral vector. Could you use recombinant DNA techniques to reduce the length (intron removal, partial splicing, etc.)?

Bone marrow transplantation is unlikely to work unless you can develop a method for expressing the Factor VIII protein in the marrow cells after transplantation—even then, it would not necessarily be well-regulated. The ideal cell target is the hepatocyte. A DNA-ligand complex might be able to introduce the gene into the liver cells. Direct DNA injection has not worked well for delivery to the liver. Liposome-mediated DNA delivery may have promise for targeting the gene to the hepatocyte, but its brief intracellular survival under current protocols is not encouraging.

Can you suggest other approaches? Many genes are quite large and since direct recombination between the introduced gene and the host gene has not been demonstrated reliably in humans, it will be necessary to introduce the entire (or, at least, most of the) coding region.

Index

Page numbers followed by the letters *f* and *t* indicate figures and tables, respectively.

ISBN 0-07-057998-9
9 780070 579989 90000